T0155850

Graduate Texts in Physics

Graduate Texts in Physics

Graduate Texts in Physics publishes core learning/teaching material for graduate- and advanced-level undergraduate courses on topics of current and emerging fields within physics, both pure and applied. These textbooks serve students at the MS- or PhD-level and their instructors as comprehensive sources of principles, definitions, derivations, experiments and applications (as relevant) for their mastery and teaching, respectively. International in scope and relevance, the textbooks correspond to course syllabi sufficiently to serve as required reading. Their didactic style, comprehensiveness and coverage of fundamental material also make them suitable as introductions or references for scientists entering, or requiring timely knowledge of, a research field.

More information about this series at http://www.springer.com/series/8431

Henryk Arodź · Leszek Hadasz

Lectures on Classical and Quantum Theory of Fields

Second Edition

Springer

Henryk Arodź
Institute of Physics
Jagiellonian University
Kraków
Poland

Leszek Hadasz
Institute of Physics
Jagiellonian University
Kraków
Poland

ISSN 1868-4513 ISSN 1868-4521 (electronic)
Graduate Texts in Physics
ISBN 978-3-319-85710-7 ISBN 978-3-319-55619-2 (eBook)
DOI 10.1007/978-3-319-55619-2

Softcover reprint of the hardcover 2nd edition 2017

Printed on acid-free paper

This Springer imprint is published by Springer Nature
The registered company is Springer International Publishing AG
The registered company address is: Gewerbestrasse 11, 6330 Cham, Switzerland

Preface to the Second Edition

Preparing the new edition we have preferred first of all to improve the existing text. Expanding it was not our priority because we would like to preserve the compact size of the book. We have removed a number of misprints and mistakes. Sections devoted to the Majorana field and to path integral approach to $U(1)$ anomaly have been revised. There are two new sections: on anticommuting (bi)spinor fields, and on more advanced supersymmetric models. Several new problems are added.

We would like to thank Thomas Williams for careful reading of the manuscript and helpful comments, Łukasz Marszałek for pointing out several mistakes, and Anna Gagatek for the help in preparation of figures.

The picture on the front cover is by courtesy of Dr. Tomasz Romańczukiewicz. It illustrates scattering of two solitonic objects in 1+1 dimensional space-time.The horizontal direction represents the time, and the vertical one the one dimensional space. The two objects are the kink, discussed in Chap. 3, and an oscillon which is a kind of breather, see Exercise 1.1(c). Before the scattering, the oscillon is at rest (then its world-line coincides with the horizontal dark dashed line), while the kink approaches it with a constant velocity (then its world-line coincides with the white continuous line).

Kraków, Poland

Henryk Arodź
Leszek Hadasz

Preface to the First Edition

This textbook on field theory is based on our lectures which we delivered to students beginning their specialization in theoretical physics at Jagiellonian University in Cracow. The lectures were accompanied by problem-solving classes. The goal was to give a presentation of the basics of field theory.

Field theory plays a fundamental role in many branches of contemporary physics, from cosmology, to particle physics, and condensed matter physics. Plenty of successful applications testify to its importance. On the other hand, there still remain unanswered questions about its foundations. For example, it is not clear what is the proper mathematical framework for its formulation. Nor do we know how to exactly solve its equations in the case of interacting fields. This state of field theory—many successful applications vs. hidden in a mist foundations—makes the task of preparing an introductory course rather challenging.

Before attending our course, the students had taken theoretical physics courses on classical mechanics, non relativistic quantum mechanics, classical electrodynamics, statistical physics, as well as mathematical courses on algebra, calculus, and differential equations. They also had a general introduction to particle physics. Concurrently with our lectures, or subsequently, they attended specialized lectures on advanced quantum mechanics including the relativistic formulation, the standard model of particle physics, statistical field theory, and the quantum theory of condensed matter. Such a curriculum has of course influenced the content of our lectures. We have entirely omitted applications of field theory, and the emphasis has been put on basic ideas. Furthermore, because of the limited time available both for the lectures and for the students, we have not at all attempted to make the course comprehensive. Our intention has been to offer a slow, step by step introduction to the main concepts of field theory. The method we have chosen consists of a carefully detailed explanation of the selected material. We hope that such a textbook can be useful, and that it is a helpful supplement to the vast amount of existing literature.

This textbook consists of three parts: classical fields are discussed in Chapters 1 to 5, an introduction to the quantum theory of fields is given in Chapters 6 to 10, and a selection of relatively modern developments is presented in Chapters 11 to 14.

We presented most of this material in three semesters using traditional tools: chalk and a blackboard. At the end of each chapter there are exercises with hints for solutions. Some are strictly tied up with the lectures, others deal with topics which were discussed at length only during the problem-solving classes. We have also included a short Appendix in which we have collected some basic facts about generalized functions. Interested students can find hundreds of books on field theory. Our list of literature includes only those books or original papers which are explicitly mentioned in this text.

Many students commented on parts of our lecture notes. We are very grateful to them all. We are particularly indebted to P. Balwierz, M. Eckstein, T. Rembiasz and P. Witaszczyk for providing lists of mistakes and unclear points. Needless to say, the full responsibility for mistakes and shortcomings still present lies entirely with us. Errata, very likely necessary in spite of our efforts, will be posted on the web page http://th-www.if.uj.edu.pl/ztp/Edukacja/index.php belonging to the Department of Field Theory of the Marian Smoluchowski Institute of Physics, Jagiellonian University.

Kraków, Poland
June 2010

Henryk Arodź
Leszek Hadasz

Contents

Chapter 1
Introduction

Abstract Sine-Gordon field as an effective description of a system of coupled pendulums in a constant gravitational field. Sine-Gordon solitons. The electromagnetic field, gauge potentials and gauge transformations. The Klein–Gordon equation and its solutions.

By definition, any physical system which has infinitely many degrees of freedom can be called a field. Systems with a finite number of degrees of freedom are called particles or sets of particles. The kinematics and dynamics of particles is the subject of classical and quantum mechanics. In parallel with these theories of particles there exist classical and quantum theories of fields. In this chapter we present two important examples of classical fields: the sine-Gordon effective field and the electromagnetic field.

Statistical mechanics deals with large ensembles of particles interacting with a thermal bath. If the particles are replaced by a field or a set of fields, the corresponding theory is called statistical field theory. This branch of field theory is not presented in our lecture notes.

1.1 Example A: Sine-Gordon Effective Field

Let us take a rectilinear, horizontal wire with $M + N + 1$ pendulums hanging from it at equally spaced points labeled by x_i. Here $i = -M, \ldots, N$, where M, N are natural numbers. The points x_i are separated by a constant distance a. The length of that part of the wire from which the pendulums are hanging is equal to $(M + N)a$. Each pendulum has a very light arm of length R, and a point mass m at the free end. It can swing only in the plane perpendicular to the wire. All pendulums are fastened to the wire stiffly, hence their swinging twists the wire (accordingly). The wire is elastic with respect to such twists. Each pendulum has one degree of freedom which may be represented by the angle $\phi(x_i)$ between the vertical direction and the arm of the pendulum. All pendulums are subject to the constant gravitational force. In the configuration with the least energy all pendulums point downward and the wire is not

H. Arodź and L. Hadasz, *Lectures on Classical and Quantum Theory of Fields*, Graduate Texts in Physics, DOI 10.1007/978-3-319-55619-2_1

twisted. We adopt the convention that in this case the angles $\phi(x_i)$ are equal to zero. Because of the presence of the wire $\phi(x_i) = 0$ is not the same as $\phi(x_i) = 2\pi k$, where $k = \pm 1, \pm 2, \ldots$—in the latter case the pendulum points downward but the wire is twisted, hence there is a non vanishing elastic energy. Therefore, the physically relevant range of $\phi(x_i)$ is from minus to plus infinity.

The equation of motion for each pendulum, except for the first and the last ones, has the following form

$$mR^2 \frac{d^2\phi(x_i,t)}{dt^2} = -mgR\sin\phi(x_i,t) + \kappa \frac{\phi(x_i - a, t) + \phi(x_i + a, t) - 2\phi(x_i,t)}{a}, \tag{1.1}$$

where κ is a constant which characterizes the elasticity of the wire with respect to twisting. The l.h.s. of this equation is the rate of change of the angular momentum of the i-th pendulum. The r.h.s. is the sum of all torques acting on the i-th pendulum: the first term is related to the gravitational force acting on the mass m, the second term represents the elastic torque due to the twist of the wire.

The equations of motion for the two outermost pendulums differ from (1.1) in a rather obvious way. In the following we shall assume that these two pendulums are kept motionless by some external force in the downward position, that is that

$$\phi(x_{-M}, t) = 0, \quad \phi(x_N, t) = 2\pi n, \tag{1.2}$$

where n is an integer. If we had put $\phi(x_{-M}) = 2\pi l$ with integer l we could stiffly rotate the wire and all of the pendulums l times by the angle -2π in order to obtain $l = 0$. Therefore, the conditions (1.2) are the most general ones in the case of motionless, downward pointing outermost pendulums. In fact, these two pendulums can be removed altogether—we may imagine that the ends of the wire are tightly held in vices.

In order to predict the evolution of the system we have to solve (1.1) assuming certain initial data for the angles $\phi(x_i, t)$, $i = -M + 1, \ldots, N - 1$, and for the corresponding velocities $\dot{\phi}(x_i, t)$. This is a rather difficult task. Practical tools to be used here are numerical methods and computers. Numerical computations are useful if we ask for the solution of the equations of motion for a finite, and not too large, time interval. If we let the number of pendulums increase, sooner or later we will be incapable of predicting the evolution of the system except for very short time intervals, unless we restrict initial data in a special way. One such special case is in the limit of small oscillations around the least energy configuration, $\phi(x_i) = 0$. In this case we can linearize the equations of motion (1.1) using the approximation $\sin\phi \approx \phi$. The resulting equations are of the same type as those obtained for a system of coupled harmonic oscillators, treatments of which can be found in textbooks on classical mechanics.

It turns out that there is another special case which can be treated analytically. We call it the field theoretical limit because, as is explained below, we pass to an auxiliary system with an infinite number of degrees of freedom. Let us introduce a function $\phi(x, t)$, where x is a new real continuous variable (a coordinate along the wire). By assumption, this function is at least twice differentiable with respect to x, and is such that its values at the points $x = x_i$ are equal to the angles $\phi(x_i, t)$ introduced earlier. Hence, $\phi(x, t)$ smoothly interpolates between $\phi(x_i, t)$ for each i. Of course, for a given set of the angles one can find infinitely many such functions. For any such function the following identity holds

$$\phi(x_i - a, t) + \phi(x_i + a, t) - 2\phi(x_i, t) = \int_0^a ds_1 \int_{-a}^0 ds_2 \frac{\partial^2 \phi(s_1 + s_2 + x, t)}{\partial x^2}\bigg|_{x=x_i}.$$

Now comes the crucial assumption: we restrict our considerations to those motions of the pendulums for which there exists an interpolating function $\phi(x, t)$ of continuous variables x and t which satisfies

$$\int_0^a ds_1 \int_{-a}^0 ds_2 \frac{\partial^2 \phi(s_1 + s_2 + x, t)}{\partial x^2}\bigg|_{x=x_i} \approx a^2 \frac{\partial^2 \phi(x, t)}{\partial x^2}\bigg|_{x=x_i} \tag{1.3}$$

for all times t and at all points x_i. For example, this is the case when the second derivative of ϕ with respect to x is almost constant as x runs through the interval $[x_i - a, x_i + a]$, for all times t. With approximation (1.3) the identity written above can be replaced by the following approximate one

$$\phi(x_i - a, t) + \phi(x_i + a, t) - 2\phi(x_i, t) \approx a^2 \frac{\partial^2 \phi(x, t)}{\partial x^2}\bigg|_{x=x_i}.$$

Using this formula in (1.1) we obtain

$$mR^2 \frac{d^2 \phi(x_i, t)}{dt^2} \approx -mgR \sin \phi(x_i, t) + \kappa a \frac{\partial^2 \phi(x, t)}{\partial x^2}\bigg|_{x=x_i}. \tag{1.4}$$

Let us now suppose that our function $\phi(x, t)$ obeys the following partial differential equation,

$$mR^2 \frac{\partial^2 \phi(x, t)}{\partial t^2} = -mgR \sin \phi(x, t) + \kappa a \frac{\partial^2 \phi(x, t)}{\partial x^2}, \tag{1.5}$$

where $x \in [-Ma, Na]$, and

$$\phi(-Ma, t) = 0, \quad \phi(Na, t) = 2\pi n, \tag{1.6}$$

where n is the same integer as in (1.2). Then, it is clear that $\phi(x_i, t)$, $i = -M + 1, \ldots, N - 1$, obey (1.4). Also the boundary conditions (1.2) are satisfied. Hence, if condition (1.3) is satisfied we obtain an approximate solution of the initial Newtonian equations (1.1).

The nonlinear partial differential equation (1.5) is well-known in mathematical physics by the jocular name 'sine-Gordon equation' which alludes to the Klein–Gordon equation. This latter equation is a cornerstone of relativistic field theory—we shall discuss it in Sect. 1.3. The sine-Gordon equation can be transformed into its standard form by dividing by mgR, and by rewriting it with the new, dimensionless variables

$$\tau = \sqrt{\frac{g}{R}}t, \quad \xi = \sqrt{\frac{mgR}{\kappa a}}x, \quad \Phi(\xi, \tau) = \phi(x, t).$$

The resulting standard form of the sine-Gordon equation reads

$$\frac{\partial^2 \Phi(\xi, \tau)}{\partial \tau^2} - \frac{\partial^2 \Phi(\xi, \tau)}{\partial \xi^2} + \sin \Phi(\xi, \tau) = 0. \tag{1.7}$$

There are many mathematical theorems about (1.7) and its solutions. One of them says that in order to determine a unique solution uniquely, one must specify the initial data, that is, one must fix the values of $\Phi(\xi, \tau)$, and $\partial\Phi(\xi, \tau)/\partial\tau$ for a chosen instant of the rescaled time $\tau = \tau_0$ and for all ξ in the interval $[\xi_{-M}, \xi_N]$ (which corresponds to the interval $[x_{-M}, x_N]$). One must also specify the so-called boundary conditions, that is the values of Φ at the boundaries $\xi = \xi_{-M}$ and $\xi = \xi_N$ of the allowed range of ξ for all values of τ. In our case their form follows from conditions (1.2),

$$\Phi(\xi_{-M}, \tau) = 0, \quad \Phi(\xi_N, \tau) = 2\pi n. \tag{1.8}$$

In order to specify the initial data we have to provide an infinite set of real numbers (to define the values of $\Phi(\xi, \tau_0)$, $\partial\Phi(\xi, \tau)/\partial\tau|_{\tau=\tau_0}$) because ξ is a continuous variable. For this reason the dynamical system defined by the sine-Gordon equation has an infinite number of degrees of freedom. This system, called the sine-Gordon field, is mathematically represented by the function Φ, and the sine-Gordon equation is its equation of motion. The sine-Gordon field is said to be the effective field for the system of pendulums described above. Let us emphasize that the sine-Gordon effective field gives an accurate description of the dynamics of the original system only if condition (1.3) is satisfied. Such a reduction of the original problem to the dynamics of an effective field, or to a set of effective fields in other cases, has become an extremely efficient tool in theoretical investigations of many physical systems considered in condensed matter physics or particle physics.

Let us end this section with a few examples of nontrivial solutions of the sine-Gordon equation in its standard form (1.7). Let us assume that Φ does not depend on the rescaled time τ, that is, that $\Phi = \Phi(\xi)$—such solutions are referred to as static. Then, (1.7) reduces to the following ordinary differential equation

$$\Phi''(\xi) = \sin \Phi(\xi), \tag{1.9}$$

where $'$ denotes differentiation with respect to ξ. Multiplying this equation by Φ' we obtain

$$\frac{1}{2}(\Phi'^2)' = -(\cos\Phi)',$$

and after integration,

$$\frac{1}{2}\Phi'^2 = c_0 - \cos\Phi, \tag{1.10}$$

where c_0 is a constant. The boundary conditions (1.8) imply that

$$c_0 = 1 + \frac{1}{2}\Phi'^2(\xi_{-M}) = 1 + \frac{1}{2}\Phi'^2(\xi_N). \tag{1.11}$$

It follows that $c_0 \geq 1$, and that $\Phi'(\xi_{-M}) = \pm\Phi'(\xi_N)$.

Let us first consider the case $c_0 = 1$. The square root of (1.10) with $c_0 = 1$ gives

$$\Phi' = 2\sin(\frac{\Phi}{2}), \tag{1.12}$$

or

$$\Phi' = -2\sin(\frac{\Phi}{2}), \tag{1.13}$$

which can be easily integrated. Apart from the trivial solution $\Phi = 0$, there exist nontrivial solutions, denoted below by Φ_+ and Φ_-. Integrating (1.12) and (1.13) we find that

$$\ln|\tan(\frac{\Phi}{4})| = \pm(\xi - \xi_0),$$

where ξ_0 is an arbitrary constant, and the signs $+$ and $-$ correspond to (1.12) and (1.13), respectively. It follows that

$$\Phi_\pm(\xi) = \pm 4\arctan[\exp(\pm(\xi - \xi_0))] \mod 4\pi. \tag{1.14}$$

Formula (1.14) implies that $\Phi'_\pm(\xi) \neq 0$ for all finite ξ, and $\Phi'_\pm(\xi) \to 0$ if $\xi \to \infty$ or $\xi \to -\infty$. Therefore, conditions (1.11) can only be satisfied if

$$\xi_{-M} = -\infty, \quad \xi_N = +\infty.$$

With the help of the identity

$$\arctan(1/x) = \pi/2 - \arctan x,$$

one can show that formula (1.14) in fact gives two solutions which obey the conditions (1.8):

$$\Phi_\pm(\xi) = \pm 4\arctan(\exp(\xi - \xi_0)). \tag{1.15}$$

It is clear that

$$\lim_{\xi \to +\infty} \Phi_{\pm}(\xi) = \pm 2\pi.$$

Hence, the integer n in (1.8) can be equal to 0 or ± 1 ($n = 0$ corresponds to the trivial solution $\Phi = 0$).

Let us summarize the case for which $c_0 = 1$. Static solutions obeying the boundary conditions (1.8) exist only if the range of ξ is infinite, from $-\infty$ to $+\infty$, and then the nontrivial solutions have the form (1.15). The solution Φ_+ is called the soliton, and Φ_- the antisoliton. ξ_0 is called the location of the (anti-)soliton. There are no static solutions with $|n| > 1$.

Coming back to our system of pendulums, the solitonic solutions (1.15) are relevant if the condition (1.3) is satisfied. The two integrals on the l.h.s. of condition (1.3) can be rewritten as integrals of $\Phi''_{\pm}$ with respect to the dimensionless variables

$$\xi_{1,2} = \sqrt{\frac{mgR}{\kappa a}} x_{1,2}.$$

Then, the integration limits are given by 0 and $\pm\alpha$, where

$$\alpha = \sqrt{\frac{mgRa}{\kappa}}.$$

We see that condition (1.3) is certainly satisfied if

$$\alpha \to 0,$$

because in this limit the range of integration shrinks to a point. The value of the dimensionless parameter α can be made small by, e.g., choosing a wire with large κ or by putting the pendulums close to each other (small a). Furthermore, note that $\xi_N = \alpha x_N/a$, $\xi_{-M} = \alpha x_{-M}/a$, $x_{-M} = -Ma$ and; $x_n = Na$. It follows that ξ_N, ξ_{-M} can tend to $\pm\infty$, respectively, in the limit $\alpha \to 0$ only if $N, M \to \infty$. Thus, the number of pendulums has to be very large.

The case $c_0 > 1$ is a little bit more complicated. Equation (1.10) is equivalent to the following equations

$$\Phi' = \pm\sqrt{2c_0 - 2\cos\Phi}, \tag{1.16}$$

which give the following relations

$$\int_0^{\Phi(\xi)} ds \frac{1}{\sqrt{1 - c_0^{-1}\cos s}} = \pm\sqrt{2c_0}(\xi - \xi_{-M}). \tag{1.17}$$

These relations implicitly define the functions $\Phi(\xi)$ which obey (1.9). The integral on the l.h.s. of (1.17) can be related to an elliptic integral of the first kind (see, e.g.

[1]), and $\Phi(\xi)$ is then given by the inverse of the elliptic function. The constant c_0 is determined from the following equation, obtained by inserting the second of the boundary conditions (1.8) into formula (1.17):

$$\int_0^{2\pi n} ds \frac{1}{\sqrt{1 - c_0^{-1} \cos s}} = \pm\sqrt{2c_0}\,(\xi_N - \xi_{-M}). \tag{1.18}$$

Note that now ξ_{-M} and ξ_N have to be finite, otherwise the r.h.s. of this equation would be meaningless.

One may also solve (1.16) numerically. These equations are rather simple and can easily be tackled by computer algebra systems like *Maple*© or *Mathematica*©. Equations (1.16) are considered on the interval (ξ_{-M}, ξ_N). They are formally regarded as evolution equations with ξ playing the role of time. The boundary condition $\Phi(\xi_{-M}) = 0$ is now regarded as the initial condition for $\Phi(\xi)$. The constant c_0 is adjusted by trial and error until the calculation gives $\Phi(\xi_N) \approx 2\pi n$ to the desired accuracy. For example, choosing $\xi_{-M} = -10$ and $\xi_N = 10$ we have obtained $c_0 \approx 1.00000008$ for $n = \pm 1$, $c_0 \approx 1.0014$ for $n = \pm 2$, and $c_0 \approx 1.0398$ for $n = \pm 3$.

These solutions of the sine-Gordon equation with $c_0 > 1$ are pertinent to the physics of the system of pendulums when the parameter α has a sufficiently small value, as in the case $c_0 = 1$. For given natural numbers N and M, the values of ξ_{-M} and ξ_N are calculated from formulas $\xi_{-M} = -\alpha M$ and $\xi_N = \alpha N$. In the limit $\alpha \to 0$ with ξ_{-M} and ξ_N kept non vanishing and constant, the number of pendulums has to increase indefinitely.

1.2 Example B: The Electromagnetic Field

We have just seen an example of an effective field—the sine-Gordon field $\phi(x, t)$—introduced in order to provide an approximate description of our original physical system: the set of coupled pendulums. Now we shall see an example from another class of fields, called fundamental fields. Such fields are regarded as elementary dynamical systems—according to present day physics there are no experimental indications that they are effective fields for an underlying system. The fundamental fields appear in particular in particle physics and cosmology. Later on we shall see several such fields. Here we briefly recall the classical electromagnetic field. It should be stressed that we regard this field as a physical entity, a part of the material world. Our main goal is to show that the Maxwell equations can be reduced to a set of uncoupled wave equations.

According to 19th century physics, the electromagnetic field is represented by two functions $\vec{E}(t, \vec{x})$, $\vec{B}(t, \vec{x})$, the electric and magnetic fields respectively. Here $\vec{x}$ is a position vector in the three dimensional space R^3, and t is time. The fields obey the Maxwell equations of the form (we use the rationalized Gaussian units)

$$\begin{array}{llll} \text{(a)} & \mathrm{div}\vec{E} = \rho, & \text{(c)} & \mathrm{div}\vec{B} = 0, \\ \text{(b)} & \mathrm{rot}\vec{B} - \frac{1}{c}\frac{\partial \vec{E}}{\partial t} = \frac{1}{c}\vec{j}, & \text{(d)} & \mathrm{rot}\vec{E} + \frac{1}{c}\frac{\partial \vec{B}}{\partial t} = 0, \end{array} \tag{1.19}$$

where ρ is the electric charge density, and $\vec{j}$ is the electric current density. ρ and $\vec{j}$ are functions of t and $\vec{x}$, and c is the speed of light in the vacuum.

Suppose that there exist fields $\vec{E}(t, \vec{x})$, $\vec{B}(t, \vec{x})$ obeying Maxwell equations (1.19). Acting with the div operator on (1.19b), then using the identity div(rot) $\equiv 0$ and (1.19a), we obtain the following condition on the charge and current density

$$\frac{\partial \rho}{\partial t} + \mathrm{div}\vec{j} = 0. \tag{1.20}$$

This is a well-known continuity equation. It is equivalent to conservation of electric charge. From the mathematical viewpoint, it should be regarded as a consistency condition for the Maxwell equations—if it is not satisfied they do not have any solutions.

Equation (1.19c) is satisfied by any field $\vec{B}$ of the form

$$\vec{B} = \mathrm{rot}\vec{A}, \tag{1.21}$$

where $\vec{A}(t, \vec{x})$ is a (sufficiently smooth) function of $\vec{x}$. Conversely, one can prove that any field $\vec{B}$ which obeys (1.19c) has the form (1.21). From (1.21) and (1.19d) follows the identity

$$\mathrm{rot}(\vec{E} + \frac{1}{c}\frac{\partial \vec{A}}{\partial t}) = 0.$$

There is a mathematical theorem (the Poincaré lemma) which says that an identity of the form rot $\vec{X} = 0$ implies that the vector function $\vec{X}$ is the gradient of a scalar function σ, i.e. $\vec{X} = \nabla\sigma$. Therefore, there exists a function A_0 such that

$$\vec{E} + \frac{1}{c}\frac{\partial \vec{A}}{\partial t} = -\nabla A_0$$

(the minus sign is dictated by tradition). Thus,

$$\vec{E} = -\frac{1}{c}\frac{\partial \vec{A}}{\partial t} - \nabla A_0. \tag{1.22}$$

The functions A_0 and $\vec{A}$ are called gauge potentials for the electromagnetic field. Note that the choice of A_0 and $\vec{A}$ for a given electric and magnetic fields is not unique—instead of A_0 and $\vec{A}$ one may just as well take

$$\vec{A}'(t, \vec{x}) = \vec{A}(t, \vec{x}) - \nabla\chi(t, \vec{x}), \quad A_0'(t, \vec{x}) = A_0(t, \vec{x}) + \frac{1}{c}\frac{\partial \chi(t, \vec{x})}{\partial t}, \tag{1.23}$$

where $\chi(t, \vec{x})$ is a sufficiently smooth but otherwise arbitrary function of the indicated variables. This freedom of choosing the gauge potentials is called the gauge symmetry. Formulas (1.23) can be regarded as transformations of the gauge potentials, and are called the gauge transformations. Often they are called local gauge transformations in order to emphasize the fact that the function χ is space and time dependent. One should keep in mind the fact that the gauge transformations appear because we adopt the mathematical description of the electromagnetic field in terms of the potentials. The fields $\vec{E}$ and $\vec{B}$ do not change under these transformations. The potentials A_0, $\vec{A}$ and A_0', $\vec{A}'$ from formulas (1.23) describe the same physical situation. The freedom of performing the gauge transformations means that the potentials form a larger than necessary set of functions for describing a given physical configuration of the electromagnetic field. Nevertheless, it turns out that the description in terms of the potentials is a most economical one, especially in quantum theories of particles or fields interacting with the electromagnetic field. In fact, it has been commonly accepted that the best mathematical representation of the electromagnetic field—one of the basic components of the material world—is given by the gauge potentials A_0 and $\vec{A}$.

Expressing $\vec{E}$ and $\vec{B}$ by the gauge potentials we have explicitly solved (1.19c, d). Now let us turn to (1.19a, b). First, we use the gauge transformations to adjust the vector potential $\vec{A}$ in such a way that

$$\mathrm{div}\vec{A} = 0. \tag{1.24}$$

This condition is known as the Coulomb gauge condition. One can easily check that for any given $\vec{A}$ one can find a gauge function χ such that $\vec{A}'$ obeys the Coulomb condition, provided that $\mathrm{div}\vec{A}$ vanishes sufficiently quickly at the spatial infinity. For that matter, let us note that from a physical viewpoint it is sufficient to consider electric and magnetic fields which smoothly[1] vanish at the spatial infinity. For such fields there exist potentials A_0 and $\vec{A}$ which also smoothly vanish as $|\vec{x}| \to \infty$. It is quite natural to assume that the gauge transformations leave the potentials within this class. Therefore, we assume that the gauge function χ also smoothly vanishes at the spatial infinity. We might have assumed that it could approach a non vanishing constant in that limit. However, such a constant leads to a trivial gauge transformation because then the derivatives present in formulas (1.23) vanish. For this reason it is natural to choose this constant equal to zero. Note that now the Coulomb gauge condition determines the gauge completely. By this we mean that if both $\vec{A}$ and $\vec{A}'$, which are related by the local gauge transformation (1.23), obey the Coulomb gauge condition, then $\chi = 0$, that is the two potentials coincide. This follows from the facts that if (1.24) is satisfied by $\vec{A}$ and $\vec{A}'$ then χ obeys the Laplace equation, $\Delta\chi = 0$, and the only nonsingular solution of this equation which vanishes at the spatial infinity is $\chi = 0$.

[1]Here this means that all derivatives of the fields with respect to the Cartesian coordinates x^i also vanish at the spatial infinity.

The condition that χ vanishes at the spatial infinity is also welcome for another reason—it makes a clear distinction between (local) gauge transformations and global transformations. Global transformations will be introduced in Chap. 3. They are given by χ which are constant in time and space. Such transformations can act nontrivially on fields other than the electromagnetic field. With the definitions we have adopted, the global transformations are not contained in the set of gauge transformations.

Equations (1.19a, b) are reduced in the Coulomb gauge to the following equations

$$\Delta A_0 = -\rho, \qquad \frac{1}{c^2}\frac{\partial^2 \vec{A}}{\partial t^2} - \Delta \vec{A} + \frac{1}{c}\nabla\frac{\partial A_0}{\partial t} = \frac{1}{c}\vec{j}. \tag{1.25}$$

The solution of the first equation has the form

$$A_0(t, \vec{x}) = \frac{1}{4\pi}\int d^3x' \frac{\rho(t, \vec{x}\,')}{|\vec{x} - \vec{x}\,'|}, \tag{1.26}$$

provided that ρ vanishes sufficiently quickly at the spatial infinity to ensure that the integral is convergent. The r.h.s. of formula (1.26) is often denoted by $-\Delta^{-1}\rho$. Because the potential A_0 is just given by integral (1.26)—there is not any evolution equation for it to be solved—it is not a dynamical variable. In the final step, formula (1.26) is used to eliminate A_0 from the second of the equations (1.25). We also eliminate $\partial\rho/\partial t$ with the help of continuity equation (1.20). The resulting equation for $\vec{A}$ can be written in the form

$$\frac{1}{c^2}\frac{\partial^2 \vec{A}}{\partial t^2} - \Delta \vec{A} = \frac{1}{c}\vec{j}_T, \tag{1.27}$$

where

$$\vec{j}_T = \vec{j} - \nabla(\Delta^{-1}\mathrm{div}\vec{j}), \tag{1.28}$$

and

$$\Delta^{-1}\mathrm{div}\vec{j}(t, \vec{x}) = -\frac{1}{4\pi}\int d^3x' \frac{\mathrm{div}\vec{j}(t, \vec{x}\,')}{|\vec{x} - \vec{x}\,'|}.$$

Of course, we assume that $\mathrm{div}\vec{j}$ vanishes sufficiently quickly at the spatial infinity. $\vec{j}_T$ is called the transverse part of the external current $\vec{j}$. The reason for such a name is that

$$\mathrm{div}\vec{j}_T \equiv 0, \tag{1.29}$$

as it immediately follows from the definition of $\vec{j}_T$. For the same reason, the potential $\vec{A}$ which obeys the Coulomb gauge condition is called the transverse vector potential.

Note that identity (1.29) is a necessary condition for the existence of the solutions of (1.27)—applying the div operator to both sides of (1.27) and using the Coulomb condition we would obtain a contradiction if (1.29) were not true.

To summarize, the set of Maxwell equations (1.19) has been reduced to (1.27) together with the Coulomb gauge condition (1.24). Equation (1.27) determines the time evolution of the electromagnetic field. It plays the same role as Newton's equation in classical mechanics. From a mathematical viewpoint, equation (1.27) is a set of three linear, inhomogeneous, partial differential equations: one equation for each component A^i of the vector potential.[2] These equations are decoupled, that is they can be solved independently from each other. They are called wave equations.

As in the case of the sine-Gordon equation (1.7), in order to uniquely determine a solution of (1.27) we have to specify the initial data at the time t_0:

$$\vec{A}(t_0, \vec{x}) = \vec{f}_1(\vec{x}), \quad \left.\frac{\partial \vec{A}(t, \vec{x})}{\partial t}\right|_{t=t_0} = \vec{f}_2(\vec{x}), \tag{1.30}$$

where $\vec{f}_1$ and $\vec{f}_2$ are given vector fields, vanishing at the spatial infinity. Moreover, in order to ensure that the Coulomb gauge condition is satisfied at the time $t = t_0$, we assume that

$$\operatorname{div} \vec{f}_1 = 0, \quad \operatorname{div} \vec{f}_2 = 0. \tag{1.31}$$

It turns out that conditions (1.31) and equation (1.27) imply that $\operatorname{div}\vec{A} = 0$ for all times t. The point is that equation (1.27) implies that $\operatorname{div}\vec{A}$ obeys the homogeneous equation

$$\frac{1}{c^2}\frac{\partial^2 \operatorname{div}\vec{A}}{\partial t^2} - \Delta(\operatorname{div}\vec{A}) = 0.$$

Due to the assumptions (1.31) the initial data for this equation are homogeneous ones, that is

$$\operatorname{div}\vec{A}|_{t=t_0} = 0, \quad \partial_t \operatorname{div}\vec{A}|_{t=t_0} = 0,$$

where ∂_t is a short notation for the partial derivative $\partial/\partial t$. We shall see in the next section that all this implies that

$$\operatorname{div}\vec{A} = 0$$

for all times. In consequence, we do not have to worry about the Coulomb gauge condition provided that the initial data (1.30) obey the conditions (1.31)—the Coulomb gauge condition has been reduced to a constraint on the initial data.

---

[2] We adhere to the convention that vectors denoted by the arrow have components with upper indices.

1.3 Solutions of the Klein–Gordon Equation

The considerations of the electromagnetic field have led us to an evolution equation of the form

$$\Box\, \phi = \eta(t, \vec{x}), \tag{1.32}$$

where

$$\Box \equiv \Delta - \frac{1}{c^2}\frac{\partial^2}{\partial t^2},$$

ϕ is a function of $(t, \vec{x})$, and η is an a priori given function, called the source. The wave equation (1.32) is a particular case of the more general Klein–Gordon equation

$$\Box\, \phi - m^2\phi = \eta(t, \vec{x}), \tag{1.33}$$

where m^2 is a real, non-negative constant of the dimension cm^{-2}, and ϕ is a real or complex function. The Klein–Gordon equation is the basic evolution equation in relativistic field theory. It also appears in non relativistic settings. For example, sine-Gordon equation (1.7) reduces to the Klein–Gordon equation with just one spatial variable ξ if we consider Φ close to 0, because in this case $\sin \Phi$ can be approximated by Φ. Therefore, one should be acquainted with solutions of the Klein–Gordon equation.

Let us introduce concise, four-dimensional relativistic notation:

$$x = (ct, \vec{x}), \quad k = (k_0, \vec{k}), \quad kx = ck_0 t - \vec{k}\vec{x}, \quad d^4x = cd^3x dt, \quad d^4k = d^3k dk_0.$$

Here k_0 is a real variable, and $\vec{k}$ is a real 3-dimensional vector called the wave vector. k_0 and $\vec{k}$ have the dimension cm^{-1}. $\omega = ck_0$ is a frequency. Furthermore, we shall often use $x^0 = ct$ instead of the time variable t and call it time too. This notation reflects the Lorentz invariant structure of space-time. In particular, the form of kx corresponds to the diagonal metric tensor of the space-time $(\eta_{\mu\nu}) = \mathrm{diag}(1, -1, -1, -1)$, where diag denotes the diagonal matrix with the listed elements on its diagonal. Note that kx is dimensionless.

Because the Klein–Gordon equation is linear with respect to ϕ and has constant coefficients, we may use the Fourier transform technique for solving it. We denote by $\tilde{\phi}(k)$ the Fourier transform of $\phi(x)$. It is defined as follows:

$$\tilde{\phi}(k) = \int d^4x \; e^{ikx} \phi(x). \tag{1.34}$$

The inverse Fourier formula has the form

$$\phi(x) = \frac{1}{(2\pi)^4} \int d^4k \; e^{-ikx} \tilde{\phi}(k). \tag{1.35}$$

Analogously,

$$\tilde{\eta}(k) = \int d^4x \; e^{ikx} \eta(x).$$

The Klein–Gordon equation is equivalent to the following algebraic (not differential!) equation for $\tilde{\phi}$

$$(k_0^2 - \vec{k}^{\,2} - m^2)\, \tilde{\phi}(k) = \tilde{\eta}(k). \tag{1.36}$$

Its solutions should be sought in a space of generalized functions. An excellent introduction to the theory of generalized functions with its applications to linear partial differential equations can be found in, e.g., [2]. Some pertinent facts can be found in Appendix A.

One can prove that the most general solution of (1.36) has the form

$$\tilde{\phi}(k) = \text{``}\, \frac{\tilde{\eta}(k)}{k_0^2 - \vec{k}^2 - m^2} \,\text{''} + C(k_0, \vec{k})\, \delta(k_0^2 - \vec{k}^2 - m^2), \tag{1.37}$$

where $C(k_0, \vec{k})$ is an arbitrary smooth function of the indicated variables. The first term on the r.h.s. denotes a particular solution of the inhomogeneous equation (1.36). We have put the quotation marks around it because in fact that term written as it stands is not correct. We explain and solve this problem shortly. The second term on the r.h.s. gives the general solution of the homogeneous equation

$$(k_0^2 - \vec{k}^{\,2} - m^2)\, \tilde{\phi}(k) = 0.$$

Formula (1.37) is in accordance with the well-known fact that the general solution of an inhomogeneous linear equation can always be written as the sum of a particular solution of that equation and of a general solution to the corresponding homogeneous equation.

The problem with the term in quotation marks is that it is not a generalized function. In consequence, its Fourier transform, formula (1.35), does not have to exist, and indeed, it does not exist. One can see this easily by looking at the integral over k_0—there are non integrable singularities of the integrand at $k_0 = \pm\omega(\vec{k})/c$, where

$$\omega(\vec{k}) = c\,\sqrt{\vec{k}^{\,2} + m^2}. \tag{1.38}$$

In order to obtain the correct formula for the solution we first find a generalized function $\tilde{G}(k)$ which obeys the equation

$$(k_0^2 - \vec{k}^{\,2} - m^2)\, \tilde{G}(k) = 1. \tag{1.39}$$

The corresponding $G(x)$ is calculated from a formula analogous to (1.35). It obeys the following equation

$$(\Box - m^2)\, G(x) = \delta(x), \tag{1.40}$$

and is called the Green's function of the Klein–Gordon equation. Knowing $\tilde{G}(k)$, we may replace the " " term by the mathematically correct expression

$$\text{“}\frac{\tilde{\eta}(k)}{k_0^2 - \vec{k}^{\,2} - m^2}\text{”} \rightarrow \tilde{\eta}(k)\tilde{G}(k),$$

provided that $\tilde{\eta}$ is a smooth function of k^0 and $\vec{k}$.

Important Green's functions for the Klein–Gordon equation have Fourier transforms of the form

$$\tilde{G}(k) = \frac{c^2}{2\omega(\vec{k})}\left(\frac{1}{ck_0 - \omega(\vec{k}) \pm i0_+} - \frac{1}{ck_0 + \omega(\vec{k}) \pm i0_+}\right). \tag{1.41}$$

The meaning of the symbol $\pm i0_+$ is explained in the Appendix. The choice $+i0_+$ in both terms of formula (1.41) gives the so called retarded Green's function

$$G_R(x-y) = \frac{c^2}{(2\pi)^4}\int d^4k\, \frac{e^{-ik(x-y)}}{2\omega(\vec{k})}\left(\frac{1}{ck_0 - \omega(\vec{k}) + i0_+} - \frac{1}{ck_0 + \omega(\vec{k}) + i0_+}\right). \tag{1.42}$$

The integral over k_0 can be performed with the help of contour integration in the plane of complex k_0. The trick consists of completing the line of real k_0 to a closed contour by adding upper (lower) semicircle with the center at $k_0 = 0$ and infinite radius when $x^0 - y^0 < 0$ ($x^0 - y^0 > 0$). We obtain

$$G_R(x-y) = \frac{-ic}{2(2\pi)^3}\Theta(x^0 - y^0)\int \frac{d^3k}{\omega(\vec{k})}\left(e^{-ik(x-y)} - e^{ik(x-y)}\right)\Big|_{k_0=\omega(\vec{k})/c}, \tag{1.43}$$

where $\Theta(x^0 - y^0)$ denotes the Heaviside step function.[3]

The Green's function G_R is used in order to obtain a particular solution of the inhomogeneous Klein–Gordon equation, denoted below by ϕ_η. Namely,

$$\phi_\eta(x) = \int d^4y\, G_R(x-y)\eta(y). \tag{1.44}$$

This solution is causal in the classical sense: the values of $\phi_\eta(x^0, \vec{x})$ at a certain fixed instant x^0 are determined by values of the external source $\eta(y^0, \vec{y})$ at earlier times, i.e., $y^0 \leq x^0$. More detailed analysis shows that the contributions come only from the interior and boundaries of the past light-cone with its tip at the point x, that is,

[3] $\Theta(x) = 1$ for $x > 1$, $\Theta(x) = 0$ for $x < 0$. The value of $\Theta(0)$ does not have to be specified because the step function is used under the integral. Formally, the step function is a generalized function, and for such functions their values at a given single point are not defined. Therefore, the question, "what is the value of $\Theta(0)$?" is meaningless.

from y such that $(x-y)^2 \geq 0$ and $x^0 - y^0 \geq 0$. This can be seen from the following formula, see Appendix 2 in [3],

$$G_R(x) = -\frac{1}{2\pi}\Theta(x^0)\left[\delta(x^2) - \Theta(x^2)\frac{m}{2\sqrt{x^2}}J_1(m\sqrt{x^2})\right],$$

where $x^2 = (x^0)^2 - \vec{x}^{\,2}$, and J_1 is a Bessel function. Therefore, waves of the field emitted from a spatially localized source η travel with velocity not greater than the velocity of light in the vacuum c. Choosing the $-i0_+$ in formula (1.41) we would obtain the so called advanced Green's function, which is anti-causal—in this case $\phi_\eta(x)$ is determined by values of $\eta(y)$ in the future light cone, $y^0 \geq x^0$ and $(x-y)^2 \geq 0$. In general, the choice of Green's function is motivated by the underlying physical problem. On purely mathematical grounds there are infinitely many Green's functions. All have the form $G_R(x) + \phi_0(x)$, where $\phi_0(x)$ is a particular solution of the homogeneous Klein-Gordon equation.

Now that we have found a particular solution for the inhomogeneous Klein-Gordon equation, let us turn our attention to finding the general solution of the homogeneous Klein–Gordon equation. The second term in formula (1.37) gives

$$\phi_0(x) = \frac{1}{(2\pi)^4}\int d^4k\, e^{-ikx} C(k^0, \vec{k})\delta(k_0^2 - \vec{k}^2 - m^2). \tag{1.45}$$

With the help of formula

$$\delta(k_0^2 - \vec{k}^2 - m^2) = \frac{\delta(k_0 - \omega(\vec{k})/c)}{2\omega(\vec{k})/c} + \frac{\delta(k_0 + \omega(\vec{k})/c)}{2\omega(\vec{k})/c},$$

ϕ_0 can be written in the form

$$\phi_0(x) = \int \frac{d^3k}{\sqrt{2(2\pi)^3\omega(\vec{k})}}\left(a_+(\vec{k})e^{-ikx} + a_-(\vec{k})e^{ikx}\right)\Big|_{k_0=\omega(\vec{k})/c}, \tag{1.46}$$

where

$$a_\pm(\vec{k}) = \frac{C(\pm\omega(\vec{k}), \pm\vec{k})}{(2\pi)^2\sqrt{4\pi\omega(\vec{k})}}.$$

The functions $a_\pm(\vec{k})$ are called the momentum space amplitudes of the field ϕ_0. The part of $\phi_0(x)$ with a_+ (a_-) is called the positive (negative) frequency part of the Klein–Gordon field. If we require that all values of $\phi(x)$ are real, we have to restrict the amplitudes $a_\pm$ by the condition

$$a_+^*(\vec{k}) = a_-(\vec{k}), \tag{1.47}$$

where * denotes the complex conjugation.

Formula (1.46), regarded as a relation between the amplitudes and the field ϕ_0, can be inverted. It is convenient first to introduce the operator $\hat{P}_{\vec{k}}(y_0)$,

$$\hat{P}_{\vec{k}}(y^0)\phi(y^0,\vec{y}) = i\int d^3y\left(f^*_{\vec{k}}(y^0,\vec{y})\frac{\partial\phi(y^0,\vec{y})}{\partial y^0} - \frac{\partial f^*_{\vec{k}}(y^0,\vec{y})}{\partial y^0}\phi(y^0,\vec{y})\right), \tag{1.48}$$

where $f_{\vec{k}}$ is a normalized plane wave

$$f_{\vec{k}}(y^0,\vec{y}) = \frac{e^{-iky}}{\sqrt{2(2\pi)^3\omega(\vec{k})}} \tag{1.49}$$

with $k_0 = \omega(\vec{k})/c$. Simple calculations show that

$$\hat{P}_{\vec{k}}(y^0)f_{\vec{k}'}(y^0,\vec{y}) = \delta(\vec{k}-\vec{k}'),\quad \hat{P}_{\vec{k}}(y^0)f^*_{\vec{k}'}(y^0,\vec{y}) = 0, \tag{1.50}$$

for any choice of y^0. It follows that

$$\hat{P}_{\vec{k}}(y^0)\phi_0(y^0,\vec{y}) = a_+(\vec{k}). \tag{1.51}$$

Note that there is no restriction on the choice of y^0 present on the l.h.s. of this formula.

Formulas (1.51) and (1.47) inserted in formula (1.46) give the following identity

$$\phi_0(x) = \int d^3k\left(f_{\vec{k}}(x)\hat{P}_{\vec{k}}(y^0)\phi_0(y^0,\vec{y}) + \text{c.c.}\right). \tag{1.52}$$

Here c.c. stands for the complex conjugate of the preceding term. At this point it is convenient to define several new generalized functions:

$$\Delta^{(+)}(x) = -\frac{ic}{2(2\pi)^3}\int\frac{d^3k}{\omega(\vec{k})}e^{-ikx}\Big|_{k_0=\omega(\vec{k})/c},$$
$$\Delta^{(-)}(x) = (\Delta^{(+)}(x))^*,\quad \Delta(x) = \Delta^{(+)}(x) + \Delta^{(-)}(x), \tag{1.53}$$

called the Pauli–Jordan functions. They obey the homogeneous Klein–Gordon equation. After simple manipulations, identity (1.52) can be rewritten in the following form

$$\phi_0(x) = -\int d^3y\left[\Delta(x-y)\frac{\partial\phi_0(y)}{\partial y^0} + \frac{\partial\Delta(x-y)}{\partial x^0}\phi_0(y)\right]. \tag{1.54}$$

This very important formula gives an explicit solution to the homogeneous Klein–Gordon equation in terms of the initial data. We just take $y^0 = ct_0$, where t_0 is the time at which $\phi_0(y^0,\vec{y})$ and $\partial\phi_0(y^0,\vec{y})/\partial y^0|_{y^0=ct_0}$ are explicitly specified as the initial data. In particular, we see from formula (1.54) that vanishing initial data imply that $\phi_0(x) = 0$. This result was used at the end of the previous section.

The explicit formula for the Pauli–Jordan function $\Delta(x)$ has the form (Appendix 2 in [3])

$$\Delta(x) = -\frac{1}{2\pi}\mathrm{sign}(x^0)\left[\delta(x^2) - \Theta(x^2)\frac{m}{2\sqrt{x^2}}J_1(m\sqrt{x^2})\right],$$

where

$$\mathrm{sign}(x^0) = +1 \text{ if } x^0 > 0, \quad \mathrm{sign}(x^0) = -1 \text{ if } x^0 < 0.$$

One can see from this formula that the initial data are propagated in space with the velocity not greater than c. In particular, if the initial data taken at the time t_0 vanish outside a certain bounded region V in space, then $\phi_0(x)$ at later times $t > t_0$ certainly vanishes at all points $\vec{x}$ which cannot be reached by a light signal emitted from V. Another implication of formula (1.54) is the Huygens principle: the value of ϕ_0 at the point $\vec{x}$ at the time t is a linear superposition of contributions from all points in space for which the initial data do not vanish (and which do not lie too far from $\vec{x}$). This principle reflects the linearity of the Klein–Gordon equation.

Exercises

1.1 (a) Check that the functions

$$\Phi_{+,v}(\xi,\tau) = 4\arctan\left(\exp[\gamma(\xi - v\tau)]\right),$$

$$\Phi_{+,+}(\xi,\tau) = 4\arctan\left(\frac{v\sinh(\gamma\xi)}{\cosh(v\gamma\tau)}\right), \quad \Phi_{+,-}(\xi,\tau) = 4\arctan\left(\frac{\sinh(v\gamma\tau)}{v\cosh(\gamma\xi)}\right),$$

where $\gamma = 1/\sqrt{1-v^2}$ and v is a real parameter such that $0 \leq |v| < 1$, are solutions of the sine-Gordon equation (1.7). Justify their interpretation: $\Phi_{+,v}$ represents the soliton moving with constant velocity v, $\Phi_{+,+}$—two solitons, $\Phi_{+,-}$—soliton + antisoliton pair.

(b) Comparing the asymptotic forms of solutions at $\tau \to -\infty$ and $\tau \to +\infty$ show that there is a repulsive force between the two solitons, and an attractive one in the case of the soliton + antisoliton pair.

(c) Check that the substitution $v = iu$, u-real, in the $\Phi_{+,-}$ solution gives a real-valued solution of the sine-Gordon equation which is periodic in time. Interpret this solution as a bound state of the soliton with the antisoliton (called the breather).

Hints: In the cases of $\Phi_{+,+}$, $\Phi_{+,-}$ consider the limits $\tau \to \pm\infty$. Use the identity

$$\arctan\frac{x-y}{1+xy} = \arctan x - \arctan y.$$

In order to show the presence of the forces, analyze shifts of the position of the soliton and the antisoliton with respect to the trajectory of the single (anti-)soliton.

1.2 (a) The advanced Green's function G_A for the Klein–Gordon equation is obtained by choosing $-i0_+$ in both terms in formula (1.41). Obtain formula analogous to (1.43) in this case.
(b) Prove also that $G_F(x)$ defined as

$$G_F(x) = \frac{1}{(2\pi)^4} \int d^4k \, \frac{e^{-ikx}}{k^2 - m^2 + i0_+},$$

where $k^2 = k_0^2 - \vec{k}^2$, is another Green's function for the Klein–Gordon equation. G_F is related to the free propagator of the scalar field, and it plays an important role in the quantum theory of such fields. What is the choice of the signs $\pm$ in formula (1.41) in this case?

1.3 Using G_R, prove that

$$\vec{A}(t, \vec{x}) = \frac{1}{4\pi c} \int d^3y \, \frac{\vec{j}_T(\underline{t}, \vec{y})}{|\vec{x} - \vec{y}|},$$

where $\underline{t} = t - |\vec{x} - \vec{y}|/c$, is a solution of the wave equation (1.27).

Chapter 2
The Euler–Lagrange Equations and Noether's Theorem

Abstract The stationary action principle and the general form of the Euler–Lagrange equations. The notion of symmetry in classical field theory. Noether's conserved currents.

2.1 The Euler–Lagrange Equations

We know from classical mechanics that equations of motion for many systems can be derived from the stationary action principle. This fact is rather mysterious if regarded on a purely classical level. It turns out that it is actually a simple consequence of the fact that such classical systems can be regarded as classical limits of quantum models. We shall see later on in Chap. 11 how the classical action appears in the quantum theory. This situation does not change when we pass to field theory, that is if the number of degrees of freedom is infinite.

Let us recall some basic facts about the stationary action principle in classical mechanics. For simplicity, we consider the case of a particle with just one degree of freedom, that is with a one-dimensional configuration space. Let q be a coordinate on that space. The trajectory of the particle is given by the function of time $q(t)$. The action functional is defined on a space of smooth trajectories $q(t)$. By definition, it has the following form

$$S[q] = \int_{t'}^{t''} dt\; L(q(t), \dot{q}(t); t), \tag{2.1}$$

where L is called the Lagrange function. All considered trajectories $q(t)$ start from a point q' at the time t', and end at a point q'' at the time t'',

$$q(t') = q', \quad q(t'') = q''. \tag{2.2}$$

H. Arodź and L. Hadasz, *Lectures on Classical and Quantum Theory of Fields*, Graduate Texts in Physics, DOI 10.1007/978-3-319-55619-2_2

The stationary action principle says that the actual (physical) trajectory $q_{phys}(t)$ of the particle obeys the condition

$$\left.\frac{\delta S[q]}{\delta q(t)}\right|_{q(t)=q_{phys}(t)} = 0. \tag{2.3}$$

The object on the l.h.s. of this formula is called the functional, or variational, derivative of the action functional S with respect to $q(t)$. Such a derivative is defined as follows. Consider a family of trajectories of the form $q(t) + \delta q(t)$, where the trajectory $q(t)$ is fixed, and $\delta q(t)$ is an arbitrary smooth function of t such that

$$\delta q(t') = 0 = \delta q(t''). \tag{2.4}$$

Thus, the trajectory $q(t) + \delta q(t)$ obeys conditions (2.2). It is also assumed that all time derivatives of $\delta q(t)$ obey the conditions (2.4). Next, we consider the difference $S[q + \epsilon\delta q] - S[q]$, where ϵ is a real number. The functional derivative $\delta S/\delta q$ is defined by the following formula

$$\lim_{\epsilon\to 0} \frac{S[q + \epsilon\delta q] - S[q]}{\epsilon} = \int_{t'}^{t''} dt\, \frac{\delta S[q]}{\delta q(t)}\, \delta q(t). \tag{2.5}$$

In the case of the action functional (2.1) with a smooth Lagrange function[1] this definition gives

$$\frac{\delta S[q]}{\delta q(t)} = \frac{\partial L}{\partial q(t)} - \frac{d}{dt}\left(\frac{\partial L}{\partial \dot{q}(t)}\right), \tag{2.6}$$

and the condition (2.3) acquires the well-known form of the Euler–Lagrange equation for $q_{phys}(t)$.

As is known from courses on classical mechanics, this formalism can be easily generalized to the case of an arbitrary finite number of degrees of freedom, when instead of the single coordinate q we have a finite number of them, $q^i(t), i = 1 \ldots n$.

The Lagrangian formalism does not guarantee that the Euler–Lagrange equations derived from a given Lagrange function will lead to acceptable equations of motion, from which one will be able to predict the actual trajectory of the particle. For example, $L = q$ gives the Euler–Lagrange 'equation' of the form $1 = 0$. Another such example: $L = \dot{q} f(q)$ gives $0 = 0$ as the Euler–Lagrange equation for any smooth function f. In the former example there is no solution, while in the latter case an arbitrary smooth function[2] $q(t)$ is a solution, therefore the equation has no predictive power. The second example is an extreme case of degenerate Euler–Lagrange equations.

Another example of problematic Euler–Lagrange equations can appear when the number of degrees of freedom is greater than 1. In the following, we use the short

[1] L is regarded as a function of q, $\dot{q}$ and t.

[2] This assumption has been made in the derivation of the Euler–Lagrange equation (2.6).

notation $q = (q^k)$ for the full set of coordinates on the configuration space. The Euler–Lagrange equations can be written in the following form

$$H_{ik}(q, \dot{q})\ddot{q}^k = \frac{\partial L}{\partial q^i} - B_{ik}(q, \dot{q})\dot{q}^k - \frac{\partial^2 L}{\partial \dot{q}^i \partial t}, \tag{2.7}$$

where

$$H_{ik} = \frac{\partial^2 L}{\partial \dot{q}^i \partial \dot{q}^k}, \quad B_{ik} = \frac{\partial^2 L}{\partial \dot{q}^i \partial q^k}.$$

In mathematical theorems about the existence and uniqueness of solutions of a system of ordinary differential equations, it is usually assumed that the system can be written in Newtonian form, that is with extracted highest order derivatives,

$$\ddot{q}^k = F^k(q, \dot{q}). \tag{2.8}$$

This is possible if the symmetric matrix $\hat{H} = (H_{ik})$ is nonsingular, $\det\hat{H} \neq 0$. In the opposite case, there exists at least one eigenvector $e_0 = (e_0^k)$ of $\hat{H}$ with the eigenvalue equal to 0,

$$H_{ik}e_0^k = 0.$$

Let us multiply both sides of (2.7) by e_0^i and sum over i. We obtain the following condition

$$e_0^i \left(\frac{\partial L}{\partial q^i} - B_{ik}(q, \dot{q})\dot{q}^k - \frac{\partial^2 L}{\partial \dot{q}^i \partial t} \right) = 0. \tag{2.9}$$

The eigenvector e_0 is a function of $(q, \dot{q})$ because $\hat{H}$ depends on these variables. Therefore, condition (2.9) is a relation between q^i and $\dot{q}^k$. Notice that its existence follows from properties of the Lagrange function only. For this reason it is called a primary Lagrangian constraint. If there were other eigenvectors of $\hat{H}$ with zero eigenvalue we would obtain more of these constraints. The total number of non-trivial primary constraints cannot be larger than the number of linearly independent eigenvectors of $\hat{H}$ with zero eigenvalues.[3] If the matrix $\hat{H}$ has K such eigenvectors, we can extract from the Euler–Lagrange equations (2.7) only $n - K$ accelerations $\ddot{q}^i$. The existence and uniqueness of the solutions in such a case is not obvious. These problems are analyzed in a branch of classical mechanics called the theory of constrained systems. Analogously, there exist constrained field theoretic systems. We shall see examples of such systems in Chap. 4.

As a final remark about the Euler–Lagrange equations in classical mechanics, let us note that the stationary action principle, which follows from quantum mechanics, has led to the variational problem in which, by assumption, both ends of the physical trajectory q_{phys} are fixed. Such a problem is not always equivalent to the initial value problem, in which we fix the initial position and velocity. For example, if the

[3] It may happen that some of the relations (2.9) reduce to trivial identities like $0 = 0$.

configuration space of a particle is a circle, the variational problem has infinitely many solutions, while the initial value problem has just one.

Field theory is obtained when the number of degrees of freedom increases to infinity, $n \to \infty$. In this case, however, more popular is a description in terms of functions of continuous variables. Thus, $q(t)$ is replaced by a set of N functions of $x = (t, \vec{x})$, denoted in this chapter by $u_a(x)$ where $a = 1 \dots N$. We assume that $(t, \vec{x}) \in R^4$. This is sufficient for most applications in the theory of particles or condensed matter systems, but in cosmology with a strong gravitational field one has to use more general Riemann spaces with non vanishing curvature instead of R^4. A typical action functional has the following form

$$S[u] = \int_{t'}^{t''} dt \int_{R^3} d^3x \, \mathcal{L}(u_a(x), \partial_\mu u_a(x); x), \tag{2.10}$$

where $\partial_\mu u_a = \partial u_a / \partial x^\mu$. $\mathcal{L}$ is called the density of the Lagrange function, or the Lagrangian for short. In most cases it does not contain second or higher order derivatives of the fields $u_a(x)$. An explicit dependence on x usually appears when the fields u_a, which are the dynamical variables, interact with certain external fields, which are represented by explicitly given functions of x. The external fields are fixed a priori; there is no equation of motion for them to be solved.

The stationary action principle says that the physical fields u_a obey the Euler–Lagrange equations

$$\frac{\delta S[u]}{\delta u_a(x)} = 0, \tag{2.11}$$

where, again, the ends of all trajectories $u_a(t, \vec{x})$ of the fields are fixed, that is

$$u_a(t', \vec{x}) = u_a'(\vec{x}), \quad u_a(t'', \vec{x}) = u_a''(\vec{x}). \tag{2.12}$$

Here u_a' and u_a'' are a priori given functions of $\vec{x}$. Moreover, boundary conditions for u_a at the spatial infinity have to be specified, that is we assume that

$$\lim_{|\vec{x}| \to \infty} u_a(t, \vec{x}) = u_a^\infty(t, \theta, \phi), \tag{2.13}$$

where $u_a^\infty(t, \theta, \phi)$ is an a priori fixed function of time t, and of the spherical angles θ, ϕ which parameterize the sphere of infinite radius. The definition of the functional derivative in the case of the fields u_a essentially coincides with (2.5). In the new notation, it is written as

$$\lim_{\epsilon \to 0} \frac{S[u_a(x) + \epsilon \delta u_a(x)] - S[u_a]}{\epsilon} = \int_{t'}^{t''} dt \int_{R^3} d^3x \, \frac{\delta S[u_a]}{\delta u_b(t, \vec{x})} \delta u_b(t, \vec{x}), \tag{2.14}$$

where the test functions $\delta u_b(t, \vec{x})$ vanish together with all their partial derivatives when $t = t'$, $t = t''$, or when $|\vec{x}| \to \infty$. Then, the trajectories $u + \epsilon \delta u$ obey the conditions (2.12), (2.13). Definition (2.14) applied to the action functional (2.10) gives

$$\frac{\delta S[u_b]}{\delta u_a(t,\vec{x})} = \frac{\partial \mathcal{L}}{\partial u_a(t,\vec{x})} - d_\mu \left(\frac{\partial \mathcal{L}}{\partial (u_{a,\mu}(t,\vec{x}))} \right). \tag{2.15}$$

In this formula d_μ denotes the total derivative with respect to x^μ—the variables x^μ can appear in $\frac{\partial \mathcal{L}}{\partial (u_{a,\mu}(x))}$ through $u_a(x)$ and $u_{a,\mu}(x) = \partial_\mu u_a(x)$, as well as explicitly (that is through the external fields).

We shall see many examples of Euler–Lagrange equations (2.11) in field theory in the following chapters. The examples considered in the previous chapter are also of the Lagrange type:

$$\mathcal{L}_{\text{sine-Gordon}} = \frac{1}{2}(\partial_\tau \Phi)^2 - \frac{1}{2}(\partial_\xi \Phi)^2 + \cos\Phi - 1, \tag{2.16}$$

$$\mathcal{L}_{\text{Maxwell}} = -\frac{1}{4} F^{\mu\nu} F_{\mu\nu} - \frac{1}{c} j_\mu(x) A^\mu(x), \tag{2.17}$$

where

$$F_{\mu\nu} = \partial_\mu A_\nu - \partial_\nu A_\mu, \quad j^0 = c\,\rho, \quad \partial_0 = \frac{1}{c}\partial_t.$$

Let us end this section with three short remarks. First, various Lagrangians can give identical Euler–Lagrange equations. For example,

$$\mathcal{L}' = \mathcal{L} + d_\mu F^\mu(u_a(x), x) \tag{2.18}$$

gives the same Euler-Lagrange equations as $\mathcal{L}$.

Second, we have assumed in the field theory case that the Lagrangian $\mathcal{L}$ depends on $u_a(x)$ and $\partial_\mu u_a(x)$ taken at the same space-time point x. Lagrangians of this type are called local.

Third, one can generalize the formalism presented above to include Lagrangians which contain partial derivatives of u_a of the second or higher order. In fact, almost no changes are needed—only the r.h.s. of formula (2.15) should be changed appropriately. It is not difficult to compute it. Lagrangian $\mathcal{L}$ can also contain derivatives of higher order than any fixed natural number. In such a case the Lagrangian is usually regarded as a nonlocal one. The point is, that the Taylor series relates the field with shifted arguments to derivatives of all orders of the field with unshifted arguments, namely

$$u_a(x + x_0) = u_a(x) + x_0^\mu \partial_\mu u_a(x) + \frac{1}{2} x_0^\mu x_0^\nu \partial_\mu \partial_\nu u_a(x) + \cdots.$$

For example, a nonlocal Lagrangian containing the term $u_a(x) u_a(x + x_0)$ with constant non vanishing x_0 can be written as a sum of local terms with derivatives of all orders.

2.2 Noether's Theorem

Noether's theorem states that invariance of a field theoretical model under a continuous group of transformations G implies the existence of integrals of motion. Integrals of motion are functionals of the fields and their derivatives which are constant in time provided that the fields obey the corresponding equations of motion.

The transformations forming the continuous group G can act both on space-time points x and on the fields u_a. The space-time points are represented by their Cartesian coordinates, $x = (x^\mu)$, and the space-time metric in these coordinates is given by the diagonal matrix $\eta = \mathrm{diag}(1, -1 - 1 - 1)$. The fields are represented by the functions $u_a(x)$ of the coordinates. Elements of G are denoted by $(f(\omega), F(\omega))$, where $\omega = (\omega^1, \omega^2, \dots \omega^s) = (\omega^\alpha)_{\alpha=1,2,\dots s}$ is a set of continuous, real parameters (often called coordinates) on the group, s is called the dimension of the group G. In fact, for our purposes it is enough to consider only a certain vicinity of the unit element of the group (the identity transformation). For this reason we do not have to specify the range of values of the parameters ω^α. However, we adopt the usual convention that $\omega = 0$ corresponds to the identity transformation which does not change x and u_a. Furthermore, we assume that $f(\omega)$ and $F(\omega)$ depend on the parameters ω^α smoothly, that is that x' and $u'_a(x')$ given by formulas (2.19), (2.20) below, are smooth functions of ω^α in certain vicinity of $\omega = 0$.

In the present chapter we assume that the parameters ω do not depend on the space-time coordinates x^μ. Such transformations are called global,[4] to distinguish them from local symmetry transformations for which $\omega = \omega(x)$. We have already seen an example of local symmetry: the gauge transformations of the potentials $A_\mu(x)$ discussed in the previous chapter.

The transformations f and F act on x and $u_a(x)$, respectively, as follows:

$$x \to x' = f(x; \omega), \tag{2.19}$$

$$u_a(x) \to u'_a(x') = F_a(u_b(x); \omega). \tag{2.20}$$

As elements of the group, these transformations are invertible. Hence, the functions $u_b(x)$ can be expressed by the functions $u'_a(x')$, and x by x'.

In the calculations presented below we need an infinitesimal form of these transformations

$$x' = x + \omega^\alpha \xi_\alpha(x) + \dots, \tag{2.21}$$

$$u'_a(x) = u_a(x) + \omega^\alpha \mathcal{D}_\alpha u_a(x) + \dots, \tag{2.22}$$

where

$$\xi_\alpha(x) = \left. \frac{\partial f(x; \omega)}{\partial \omega^\alpha} \right|_{\omega=0}, \tag{2.23}$$

[4]Nevertheless, up to formula (2.30) below, we do not make use of the assumption that the transformations are global. Only the derivation of Noether's identity (2.31) from formula (2.30) depends on this assumption.

$$\mathcal{D}_\alpha u_a(x) = \left.\frac{\partial F_a(u(x);\omega)}{\partial \omega^\alpha}\right|_{\omega=0} - \xi^\mu_\alpha(x)\frac{\partial u_a(x)}{\partial x^\mu}. \tag{2.24}$$

The dots denote terms of second or higher order in ω^α. Formula (2.21) is obtained by taking the Taylor expansion of the r.h.s. of formula (2.19) with respect to ω^α around $\omega = 0$. It is consistent with the condition $x'(\omega = 0) = x$. Formula (2.22) follows from the Taylor expansion of both sides of formula (2.20)—on the l.h.s. of it, formula (2.21) for x' is used. The four-vectors $\xi_\alpha(x) = (\xi^\mu_\alpha(x))$, where $\alpha = 1 \ldots s$, are called Killing four-vectors. $\mathcal{D}_\alpha u_b(x)$ is called the Lie derivative of u_b in the direction ξ_α at the point x.

Let us now specify what we mean by invariance of the field theoretic model with Lagrangian $\mathcal{L}(u_a(x), \partial_\mu u_a(x); x)$ under transformations (2.19), (2.20). By $S_\Omega[u]$ we denote the action functional calculated for the fields u_a on the whole space R^3 in the time interval $[t', t'']$:

$$S_\Omega[u] = \int_\Omega d^4x\, \mathcal{L}(u_a(x), \partial_\mu u_a(x); x),$$

where

$$\Omega = \{(ct, \vec{x}) : t \in [t', t''], \vec{x} \in R^3\}.$$

Transformation (2.19) acting on Ω gives a new region Ω':

$$\Omega' = f(\Omega; \omega).$$

The action functional calculated for the new functions $u'_a(x')$ in the new region Ω' has the form

$$S_{\Omega'}[u'] = \int_{\Omega'} d^4x'\, \mathcal{L}(u'_a(x'), \frac{\partial u'_a(x')}{\partial x'^\mu}; x').$$

We say that the transformation (f, F) is a symmetry transformation of our model if

$$S_{\Omega'}[u'] = S_\Omega[u] + \int_{\partial\Omega} dS_\mu\, K^\mu(u; x; \omega), \tag{2.25}$$

for all choices of t' and t''. In condition (2.25) x' and u'_a are related to x and u_a by transformations (2.19), (2.20), and $\partial\Omega$ denotes the three-dimensional boundary of the four-dimensional region Ω.

The last term on the r.h.s. of formula (2.25), called the surface term, has the form of surface integral; $\partial\Omega$ is regarded here as a three-dimensional surface embedded in the four-dimensional space-time.[5] With the help of Stokes' theorem the surface term can also be written as the four-dimensional volume integral

[5] $\partial\Omega$ is called surface in the space-time because its dimension, equal to 3, differs from the dimension of the space-time by 1.

$$\int_{\partial\Omega} dS_\mu K^\mu(u; x; \omega) = \int_\Omega d^4x \, \frac{dK^\rho}{dx^\rho},$$

where d/dx^ρ denotes the total derivative.

Note that condition (2.25) is a relation between the action functionals computed for arbitrary functions $u_a(x)$, even those which do not obey the Euler–Lagrange equations. In field theoretical jargon, one says that (2.25) is an 'off-shell' condition. 'On-shell' would mean that the fields $u_a(x)$ were solutions of the Euler–Lagrange equations.

Postulate (2.25) might seem quite strange. As in the case of the stationary action principle, its origin lies in quantum mechanics. In particular, the surface term can be related to a change of phase factor of state vectors. Nevertheless, one can show also on purely classical grounds, that the postulate (2.25) correctly captures the idea of a symmetry of the model.[6] One expects that in such a model, symmetry transformations acting on physically admissible fields give physically admissible fields. Which fields are physically admissible? By assumption, they are those fields which are solutions to the pertinent Euler–Lagrange equations. Therefore, it is important to check whether the symmetry transformations applied to a solution of the Euler–Lagrange equations give a solution to the same equations. Below we show that indeed, this is the case.

Let us compute the functional derivative $\delta/\delta u_a(x)$ of both sides of condition (2.25). The surface term has a vanishing derivative because the test functions used in the definition (2.14) vanish on $\partial\Omega$. The derivative of $S_\Omega[u]$ also vanishes because we now consider the fields $u_a(x)$ which obey the Euler–Lagrange equations (2.11). The r.h.s. is regarded as a composite functional of u_a, and in order to compute its functional derivative we use a chain rule analogous to the one well known from calculus. Hence, if $u_a(x)$ are solutions of the Euler–Lagrange equations,

$$\frac{\delta S_\Omega[u]}{\delta u_b(y)} = \int_\Omega d^4x \, \frac{\delta S_{\Omega'}[u']}{\delta u'_a(x')}\bigg|_{u'(x')=F(u(x);\omega)} \frac{\delta F_a(u(x);\omega)}{\delta u_b(y)} = 0. \tag{2.26}$$

Let us introduce the new notation

$$\frac{\delta F_a(u(x);\omega)}{\delta u_b(y)} \equiv \frac{\delta F}{\delta u}(a, x; b, y).$$

Its purpose is to mark the fact that this functional derivative can be regarded as an integral kernel of a certain linear operator $\delta F/\delta u$. For transformations (2.20) this operator is nonsingular, that is there exists a linear operator $(\delta F/\delta u)^{-1}$ such that

$$\int_\Omega d^4y \frac{\delta F_a(u(x);\omega)}{\delta u_b(y)} (\frac{\delta F}{\delta u})^{-1}(b, y; c, z) = \delta_{ac}\delta(x-z),$$

[6] One should not confuse a symmetry of a model with a symmetry of a concrete physical state. For example, a model which is invariant under rotations can predict the existence of physical states which are not invariant under rotations.

where the first δ on the r.h.s. is Kronecker delta, while the second one is the four-dimensional Dirac delta. Therefore, (2.26) implies that

$$\left.\frac{\delta S_{\Omega'}[u']}{\delta u'_a(x')}\right|_{u'(x')=F(u(x);\omega)} = 0,$$

but this means that $u'_a(x')$ obeys the Euler–Lagrange equations in the region Ω'.

As the next step in our analysis of the invariance condition (2.25) we derive the so called Noether's identity. The l.h.s. of this identity gives an explicit formula for the integrals of motion. The main part of the derivation is just a calculation of the first two terms of the Taylor expansion of the l.h.s. of condition (2.25) with respect to ω^α. The change of the integration variable from x' to x gives

$$d^4x' = J d^4x,$$

where J is the Jacobian corresponding to transformation (2.19), that is

$$J = \det\left[\frac{\partial x'^\mu}{\partial x^\nu}\right].$$

Using formula (2.21) we may write

$$J = 1 + \frac{\partial \delta x^\mu}{\partial x^\mu} + \cdots, \tag{2.27}$$

where

$$\delta x^\mu = \omega^\alpha \xi^\mu_\alpha(x). \tag{2.28}$$

Here and in the subsequent calculations, the multi-dots denote terms of the second or higher order in ω^α. The Taylor expansion of $\mathcal{L}(u'_a(x'), \frac{\partial u'_a(x')}{\partial x'^\mu}; x')$ has the following form:

$$\begin{aligned}\mathcal{L}(u'_b(x'), \frac{\partial u'_b(x')}{\partial x'^\mu}; x') &= \mathcal{L}(u_b(x), \frac{\partial u_b(x)}{\partial x^\mu}; x) + \frac{\partial \mathcal{L}}{\partial x^\lambda}\delta x^\lambda \\ &+ \frac{\partial \mathcal{L}}{\partial u_a(x)}\left(u'_a(x') - u_a(x)\right) + \frac{\partial \mathcal{L}}{\partial (u_{a,\nu}(x))}\left(\frac{\partial u'_a(x')}{\partial x'^\nu} - \frac{\partial u_a(x)}{\partial x^\nu}\right) + \cdots .\end{aligned} \tag{2.29}$$

Next, we use formulas (2.21), (2.22):

$$u'_a(x') - u_a(x) = \bar{\delta} u_a(x) + \frac{\partial u_a(x)}{\partial x^\lambda}\delta x^\lambda + \cdots,$$

where

$$\bar{\delta} u_a(x) = \omega^\alpha \mathcal{D}_\alpha u_a(x),$$

and

$$\frac{\partial u'_a(x')}{\partial x'^\nu} - \frac{\partial u_a(x)}{\partial x^\nu} = \frac{\partial x^\mu}{\partial x'^\nu}\frac{\partial u'_a(x')}{\partial x^\mu} - \frac{\partial u_a(x)}{\partial x^\nu}$$
$$= \frac{\partial}{\partial x^\nu}\left(u'_a(x') - u_a(x)\right) - \frac{\partial(\delta x^\mu)}{\partial x^\nu}\frac{\partial u_a(x)}{\partial x^\mu} + \cdots$$
$$= \frac{\partial}{\partial x^\nu}\left(\bar{\delta}u_a(x)\right) + \frac{\partial^2 u_a(x)}{\partial x^\nu \partial x^\lambda}\delta x^\lambda + \cdots .$$

Therefore,

$$J\mathcal{L}\left(u'_a(x'), \frac{\partial u'_a(x')}{\partial x'^\mu}; x'\right) = \mathcal{L}\left(u_a(x), \partial_\mu u_a(x); x\right) + \frac{\partial \delta x^\mu}{\partial x^\mu}\mathcal{L}\left(u_a(x), \partial_\mu u_a(x); x\right)$$
$$+ \frac{\partial \mathcal{L}}{\partial x^\nu}\delta x^\nu + \frac{\partial \mathcal{L}}{\partial u_a(x)}\frac{\partial u_a(x)}{\partial x^\lambda}\delta x^\lambda + \frac{\partial \mathcal{L}}{\partial(u_{a,\nu}(x))}\frac{\partial^2 u_a(x)}{\partial x^\nu \partial x^\lambda}\delta x^\lambda$$
$$+ \frac{\partial \mathcal{L}}{\partial u_a(x)}\bar{\delta}u_a(x) + \frac{\partial \mathcal{L}}{\partial(u_{a,\nu}(x))}\frac{\partial}{\partial x^\nu}\bar{\delta}u_a(x) + \cdots$$
$$= \mathcal{L}\left(u_a(x), \partial_\mu u_a(x); x\right) + \frac{d\left(\delta x^\mu \mathcal{L}\left(u_a(x), \partial_\mu u_a(x); x\right)\right)}{dx^\mu}$$
$$+ \left[\frac{\partial \mathcal{L}}{\partial u_a(x)} - \frac{d}{dx^\nu}\left(\frac{\partial \mathcal{L}}{\partial(u_{a,\nu}(x))}\right)\right]\bar{\delta}u_a + \frac{d}{dx^\nu}\left(\frac{\partial \mathcal{L}}{\partial(u_{a,\nu}(x))}\bar{\delta}u_a\right) + \cdots .$$

This last expression is used in $S_{\Omega'}[u']$ on the l.h.s. of condition (2.25). On the r.h.s. of that condition we have $\mathcal{L}\left(u_a(x), \partial_\mu u_a(x); x\right)$ and $K^\mu(u_a; x; \omega)$. Notice that

$$K^\mu(u_a; x; \omega = 0) = 0,$$

because $\omega = 0$ corresponds to the trivial transformation $u'_a(x') = u_a(x)$ and $x' = x$. Therefore,

$$K^\mu(u_a; x; \omega) = \omega^\alpha K^\mu_\alpha(u_a; x) + \cdots .$$

Now it is clear that condition (2.25) can be written in the following form

$$\int_\Omega d^4x \, \frac{d}{dx^\nu}\left(K^\nu_\alpha \omega^\alpha - \mathcal{L}\delta x^\nu - \frac{\partial \mathcal{L}}{\partial(u_{a,\nu}(x))}\bar{\delta}u_a\right)$$
$$= \int_\Omega d^4x \, \bar{\delta}u_a(x)\left[\frac{\partial \mathcal{L}}{\partial u_a(x)} - \frac{d}{dx^\nu}\left(\frac{\partial \mathcal{L}}{\partial(u_{a,\nu}(x))}\right)\right] + \cdots . \tag{2.30}$$

Because the parameters ω^α vary continuously in an interval around $\omega = 0$, we may take the derivative with respect to ω^α of both sides of (2.30) and put $\omega = 0$ afterwards. In this way we obtain Noether's identity

$$\int_\Omega d^4x\, \frac{dj^\nu_\alpha}{dx^\nu} = \int_\Omega d^4x\, \mathcal{D}_\alpha u_a(x) \left[\frac{\partial \mathcal{L}}{\partial u_a(x)} - \frac{d}{dx^\nu} \left(\frac{\partial \mathcal{L}}{\partial (u_{a,\nu}(x))} \right) \right], \tag{2.31}$$

where the current density j^ν_α is defined as follows

$$j^\nu_\alpha = K^\nu_\alpha(u_a; x) - \mathcal{L}\xi^\nu_\alpha - \frac{\partial \mathcal{L}}{\partial (u_{a,\nu}(x))} \mathcal{D}_\alpha u_a(x). \tag{2.32}$$

The fact that this identity exists is known as Noether's theorem.

Noether's identity (2.31) reduces to a conservation law when the fields u_a obey the Euler–Lagrange equations—then the r.h.s. of the identity vanishes, and therefore

$$\int_{t'}^{t''} dt \int_{R^3} d^3x \left(\frac{dj^0_\alpha}{dt} + \frac{dj^k_\alpha}{dx^k} \right) = 0. \tag{2.33}$$

With the help of Gauss's theorem, the second term on the l.h.s. of formula (2.33) can be written as integral over a sphere of radius increasing to infinity. Therefore, if the spatial components j^k_α of the current density vanish sufficiently quickly when $|\vec{x}| \to \infty$, this term gives a vanishing contribution. The integral with respect to time is trivial. The result can be written in the form

$$\mathcal{Q}_\alpha(t'') = \mathcal{Q}_\alpha(t'), \tag{2.34}$$

where

$$\mathcal{Q}_\alpha(t) = \int_{R^3} d^3x\, j^0_\alpha(t, \vec{x}). \tag{2.35}$$

Because t' and t'' are arbitrary, this means that the 'charges' $\mathcal{Q}_\alpha$, $\alpha = 1, \ldots, s$, are constant in time if the fields u_a obey the pertinent Euler–Lagrange equations.

One often postulates an invariance condition stronger than (2.25), obtained by omitting the integrals. In this sense, it is the local version of condition (2.25). It has the following form

$$J(x)\mathcal{L}\left(u'_a(x'), \frac{\partial u'_a(x')}{\partial x'^\mu}; x' \right) = \mathcal{L}\left(u_a(x), \frac{\partial u_a(x)}{\partial x^\mu}; x \right) + \frac{dK^\mu}{dx^\mu}. \tag{2.36}$$

This condition leads to the continuity equation

$$\frac{dj^\nu_\alpha}{dx^\nu} = 0, \tag{2.37}$$

where j^ν_α are still given by formula (2.32). Equation (2.37) is the local version of the conservation law (2.34) of the charges $\mathcal{Q}_\alpha$. The derivation of (2.37) from the condition (2.36) is essentially the same as in the case of global condition (2.25) and global conservation law (2.34).

Exercises

2.1 Let $S[\phi]$ denotes a functional which assigns (real or complex) numbers to the functions ϕ defined on R^D. The functional derivative $\frac{\delta S[\phi]}{\delta\phi(x)}$ is a generalized function defined as follows

$$\lim_{\epsilon\to 0}\frac{S[\phi+\epsilon f]-S[\phi]}{\epsilon}=\int_{R^D} d^D x\,\frac{\delta S[\phi]}{\delta\phi(x)}\,f(x),$$

for arbitrary test function $f\in S(R^D)$ (see the Appendix). Calculate $\frac{\delta S[\phi]}{\delta\phi(x)}$ for:

(a) $S[\phi]=\phi(x_0)$ with fixed x_0,

(b) $S[\phi]=\left.\frac{d^p\phi(x)}{dx^p}\right|_{x=x_0}$,

(c) $S[\phi]=\int_{R^D} d^D y\; h(y)\phi(y)$, where $h(y)$ is a fixed function of y,

(d) $S[\phi]=\exp\left\{\frac{1}{2}\int_{R^D} d^D y\int_{R^D} d^D z\;\phi(y)G(y,z)\phi(z)\right\}$.

2.2 During its propagation in space-time, a structureless, relativistic string sweeps a world-sheet $X^\mu(t,s)$ (a two-dimensional generalization of the world line of a particle). Here t is time, and $s\in[0,2\pi]$ is a parameter along the string. We consider only the closed string for which $X^\mu(t,0)=X^\mu(t,2\pi)$ at all t. We also assume that the vector $\dot{X}^\mu\equiv\partial_t X^\mu(t,s)$ is time-like and $X'^\mu\equiv\partial_s X^\mu(t,s)$ is space-like. For the simplest string, the so called Nambu–Goto string, the pertinent action is proportional to the area of the world-sheet,

$$S_{\mathrm{NG}}=\gamma\int_{t_1}^{t_2}dt\int_0^{2\pi}ds\,\sqrt{\left(\dot{X}^\mu X'_\mu\right)^2-\left(\dot{X}^\mu\dot{X}_\mu\right)\left(X'^\mu X'_\mu\right)},$$

with the dimensional constant γ.

(a) Rewrite this action in terms of the determinant of the induced world-sheet metric g_{ab}, which can be read off from the identity

$$dX^\mu(t,x)dX_\mu(t,s)=g_{ab}(s,t)d\sigma^a d\sigma^b,\qquad \sigma^0=t,\ \sigma^1=s.$$

(b) Let g^{ab} denote the inverse of the induced metric, $g^{ab}g_{bc}=\delta^a_c$, and let $g\equiv\det(g_{ab})$. Using a well-known formula for the determinant check that the variation of g that corresponds to a variation of the induced metric can be written in the form

$$\delta g=g g^{ab}\delta g_{ab}.$$

Show that the equation of motion of the closed Nambu–Goto string can be written as the Laplace equation for $X^\mu(s,t)$:

$$\Delta_g X^\mu(s,t)=0,\qquad \Delta_g(\ldots)=\frac{1}{\sqrt{-g}}\partial_a\left(\sqrt{-g}g^{ab}\partial_b(\ldots)\right).$$

2.3 Check the invariance of S_{NG} under the infinitesimal space-time translations $\delta X^\mu = \omega^\mu$ and rotations $\delta X_\mu = \omega_{\mu\nu} X^\nu$, $\omega_{\mu\nu} + \omega_{\nu\mu} = 0$, where ω^μ and $\omega_{\mu\nu}$ are constants. Show that the corresponding conserved quantities—the total energy-momentum and angular momentum of the closed Nambu–Goto string—have the form

$$P^\mu = \gamma \int_0^{2\pi} ds \, \sqrt{-g} g^{0a} \partial_a X^\mu,$$

$$M^{\mu\nu} = \frac{\gamma}{2} \int_0^{2\pi} ds \, \sqrt{-g} g^{0a} \left[X^\mu \partial_a X^\nu - X^\nu \partial_a X^\mu \right],$$

respectively.

2.4 The transformation rule for a scalar field Φ under the dilatation $x^\mu \to x'^\mu = \mathrm{e}^\sigma x^\mu$ reads:

$$\Phi'(x') = \mathrm{e}^{-\sigma d_\Phi} \Phi(x), \tag{2.38}$$

where d_Φ is the so called canonical scaling dimension of the field Φ, i.e. $\dim(\Phi) = \text{cm}^{-d_\Phi}$. The action functional for a free massless scalar field, propagating in D-dimensional space-time, has the form

$$S[\Phi] = \frac{1}{2} \int d^D x \; \partial_\mu \Phi \partial^\mu \Phi.$$

(a) In the system of units where $\hbar = 1$ the action should be dimensionless. Find the value d_Φ which follows from this requirement.
(b) Prove that the action of the massless free field Φ is invariant under the dilatation (2.38). Is the Lagrangian invariant as well?
(c) Find the form of the relevant conserved current.

2.5 Consider the action functional for an interacting massless scalar field in D-dimensional space-time,

$$S[\Phi] = \int d^D x \left(\tfrac{1}{2} \partial_\mu \Phi \partial^\mu \Phi - \lambda \Phi^n \right), \tag{2.39}$$

where $n \geq 3$ is an integer and λ is a (coupling) constant. For which values of D and n does the action (2.39) possess a dilatational invariance? What is the dimension of λ in these cases?

2.6 Consider the Lagrangian

$$\mathcal{L}_{\text{U}} = \text{tr} \left(\partial_\mu U^\dagger \partial^\mu U \right)$$

with $U(x)$ being a unitary, $N \times N$ matrix.
(a) Check that it is invariant under the transformations

$$U(x) \to A^\dagger U(x) B, \tag{2.40}$$

where A and B are arbitrary constant, $N \times N$ unitary matrices with unit determinant (i.e. $A, B \in \mathrm{SU}(N)$).
(b) For A and B close to the unit $N \times N$ matrix I_N we may write

$$A = \exp\left\{ i \sum_{a=1}^{N^2-1} \epsilon_a T^a \right\}, \qquad B = \exp\left\{ i \sum_{a=1}^{N^2-1} \eta_a T^a \right\}$$

where ϵ_a and η_a are real, infinitesimal parameters (playing the role of the ω parameters used in the derivation of Noether's current) and T^a are linearly independent over $\mathbb{R}$, Hermitian, traceless, $N \times N$ matrices.

Find the expressions for the conserved charges that exist thanks to this symmetry.

Chapter 3
Scalar Fields

Abstract The Lorentz and Poincaré groups. The equation of motion and energy-momentum tensor for a real scalar field. Domain walls in a model with spontaneously broken Z_2 symmetry. The complex scalar field with $U(1)$ symmetry and the Mexican hat potential. The Goldstone mode of the field. Global vortex and winding number.

In this and the next two chapters we review the main types of classical fields appearing in particle physics. We begin with a presentation of several models which involve only scalar fields. In Chap. 4 we discuss vector fields, and in Chap. 5 spinor fields. The main feature all of these fields have in common is the simplicity of their transformation laws under Poincaré transformations of Minkowski space-time. For this reason, they are called the relativistic fields. Moreover, Poincaré transformations are symmetries of their corresponding action functionals in the sense described in the previous chapter. Therefore, we first discuss the Lorentz and Poincaré groups.

3.1 The Lorentz and Poincaré groups

Let us endow Minkowski space-time M with a Cartesian coordinate system (x^μ), in which the metric on M has the diagonal form $\eta = \text{diag}(1, -1, -1, -1)$. Matrix elements of η are denoted as $\eta_{\mu\nu}$, where $\mu, \nu = 0, 1, 2, 3$. The inverse matrix η^{-1} coincides with η, but by convention its matrix elements have upper indices. Hence, $\eta^{\mu\nu}$ are matrix elements of η^{-1}. Minkowski space-time has a very simple structure. In particular, it can be covered by one Cartesian coordinate system, and then its points can be identified with the set of four coordinates x^μ, $x = (x^\mu)$. Poincaré transformations of M have the form

$$x'^\mu = L^\mu{}_\nu x^\nu + a^\mu, \tag{3.1}$$

where $L^\mu{}_\nu$ and a^μ do not depend on x^ν and are real. By definition, they preserve the form of the metric η, that is

H. Arodź and L. Hadasz, *Lectures on Classical and Quantum Theory of Fields*, Graduate Texts in Physics, DOI 10.1007/978-3-319-55619-2_3

$$\frac{\partial x'^{\mu}}{\partial x^{\rho}} \frac{\partial x'^{\nu}}{\partial x^{\lambda}} \eta_{\mu\nu} = \eta_{\rho\lambda}. \tag{3.2}$$

For comparison, the transformation of a general second rank covariant tensor field $a_{\mu\nu}(x)$ has the form

$$\frac{\partial x'^{\mu}}{\partial x^{\rho}} \frac{\partial x'^{\nu}}{\partial x^{\lambda}} a'_{\mu\nu}(x') = a_{\rho\lambda}(x).$$

Because η is constant on M in the Cartesian coordinates, the arguments x and x' may be omitted. It is clear that (3.2) actually means that $\eta' = \eta$. This shows that (3.2) is indeed an invariance condition. The partial derivatives in (3.2) can easily be calculated, and the condition is equivalently written as

$$L^{\mu}{}_{\rho} L^{\nu}{}_{\lambda} \eta_{\mu\nu} = \eta_{\rho\lambda}. \tag{3.3}$$

Transformations of the form

$$x'^{\mu} = L^{\mu}{}_{\nu} x^{\nu} \tag{3.4}$$

with $L^{\mu}{}_{\nu}$ obeying condition (3.3) are called general Lorentz transformations. They form a subset of Poincaré transformations, obtained by setting $a^{\mu} = 0$. One may associate with the Lorentz transformation a four by four matrix $\hat{L}$ with real elements $L^{\mu}{}_{\nu}$,

$$\hat{L} = \left(L^{\mu}{}_{\nu}\right).$$

Here the first index μ enumerates rows and the second index ν columns of this matrix. The same convention holds also for the metric tensor $\eta = (\eta_{\mu\nu})$: the first index (μ) enumerates rows and the second index (ν) columns. Condition (3.3) can be written in the matrix form

$$\hat{L}^{T} \eta \hat{L} = \eta, \tag{3.5}$$

where ^{T} denotes the transposed matrix, i.e., $(\hat{L}^{T})^{\mu}{}_{\nu} = L^{\nu}{}_{\mu}$. It follows from (3.5) that

$$(\det\hat{L})^{2} = 1,$$

hence the Lorentz transformations are represented by nonsingular matrices with determinant equal to $+1$ or -1. Another consequence of the matrix condition (3.5) is the following formula for the inverse of the Lorentz matrix

$$\hat{L}^{-1} = \eta^{-1} \hat{L}^{T} \eta.$$

For matrix elements,

$$(\hat{L}^{-1})^{\mu}{}_{\nu} = \eta^{\mu\lambda}(\hat{L}^{T})^{\lambda}{}_{\rho}\eta_{\rho\nu} = \eta^{\mu\lambda}L^{\rho}{}_{\lambda}\eta_{\rho\nu} = L_{\nu}{}^{\mu},$$

where we have used the standard conventions about raising and lowering indices by the metric tensor and its inverse.

Condition (3.5) implies that the four by four unit matrix I_4, the inverse matrix $\hat{L}^{-1}$, and the matrix product $\hat{L}_1\hat{L}_2$, are all Lorentz transformations if $\hat{L}$, $\hat{L}_1$, $\hat{L}_2$ are. Therefore, the set of all Lorentz transformations forms a matrix group, called the general Lorentz group. It can be regarded as a subset of the 16 dimensional space of all four by four real matrices, determined by conditions (3.3) or (3.5), which are constraints on the 16 elements of a general four by four real matrix. There are 10 independent constraints because the matrix on the l.h.s. of condition (3.5) is automatically symmetric, hence elements lying above its diagonal are identical with the ones placed symmetrically below the diagonal. The 10 constraints allow us to express 10 of the matrix elements by the remaining 6. Therefore, the general Lorentz group is six dimensional.

The general Lorentz group regarded as a set is not connected. We have seen that we can have either $\det\hat{L} = +1$ or $\det\hat{L} = -1$. Moreover, condition (3.3) considered for $\mu = \nu = 0$ can be written in the form

$$(L^{0}{}_{0})^2 = 1 + L^{i}{}_{0}L^{i}{}_{0},$$

which shows that either $L^{0}{}_{0} \geq 1$ or $L^{0}{}_{0} \leq -1$. It turns out that the general Lorentz group has four connected components (maximal connected subsets) which differ by the signs of $\det\hat{L}$ and $L^{0}{}_{0}$. Only one of them, namely that which is characterized by

$$\det\hat{L} = +1, \quad L^{0}{}_{0} \geq 1, \tag{3.6}$$

is also a group; a subgroup of the general Lorentz group. It is called the proper orthochronous Lorentz group, or Lorentz group for short, and is denoted by $L^{\uparrow}_{+}$. This is the only connected component of the general Lorentz group which contains the unit matrix. The other connected components can be obtained by taking products of matrices from $L^{\uparrow}_{+}$ with one of the three matrices

$$T = \text{diag}(-1, 1, 1, 1), \quad P = \text{diag}(1, -1 - 1 - 1), \quad TP.$$

Lorentz transformations corresponding to matrices T, P and TP change direction of time, give spatial reflection $\vec{x} \rightarrow -\vec{x}$, or both, respectively. Definitions of relativistic fields given below refer only to the Lorentz group $L^{\uparrow}_{+}$. The transformations T, P and TP are usually included at a later stage. In our lecture notes we shall not discuss them.

By definition, the Poincaré group $\mathcal{P}$ consists of transformations (3.1) such that $\hat{L} \in L_+^\uparrow$. Elements of $\mathcal{P}$ are denoted as $(\hat{L}, a)$, where $a = (a^\mu)$. The group multiplication in $\mathcal{P}$ follows from the superposition of two transformations (3.1):

$$(\hat{L}_2, a_2)(\hat{L}_1, a_1) = (\hat{L}_2\hat{L}_1, \hat{L}_2 a_1 + a_2). \tag{3.7}$$

The unit element has the form $(I_4, 0)$, where I_4 denotes four by four unit matrix. Furthermore, $(\hat{L}, a)^{-1} = (\hat{L}^{-1}, -\hat{L}^{-1}a)$. The Poincaré group is ten dimensional.

The Poincaré group has many subgroups. One of them consists of all transformations of the form $(\hat{L}, 0)$. It is isomorphic to the Lorentz group $L_+^\uparrow$. Another subgroup is isomorphic to the group of all translations in Minkowski space-time denoted by T_4. That subgroup consists of all transformations of the form (I_4, a). Each element of the Poincaré group can be uniquely written as the product of a Lorentz transformation and a translation,

$$(\hat{L}, a) = (I_4, a)(\hat{L}, 0).$$

Moreover, using the multiplication rule (3.7) one can check that

$$(\hat{L}, 0)(I_4, a)(\hat{L}, 0)^{-1} = (I_4, \hat{L}a).$$

The last two properties together with the multiplication rule (3.7) are summarized in the statement that the Poincaré group is a semidirect product of the group T_4 of all translations in Minkowski space-time and of the Lorentz group $L_+^\uparrow$.[1]

The translations in Minkowski space-time have the form

$$x'^\mu = x^\mu + a^\mu.$$

It is clear that a parametrization of the translations which is convenient for applications of Noether's theorem is provided by a^μ themselves. The Cartesian components of the corresponding Killing vectors have the form

$$\xi^\mu_\alpha = \left.\frac{\partial x'^\mu}{\partial a^\alpha}\right|_{a=0} = \delta^\mu_\alpha, \tag{3.8}$$

where $\alpha = 0, 1, 2, 3$.

Finding a suitable parametrization of the Lorentz group is more cumbersome. We use the mathematical theorem which says that with the help of the exponential mapping one can parameterize a vicinity of the unit matrix by certain matrices from a neighborhood of the zero matrix. In the case of the Lorentz group, this means that for each $\hat{L}$ from such a vicinity of the unit matrix I_4 there exists just one real matrix $\hat{\epsilon}$ such that

$$\hat{L} = \exp\hat{\epsilon}. \tag{3.9}$$

[1] In the case of a direct product, the multiplication rule would have the form $(\hat{L}_1, a_1)(\hat{L}_2, a_2) = (\hat{L}_1\hat{L}_2, a_1 + a_2)$.

It is clear that $\hat{L} = I_4$ is obtained for $\hat{\epsilon} = 0$. Let us write condition (3.5) in the following form

$$\eta^{-1}\hat{L}^T\eta = \hat{L}^{-1}.$$

By inserting formula (3.9) we obtain the condition

$$\eta^{-1}\hat{\epsilon}^T\eta = -\hat{\epsilon}, \tag{3.10}$$

which in fact says that the matrix $\eta\hat{\epsilon}$ is antisymmetric. With our conventions for indices, we have $\hat{\epsilon} = (\epsilon^{\mu}{}_{\nu})$ and $\eta\hat{\epsilon} = (\epsilon_{\mu\nu})$, and therefore,

$$\epsilon_{\mu\nu} = -\epsilon_{\nu\mu}.$$

In consequence,

$$\epsilon^{0}{}_{i} = \epsilon^{i}{}_{0}, \;\; \epsilon^{i}{}_{k} = -\epsilon^{k}{}_{i}, \epsilon^{0}{}_{0} = \epsilon^{1}{}_{1} = \epsilon^{2}{}_{2} = \epsilon^{3}{}_{3} = 0. \tag{3.11}$$

Note that $\hat{\epsilon}$ is not antisymmetric. As parameters on the Lorentz group in a neighborhood of the unit matrix we take $\epsilon^{\mu\nu}$ with $\mu < \nu$, that is those elements of the matrix $\hat{\epsilon}\hat{\eta}^{-1}$ which lie above its diagonal. This last matrix is antisymmetric. The corresponding Killing vectors are calculated from the formula

$$\xi^{\mu}_{\alpha\beta} = \left.\frac{\partial x'^{\mu}}{\partial\epsilon^{\alpha\beta}}\right|_{\epsilon=0},$$

where $x'^{\mu} = L^{\mu}{}_{\nu}x^{\nu}$ and $\alpha < \beta$. Because

$$x'^{\mu} = (\hat{L}\hat{\eta}^{-1})^{\mu\nu}(\hat{\eta}x)_{\nu} \equiv L^{\mu\nu}x_{\nu},$$

and

$$\left.\frac{\partial L^{\mu\nu}}{\partial\epsilon^{\alpha\beta}}\right|_{\epsilon=0} = \delta^{\mu}_{\alpha}\delta^{\nu}_{\beta} - \delta^{\mu}_{\beta}\delta^{\nu}_{\alpha},$$

we obtain

$$\xi^{\mu}_{\alpha\beta} = (\delta^{\mu}_{\alpha}\eta_{\beta\nu} - \delta^{\mu}_{\beta}\eta_{\alpha\nu})x^{\nu}. \tag{3.12}$$

Let us recall that we regard the Poincaré transformations as transformations of points in Minkowski space-time. Therefore, x^{μ} and x'^{μ} are coordinates of two points with respect to a single, fixed Cartesian reference frame in the space-time. The parameters ϵ^{12}, ϵ^{13}, ϵ^{23} correspond to transformations which do not change x^0, that is to spatial rotations. For example, when all $\epsilon^{\alpha\beta}$ except ϵ^{12} are equal to zero and ϵ^{12} is infinitesimally small, we obtain an infinitesimal rotation around the x^3 axis by the angle ϵ^{12}:

$$x'^0 = x^0, \;\; x'^3 = x^3, \;\; x'^1 = x^1 - \epsilon^{12}x^2, \;\; x'^2 = x^2 + \epsilon^{12}x^3,$$

where all terms with second and higher powers of ϵ^{12} have been neglected. The parameters ϵ^{01}, ϵ^{02}, ϵ^{03} give the so called Lorentz boosts. This name is justified by the fact that boosts transform a particle at rest into a particle moving with non zero velocity. For example, if only ϵ^{01} is not equal to zero, then for ϵ^{01} infinitesimally small,

$$x'^0 = x^0 - \epsilon^{01} x^1, \ x'^1 = x^1 - \epsilon^{01} x^0, \ x'^2 = x^2, \ x'^3 = x^3,$$

where again we have kept only the terms constant or linear in ϵ^{01}. These formulas imply that this boost, acting on a particle which is at rest at the point $\vec{x}_0$ and which has the world-line $x(t) = (ct, \vec{x}_0)$, gives a particle moving with the infinitesimal velocity $-\epsilon^{01}$ along the x^1 axis in the negative direction.

Formulas (3.8) and (3.12) are used in this and the next chapters, where we apply Noether's theorem to relativistic fields. We adopt the stronger, local form (2.36) of the invariance condition. Note that in the case of the Poincaré transformations (3.1) the Jacobian $J = \det\hat{L}$ is equal to +1 because $\hat{L} \in L_+^\uparrow$.

3.2 The Real Scalar Field

The configuration space of the relativistic real scalar field is a space of real functions $\phi(\vec{x})$ on R^3, and trajectories of the field are described by real functions $\phi(x)$, $x = (ct, \vec{x})$, on Minkowski space-time. By definition, the scalar field ϕ has the following transformation law under the Poincaré transformations

$$\phi'(x') = \phi(x), \tag{3.13}$$

where

$$x'^\mu = L^\mu{}_\nu x^\nu + a^\mu, \quad \hat{L} \in L_+^\uparrow.$$

This definition implies that

$$\phi'(x) = \phi(\hat{L}^{-1}(x - a)). \tag{3.14}$$

Comparing (3.13) with the general formula (2.20) we see that in the present case F is trivial, $F(\phi(x); \omega) = \phi(x)$. As the parameters ω we choose a^μ and $\epsilon^{\mu\nu}$ introduced in the last section. Therefore, the first term in definition (2.24) of the Lie derivative vanishes, and

$$\mathcal{D}\phi(x) = -\xi^\rho(x)\frac{\partial\phi(x)}{\partial x^\rho},$$

where as the Killing vector ξ we now take ξ_α or $\xi_{\alpha\beta}$ given by formulas (3.8) and (3.12), respectively.

The invariance condition in the local form, with vanishing surface term and in absence of external fields, has the form

$$\mathcal{L}\left(\phi'(x'), \frac{\partial \phi'(x')}{\partial x'^{\nu}}\right) = \mathcal{L}\left(\phi(x), \frac{\partial \phi(x)}{\partial x^{\nu}}\right). \tag{3.15}$$

Using formula (3.13) we find that

$$\frac{\partial \phi'(x')}{\partial x'^{\nu}} = \frac{\partial x^{\rho}}{\partial x'^{\nu}} \frac{\partial \phi(x)}{\partial x^{\rho}} = L_{\nu}{}^{\rho} \frac{\partial \phi(x)}{\partial x^{\rho}}$$

(recall that $L_{\nu}{}^{\rho} = (\hat{L}^{-1})^{\rho}{}_{\nu}$). Therefore, condition (3.15) acquires the form

$$\mathcal{L}\left(\phi(x), L_{\nu}{}^{\rho} \frac{\partial \phi(x)}{\partial x^{\rho}}\right) = \mathcal{L}\left(\phi(x), \frac{\partial \phi(x)}{\partial x^{\nu}}\right).$$

It is clear that this condition does not impose any restriction on the dependence of the Lagrangian on the field ϕ, and that the derivatives $\partial_{\nu}\phi$ can appear only in Lorentz invariant combinations.

In almost all applications of the real scalar field, the pertinent Lagrangian has the form

$$\mathcal{L} = \frac{1}{2}\eta^{\mu\nu}\partial_{\mu}\phi(x)\partial_{\nu}\phi(x) - \frac{1}{2}m^2\phi^2 - V(\phi(x)), \tag{3.16}$$

where m^2 is a real constant, and $V(\phi)$ is a simple function of ϕ—a polynomial in most cases—called the interaction potential[2] of the field ϕ. Also, non-polynomial $V(\phi)$ are considered, e.g., exponential, logarithmic or trigonometric functions. The Euler–Lagrange equation corresponding to Lagrangian (3.16) has the form

$$\partial_{\mu}\partial^{\mu}\phi(x) + m^2\phi(x) + V'(\phi(x)) = 0, \tag{3.17}$$

where $V' = dV/d\phi$. Simple calculation shows that $V(\phi) = c_2\phi^2 + c_1\phi + c_0$ leads to the Euler–Lagrange equation of the Klein–Gordon type, namely

$$\partial_{\mu}\partial^{\mu}\phi(x) + (m^2 + 2c_2)\phi(x) = -c_1,$$

which can be reduced to the homogeneous Klein–Gordon equation by a constant shift of the field ϕ when $m^2 + 2c_2 \neq 0$, or by $-c_1 x^{\mu}x_{\mu}/8$ if $m^2 + 2c_2 = 0$.

The first really new Euler–Lagrange equation, with a term quadratic in ϕ, is obtained when $V(\phi) = \lambda\phi^3$ with constant λ. In mathematical terminology, it is a nonlinear partial differential evolution equation of hyperbolic type. At present, there are no methods which would allow us to construct a general solution for such equa-

---

[2] The term 'potential' is reserved for the sum $m^2\phi^2/2 + V(\phi)$.

tions. Particular examples of solutions can be obtained with the help of approximation methods, which include numerical calculations performed by computers. Sometimes one can find analytic solutions, especially when one is interested in particularly symmetric ones. In general, nonlinear partial differential equations of the hyperbolic type can lead to quite complicated and surprising time evolution of the field. Coming back to our Euler–Lagrange equation, it turns out that the above introduced cubic $V(\phi)$ is not quite satisfactory, because, as we show in the next paragraph, the corresponding energy is not bounded from below. This fact does not mean that some mathematical inconsistency is present. The point is that all physical objects in Nature discovered until now seem to have energy bounded from below. In consequence, models in which the energy is not bounded from below are regarded as less interesting.

The energy and momentum of the field are identified with the integrals of motion obtained from Noether's theorem applied to time and space translations, respectively. We already know the Killing vectors for the translations and the Lie derivatives of the scalar field. The surface term in formula (2.32) is absent. Simple calculations give the currents corresponding to the four independent translations,

$$j^\mu_\alpha = -\mathcal{L}\delta^\mu_\alpha + \partial^\mu\phi(x)\partial_\alpha\phi(x). \tag{3.18}$$

Often one introduces the so called energy-momentum tensor $T^\mu{}_\nu$. It is defined by the following formula

$$j^\mu_\alpha = T^\mu{}_\nu \xi^\nu_\alpha. \tag{3.19}$$

Thus,

$$T^\mu{}_\nu = \partial^\mu\phi(x)\partial_\nu\phi(x) - \mathcal{L}\delta^\mu_\nu = j^\mu{}_\nu. \tag{3.20}$$

The continuity equations $d_\mu j^\mu_\alpha = 0$ imply that

$$\partial_\nu T^\nu{}_\alpha = 0.$$

The total energy E and momentum P^i of the field are defined as

$$E = \int_{R^3} d^3x \; j^0_0, \quad P^i = -\int_{R^3} d^3x \; j^0_i. \tag{3.21}$$

The minus sign in the formula for P^i is due to the metric tensor $\eta^{\mu\nu}$ used here to raise the index i. Using formulas (3.16) and (3.18) we obtain

$$j^0_0 = \frac{1}{2}\partial_0\phi\partial_0\phi + \frac{1}{2}\partial_i\phi\partial_i\phi + \frac{1}{2}m^2\phi^2 + \lambda\phi^3, \tag{3.22}$$

and

$$j^0_i = \partial_0\phi \, \partial_i\phi. \tag{3.23}$$

Note that non-vanishing momentum is possible only when the field varies in time and space. Moreover, the momentum does not depend on the potential $m^2\phi^2/2 + V(\phi)$.

Because $\phi(x)$ can take arbitrary real values, the cubic term $\lambda\phi^3$ in j_0^0 can also have arbitrary values, from minus to plus infinity. Arbitrarily large positive values of energy are regarded as physically acceptable, but at the same time one does expect that values of energy should be bounded from below. Therefore, the model with cubic interaction potential is used mainly as a relatively simple example of a field theory with interactions, convenient for illustrating methods of field theory. Also note that the Klein–Gordon model ($V(\phi) = 0$) would have the energy unbounded from below if $m^2 < 0$. Precisely for this reason, we have assumed in the Klein–Gordon equation (1.33) that $m^2 \geq 0$.

Much more interesting is the model with quartic interaction energy

$$V(\phi) = \frac{\lambda}{4!}\phi^4(x), \tag{3.24}$$

where $\lambda > 0$ in order to ensure that the corresponding total energy E is bounded from below, and the factor $1/4!$ is included for later convenience. Now the Euler–Lagrange equation (3.17) has the form

$$\partial_\mu\partial^\mu\phi(x) + m^2\phi(x) + \frac{\lambda}{3!}\phi^3(x) = 0. \tag{3.25}$$

The total energy E is given by the following formula

$$E = \int_{R^3} d^3x \left(\frac{1}{2}\partial_0\phi\partial_0\phi + \frac{1}{2}\partial_i\phi\partial_i\phi + \frac{1}{2}m^2\phi^2 + \frac{\lambda}{4!}\phi^4\right). \tag{3.26}$$

Due to the presence of the positive quartic term, the energy is bounded from below also for negative m^2.

It turns out that physical predictions of the model crucially depend on the sign of m^2. Let us first consider the case $m^2 \geq 0$. It is obvious that the minimum value of the total energy $E = 0$ is obtained for $\phi(x) = 0$. This trivial trajectory of the field is called the classical ground state[3] of the field. If the field is close to the ground state, then we may neglect the interaction term $\lambda\phi^3/3!$ in (3.25), and we obtain the familiar Klein–Gordon equation. The fields which are close to the ground state form a so called ground state sector in the space of solutions of the Euler–Lagrange equations. Fields from this sector can be written as superpositions of the plane waves $f_{\vec{k}}$ introduced in Sect. 1.3 with the amplitudes $a_\pm(\vec{k})$ which are approximately constant in time as long as the interaction term is small.

The model (3.24) with $m^2 < 0$ is a little bit more intricate. It exhibits spontaneous symmetry breaking, and it has sectors characterized by a topological charge. First, let us notice that the Lagrangian can be rewritten in the form

[3]Often another term is used, namely the classical vacuum.

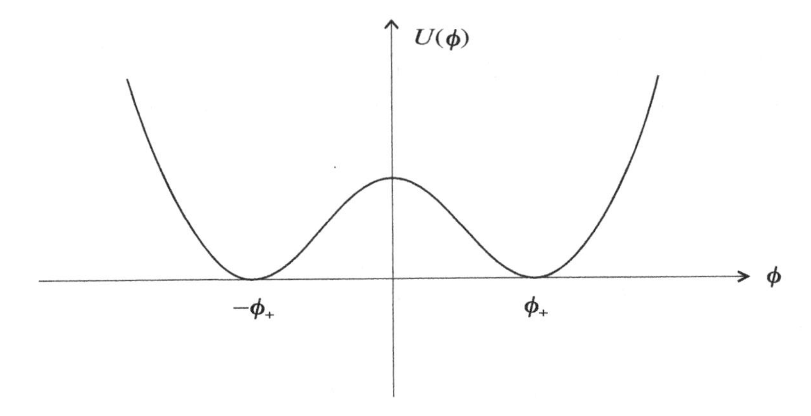

Fig. 3.1 The shape of the potential $U(\phi)$ given by formula (3.27)

$$\mathcal{L} = \frac{1}{2}\partial^{\mu}\phi(x)\partial_{\mu}\phi(x) - U(\phi) + \frac{3m^4}{2\lambda},$$

where

$$U(\phi) = \frac{\lambda}{4!}\left(\phi^2 - \frac{6|m^2|}{\lambda}\right)^2 \qquad (3.27)$$

($|m^2|$ denotes the modulus of m^2). Furthermore, we omit the last term in $\mathcal{L}$ because it does not contribute to the Euler-Lagrange equation and gives a trivial constant in $T^0_{\ 0}$. The energy density $T^0_{\ 0}$ calculated from the new Lagrangian has the form

$$T^0_{\ 0} = \frac{1}{2}\partial_0\phi(x)\partial_0\phi(x) + \frac{1}{2}\partial_i\phi(x)\partial_i\phi(x) + U(\phi). \qquad (3.28)$$

We see that it is bounded from below by 0. It reaches its minimum value 0 for constant $\phi = \pm\phi_+$, where

$$\phi_+ = \sqrt{\frac{6|m^2|}{\lambda}}, \qquad (3.29)$$

see Fig. 3.1.

Thus, there are two classical ground states $\pm\phi_+$. They are transformed into each other by the transformation S

$$S : \phi(x) \to S\phi(x) = -\phi(x).$$

Actually, this transformation is a symmetry of the model: the Lagrangian does not change its form if we write it as a function of $S\phi$. It follows that $S\phi(x)$ is a solution of the Euler–Lagrange equation together with $\phi(x)$. The fact that the classical ground states of the model are not invariant under such a symmetry transformation is called the spontaneous symmetry breaking (SSB). In the case $m^2 > 0$ the symmetry S is also present, but the ground state $\phi = 0$ is invariant under it.

Examples of the spontaneous symmetry breaking are already ubiquitous in classical mechanics. For example, consider a point particle with a one dimensional configuration space and with the energy $m\dot{q}^2/2 + V(q)$, where $V(q) = a(q^2 - b^2)^2$, and a and b are positive constants. There are two classical ground states $q = \pm b$ with the same energy (equal to zero), and neither of them is invariant under the symmetry transformation $q \to -q$. The quantum mechanical counterpart of this model has the Hamiltonian $\hat{H} = \hat{p}^2/(2m) + V(\hat{q})$, where $\hat{p}, \hat{q}$ are momentum and position operators, respectively. It turns out that this quantum Hamiltonian has a single ground state (with the energy >0)—the degeneracy of the ground state is absent. The corresponding wave function $\psi(q)$ is a symmetric function of q, hence it is invariant under the symmetry transformation.

This lack of SSB in the quantum case can be explained with the help of Heisenberg's uncertainty relation. If the quantum particle is confined to a finite segment of the q-axis with length Δq, then it can not have any fixed value of momentum p—all momenta from a band of width $\Delta p \approx \hbar/\Delta q$ are present. Because we look for the least energy state, we assume that this band contains the momenta with modulus from 0 up to Δp—a shift of the band towards higher momenta would give higher expectation values of the kinetic energy $\hat{p}^2/(2m)$. Thus, we may estimate that the expectation value of the kinetic energy is not larger than $(\Delta p)^2/(2m)$. It is clear that this contribution is minimized when in the ground state the particle occupies as large an interval Δq as possible. The only limitation is that the particle should avoid the regions where the potential V has large values, otherwise the gain in the kinetic energy would be overwhelmed by an increase of the expectation value of the potential energy. This means that the values of the ground state wave function should be as close to zero as possible in such regions. Therefore, we expect that the normalized ground state wave function does not vanish close to the two minima of the potential $V(q)$, while in all other regions it is close to zero. Then Δq is as large as possible, and the expectation value of the potential energy is small. The quantum particle adjusts its wave function globally in space taking into account all minima of the potential. Not surprisingly, there exists just one state that has the least energy.

In the heuristic reasoning presented above, we have been concerned directly with energy eigenfunctions. The complementary view is obtained by inspecting the time evolution of a wave packet which initially is localized around one of the minima of the potential, say $q = -b$. Even if the initial wave function vanishes in the region $q \geq 0$ at the initial instant $t = 0$, due to quantum tunneling through the potential barrier which separates the two minima of $V(q)$ it will not vanish in that region when $t > 0$. Actually, it turns out that the wave packet oscillates between the two minima. If we switch on a 'cooling procedure', that is if we gradually take away some energy from the particle, it will finally reach the ground state with the corresponding wave function evenly distributed around each of the two minima.

Similar results are obtained for systems with an arbitrary finite number of degrees of freedom, for instance, for several particles. Of course, the probability that all particles will tunnel decreases with the number of particles. For example, in the case of N mutually noninteracting, distinct particles the total probability is equal to the product $p_1 p_2 \dots p_N$, where p_i is the probability of tunneling for the i-th particle, $i = 1, 2 \dots, N$. Of course $p_i < 1$, so it follows that in the field theoretical limit, when the number of degrees of freedom is infinite, the total tunneling probability vanishes. In particular, the 'wave function' of our scalar field ϕ will stay close to one of the vacuum fields ϕ_+ or $-\phi_+$ forever if it is localized around it at a certain initial time. Therefore, we expect that the degeneracy of the ground state can be present also in a quantum version of our model (3.27). The spontaneous symmetry breaking in field theory does not have to disappear when we pass to a quantum version of the classical model as it did in the case of passing from classical mechanics to quantum mechanics.

Let us now have a look at small perturbations of the classical ground states. For concreteness, we consider perturbations of ϕ_+, that is the fields of the form

$$\phi(x) = \phi_+ + \epsilon(x), \tag{3.30}$$

where $\epsilon(x)$ is small in comparison with ϕ_+. Substituting formula (3.30) into (3.25) (in which $m^2 = -|m^2| < 0$), expanding with respect to ϵ and keeping only the terms linear in ϵ we again obtain the Klein–Gordon equation, namely

$$(\partial_\mu \partial^\mu + 2|m^2|)\, \epsilon(x) = 0. \tag{3.31}$$

We see that $\epsilon(x)$ has an effective mass coefficient $m_{\text{eff}}^2 = 2|m^2|$ which is positive. In consequence, $\epsilon(x)$ can be written as a superposition of the normalized plane waves $f_{\vec{k}}(x)$ with the frequencies $k_0 = \pm\omega(\vec{k})/c$, where

$$\omega(\vec{k}) = c\sqrt{\vec{k}^{\,2} + m_{\text{eff}}^2}$$

is positive. Small perturbations around the other ground state ϕ_- have the same effective mass coefficient.

Analogous expansion around $\phi = 0$ leads to the following equation

$$(\partial_\mu \partial^\mu - |m^2|)\epsilon(x) = 0.$$

It also has the plane wave solutions $f_{\vec{k}}(x)$, but now

$$\omega(\vec{k}) = c\sqrt{\vec{k}^{\,2} - |m^2|}.$$

We see that the modes with the wave vectors $\vec{k}$ such that $\vec{k}^{\,2} < |m^2|$ have imaginary frequencies. They do not oscillate in time, but monotonically increase if Im $k_0 < 0$ or decreases if Im $k_0 > 0$. The increasing amplitude means that after some time, $\epsilon(x)$ is no longer a small correction to ϕ_+ and one has to include the terms quadratic and cubic in $\epsilon(x)$. Therefore, the linear approximation around $\phi = 0$ is of limited use. One says that the constant field $\phi = 0$ is unstable with respect to the small oscillations, as opposed to the constant fields $\pm\phi_+$. It is clear that difference in the behavior of the small perturbations is due to the fact that the field potential $U(\phi)$ has minima at $\pm\phi_+$, while at $\phi = 0$ it has a local maximum.

The presence of SSB is a special property that can have rather interesting consequences. Related to the presence of the two ground states in the model (3.27) is the existence of a particular class of static solutions of the field equation (3.25). These solutions, called planar domain walls, smoothly interpolate between the two ground states in the following sense. Let us choose a plane in R^3 space. Without any loss of generality it can be the $x^3 = 0$ plane. Then, the coordinates x^1 and x^2 parameterize the plane, and x^3 varies in the direction perpendicular to the plane. Let us assume that the field ϕ is constant in each plane parallel to the $x^3 = 0$ plane, i.e., that ϕ can depend only on x^3: $\phi = \phi(x^3)$. Planar domain walls are solutions which merge with the two ground states when $x^3 \to \pm\infty$, that is, by definition, they obey the following boundary conditions

$$\lim_{x^3\to-\infty} \phi(x^3) = -\phi_+, \quad \lim_{x^3\to+\infty} \phi(x^3) = \phi_+. \tag{3.32}$$

When ϕ depends only on x^3, (3.25) is reduced to the following ordinary differential equation

$$\partial_3^2\phi + |m^2|\phi - \frac{\lambda}{3!}\phi^3 = 0. \tag{3.33}$$

Multiplying it by $2\partial_3\phi$, and integrating we obtain the equation

$$(\partial_3\phi)^2 + |m^2|\phi^2 - \frac{2\lambda}{4!}\phi^4 = \text{const}. \tag{3.34}$$

The boundary conditions (3.32) determine the integration constant

$$\text{const} = |m^2|\phi_+^2 - \frac{2\lambda}{4!}\phi_+^4 = \frac{3|m^2|^2}{\lambda}.$$

Equation (3.34) can be written in the form

$$\frac{1}{2}(\partial_3\phi)^2 - U(\phi) = 0. \tag{3.35}$$

It is easy to check that

$$\phi_d(x^3) = \phi_+ \tanh\frac{\sqrt{|m^2|}(x^3 - x_0^3)}{\sqrt{2}} \tag{3.36}$$

obeys (3.35) and the boundary conditions (3.32). In the solution (3.36), x_0^3 is another integration constant. Its value is arbitrary. Physically, it gives the position of the

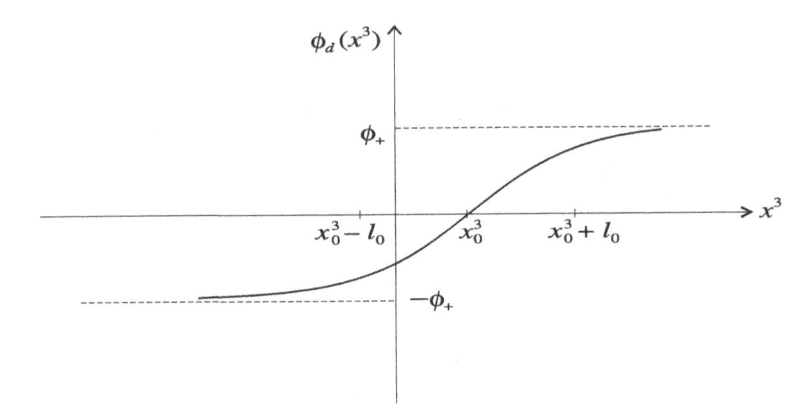

Fig. 3.2 The plot of the function $\phi_d(x^3)$

domain wall along the x^3-axis, and its change corresponds to a translation of the domain wall parallel to the (x^1, x^2)-plane. Note that $\phi_d(x^3)$ vanishes at $x^3 = x_0^3$, and the potential energy $U(\phi_d(x^3))$ has the largest value there. For large positive x^3

$$\phi_d(x^3) \cong \phi_+ - 2\phi_+ \exp(-\sqrt{2|m^2|}(x^3 - x_0^3)),$$

and for large negative x^3

$$\phi_d(x^3) \cong -\phi_+ + 2\phi_+ \exp(\sqrt{2|m^2|}(x^3 - x_0^3)).$$

Thus, for $|x^3 - x_0^3| \gg l_0$, where $l_0 = 1/\sqrt{2|m^2|}$, the domain wall solutions practically merge with the classical ground states. The constant l_0 is equal to the inverse of the effective mass coefficient m_{eff}. It essentially gives the thickness of the planar domain walls. The function $\phi_d(x^3)$ is plotted in Fig. 3.2.

The energy density for the domain wall is given by

$$T^0_{\ 0} = \frac{1}{2}(\partial_3\phi_d)^2 + U(\phi_d) = \frac{3m^4}{\lambda \cosh^4(\sqrt{|m^2|}(x^3 - x_0^3)/\sqrt{2})}.$$

It has maximal value at $x^3 = x_0^3$, and it exponentially approaches 0 when $x^3 \to \pm\infty$, see Fig. 3.3. Recall that ϕ_d vanishes precisely at $x^3 = x_0^3$. Existence of at least one zero is implied by the boundary conditions (3.32) because $\phi(x^3)$ is by assumption a continuous function of x^3.

The total energy of the domain wall is of course infinite, because of the integration over x^1 and x^2. The energy density per unit area, denoted below by σ, is finite and constant along the domain wall. It is given by the integral

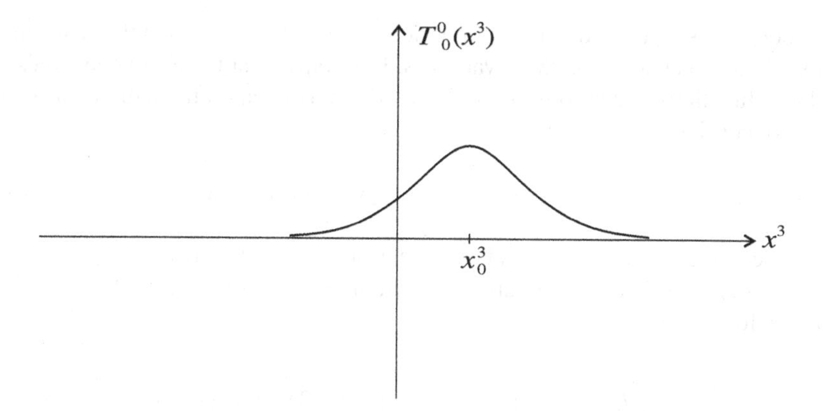

Fig. 3.3 The energy density across the domain wall

$$\sigma = \int dx^3 \, T^0_{\ 0} = \frac{3m_{\text{eff}}^3}{2\lambda} \int_{-\infty}^{+\infty} \frac{ds}{\cosh^4 s} = \frac{2m_{\text{eff}}^3}{\lambda}. \tag{3.37}$$

Note that σ becomes infinite when the coupling constant λ decreases to zero. Such singularity at $\lambda = 0$ means that the domain wall cannot be obtained as a perturbative expansion in positive powers of λ. In this sense, the presence of the domain walls is a non perturbative phenomenon.

Because of their infinite energy, strictly planar infinite domain walls are not physically possible. Nevertheless, they are quite useful in the theoretical analysis of other domain walls which have finite energy. There exist closed domain walls, e.g., spherical ones, which have finite total energy. The corresponding solutions of (3.25) are not static—ϕ depends also on time. Such solutions approach $\pm\phi_+$ in the directions perpendicular to the domain wall, but the domain wall shrinks or expands. The total potential energy is approximately equal to σS, where S denotes the area of the domain wall, regarded (approximately) as an infinitely thin surface. The total energy also contains a finite kinetic energy which does not vanish because of the time dependence of ϕ. If, at a given point of the domain wall, its curvature is not very large, one may expect that the solution will not be very different from $\phi_d(x^3)$ around that point, except that now x^3 is replaced by a coordinate perpendicular to the domain wall. Another class of finite energy domain walls appears in condensed matter physics. They just end on the boundaries of the material in which they are created. If the bulk of the material is sufficiently large, the surface effects can be neglected, and again, our infinite planar domain wall can be quite a reasonable first approximation.

Let us investigate small perturbations of the planar domain wall. Substituting $\phi(x) = \phi_d(x^3) + \epsilon(x)$ into (3.25) and neglecting terms which are quadratic or cubic in ϵ we obtain the following linear equation for $\epsilon(x)$

$$\partial_\mu \partial^\mu \epsilon - |m^2|\epsilon + \frac{\lambda}{2}\phi_d^2(x^3)\epsilon = 0. \tag{3.38}$$

The coefficients in this equation do not depend on x^0, x^1, x^2. Therefore, we may factorize the dependence on these variables. It is convenient to take exponentials as the basis functions, other solutions can be written as linear combinations of them. Thus, we consider $\epsilon(x)$ of the form

$$\epsilon(x) = \exp(-ik_0x^0)\exp(ik_1x^1 + ik_2x^2)\psi(x^3).$$

It is understood that in fact we take the real or imaginary part of this expression because $\epsilon(x)$ should have real values. Equation (3.38) is reduced to the following equation for $\psi(x^3)$

$$k_0^2\psi = [(k_1)^2 + (k_2)^2]\psi - \partial_3^2\psi + |m^2|(3\tanh^2\frac{x^3}{2l_0} - 1)\psi, \tag{3.39}$$

where we have put $x_0^3 = 0$ for simplicity. Suppose that we know the solutions of the auxiliary eigenvalue problem

$$-\frac{1}{2}\partial_3^2\psi + \frac{|m^2|}{2}(3\tanh^2\frac{x^3}{2l_0} - 1)\psi = \kappa\psi, \tag{3.40}$$

where κ is the eigenvalue and ψ the eigenfunction. Then,

$$k_0^2 = (k_1)^2 + (k_2)^2 + 2\kappa.$$

It is clear that if there exists a negative eigenvalue κ then we can have $k_0^2 < 0$ if k_1^2 and k_2^2 are sufficiently small. This would imply that exponentially growing modes are present. In physical realizations of the domain wall, we can never exactly construct the one given by ϕ_d—small perturbations are always present. If there is a growing mode having finite energy per unit square it will significantly modify the domain wall or even destroy it. Therefore, it is important to check the sign of the eigenvalues κ.

When looking for eigenvalues it is important to specify which eigenfunctions we allow. In our case, relevant eigenfunctions are those which can give the perturbations $\epsilon(x)$ with finite energy per unit area. Thus, apart from ψ vanishing when $x_3 \to \pm\infty$, we also admit eigenfunctions which become plane waves in these limits because one can construct from them wave packets with finite energy.

It turns out that there exists just one eigenfunction with $\kappa = 0$, called the translational zero mode, and all other eigenfunctions have strictly positive eigenvalues ($\kappa > 0$). The existence of the zero mode is related to the translational invariance of the model, which is responsible for the presence of the arbitrary constant x_0^3 in the domain wall solution (3.36). This solution inserted on the l.h.s. of (3.33) gives an identity. Let us differentiate both sides of this identity with respect to x_0^3 and put $x_0^3 = 0$ afterwards. The resulting identity has the form (3.40) with the eigenfunction

$$\psi_0(x^3) = \frac{\partial \psi_d(x^3 - x_0^3)}{\partial x_0^3}\bigg|_{x_0^3=0} = -\frac{\partial \psi_d(x^3)}{\partial x^3} = \sqrt{\frac{3}{4\lambda}}\,\frac{m_{\text{eff}}^2}{\cosh^2(m_{\text{eff}}\,x^3/2)}, \tag{3.41}$$

and $\kappa = 0$. The zero mode $\psi_0(x^3)$ does not vanish for any finite x^3. There is a theorem, discussed in textbooks on quantum mechanics, which says that the eigenvalue corresponding to such non vanishing eigenfunction is the smallest one. Therefore, all other eigenvalues κ are positive.[4] In conclusion, the planar domain wall is stable with respect to the small perturbations.

The perturbation of the planar domain wall given by $\epsilon(x) = a\psi_0(x^3)$, where a is a small number, is time independent. It results in the uniform, parallel shift of the domain wall along the x^3-axis,

$$\phi_d(x^3) + a\psi_0(x^3) \cong \phi_d(x^3 + a).$$

The perturbations with $(k_1)^2 + (k_2)^2 > 0$ give waves traveling along the planar domain wall.

3.3 The Complex Scalar Field

The complex scalar field is mathematically represented by a function $\phi(x)$, $x \in M$, which can have complex values. Similarly as in the case of real scalar field, we require that under the Poincaré transformations $x' = \hat{L}x + a$

$$\phi'(x') = \phi(x). \tag{3.42}$$

A typical Lagrangian for the complex scalar field has the form

$$\mathcal{L} = \partial_\mu \phi^*(x)\partial^\mu \phi(x) - m^2\phi^*(x)\phi(x) - V(\phi^*(x)\phi(x)), \tag{3.43}$$

where * denotes the complex conjugation. Equivalently, one may replace the complex scalar field by two real scalar fields $\phi_1(x)$ and $\phi_2(x)$,

$$\phi(x) = \frac{1}{\sqrt{2}}(\phi_1(x) + i\phi_2(x)).$$

Lagrangian (3.43) is equal to

$$\mathcal{L} = \frac{1}{2}\partial_\mu\phi_1\partial^\mu\phi_1 + \frac{1}{2}\partial_\mu\phi_2\partial^\mu\phi_2 - \frac{1}{2}m^2\phi_1^2 - \frac{1}{2}m^2\phi_2^2 - V\left(\frac{\phi_1^2 + \phi_2^2}{2}\right).$$

[4] Actually, the eigenvalue problem (3.40) is explicitly solved in textbooks on quantum mechanics. It turns out that apart from the zero mode there is one bound state with $0 < \kappa < |m^2|$ and a continuum of eigenfunctions with $\kappa \geq |m^2|$.

This form of the Lagrangian suggests a generalization to the so called $O(N)$ models. Let $\vec{\phi}$ denote a multiplet of N real scalar fields

$$\vec{\phi}(x) = \begin{pmatrix} \phi_1(x) \\ \phi_2(x) \\ \vdots \\ \phi_N(x) \end{pmatrix},$$

and $\vec{\phi}^2 = \sum_{i=1}^{N} \phi_i \phi_i$. The Lagrangian of the $O(N)$ model has the form

$$\mathcal{L} = \frac{1}{2}\partial_\mu \vec{\phi} \partial^\mu \vec{\phi} - \frac{1}{2} m^2 \vec{\phi}^2 - V\left(\vec{\phi}^2\right).$$

It has a global $O(N)$ symmetry which consists of transformations

$$\vec{\phi}'(x) = \mathcal{O}\vec{\phi}(x),$$

where $\mathcal{O}$ denotes an arbitrary $N \times N$ real matrix which obeys the condition $\mathcal{O}^T\mathcal{O} = I_N$, $\mathcal{O}^T$ denotes the transposed matrix, I_N is the N by N unit matrix. Such matrices form the orthogonal matrix group $O(N)$. The $O(N)$ models play an important role in applications of field theory. They also provide a testing ground for certain mathematical techniques developed in field theory.

Lagrangian (3.43) is invariant under the Poincaré transformations. Moreover, it also possesses a $U(1)$ global symmetry. The $U(1)$ group consists of all complex numbers z such that $|z| = 1$, or equivalently, of all phase factors $\exp(i\alpha)$. The $U(1)$ transformations of the complex scalar field have the form

$$\phi'(x) = \exp(iq\alpha)\phi(x), \tag{3.44}$$

where q is an integer different from 0. For a non integer q the transformation (3.44) would be multi-valued. Of course, α does not depend on x, as expected for the global transformations. Note that the space-time points x are not transformed. In such cases one says that the symmetry is an internal one.

Let us calculate the conserved current j_ν corresponding to the $U(1)$ symmetry using the formalism developed in Chap. 2. As the parameter on the group in a vicinity of the unit element we may take α. The Lie derivative of the field ϕ has the form

$$\mathcal{D}\phi(x) = iq\phi(x). \tag{3.45}$$

Lagrangian (3.43) contains also the complex conjugate field ϕ^*. Its transformation law is obtained by taking the complex conjugate of formula (3.44), and

$$\mathcal{D}\phi^*(x) = -iq\phi^*(x). \tag{3.46}$$

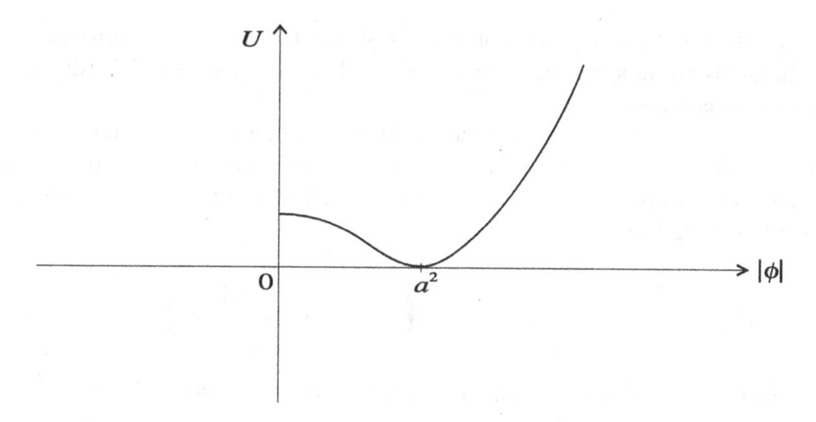

Fig. 3.4 The potential $U(|\phi|)$ given by formula (3.49)

The formula for the conserved current follows from the general formula (2.32). In the present case $K = 0$, $\xi = 0$, and as the fields u_a we take ϕ and ϕ^*. Alternatively, we could take the real fields ϕ_1, ϕ_2. Simple calculation gives

$$j_\nu = iq\left(\partial_\nu\phi\phi^* - \phi\partial_\nu\phi^*\right). \tag{3.47}$$

The corresponding conserved charge Q is given by the integral

$$Q = \int_{R^3} d^3x \; j^0.$$

Note that Q vanishes if the imaginary part of ϕ is equal to zero, or if ϕ is constant in time.

The choice $V(\phi^*\phi) = 0$ in Lagrangian (3.43) gives the free complex scalar field model. In this case the Euler–Lagrange equations are linear in ϕ. They coincide with the familiar Klein–Gordon equation for the real and imaginary parts of ϕ.

Another important particular choice of the interaction potential U gives the Goldstone model. In this case

$$\mathcal{L} = \partial_\mu\phi^*\partial^\mu\phi - U, \tag{3.48}$$

where

$$U = \frac{\lambda}{4!}\left(\phi^*(x)\phi(x) - \frac{12|m^2|}{\lambda}\right)^2 \tag{3.49}$$

($V = U - m^2\phi^*\phi$ with $m^2 < 0$). The potential U regarded as a function of $|\phi|$ is shown in Fig. 3.4. Often it is called the 'Mexican hat' potential. This name refers to the characteristic shape of the surface obtained by plotting U over the plane (ϕ_1, ϕ_2).

Calculation of the term proportional to $\phi^*\phi$ shows that $m^2 = -|m^2|$ is negative. In fact, the Goldstone model is the complex field analogue of the model (3.27) discussed in the previous section.

Let us find the classical ground states in this model, that is the fields for which the energy density $T^0_{\ 0}$ acquires its minimal value. The energy density for the field ϕ with Lagrangian (3.48) is easily obtained from the general formulas given in Chap. 2. It has the following form

$$T^0_{\ 0} = \partial_0\phi^*\partial_0\phi + \partial_i\phi^*\partial_i\phi + \frac{\lambda}{4!}\left(\phi^*(x)\phi(x) - \frac{12|m^2|}{\lambda}\right)^2. \tag{3.50}$$

It is clear that the least energy density is obtained for any constant ϕ such that

$$\phi^*\phi = a^2, \tag{3.51}$$

where

$$a = \sqrt{\frac{12|m^2|}{\lambda}}. \tag{3.52}$$

The set of all classical ground states is called the vacuum manifold. We denote it by $\mathcal{V}$. In the model discussed in the previous section it consists of just two points $\pm\phi_+$. In the Goldstone model, the vacuum manifold is defined by condition (3.51), hence it consists of all constant fields of the form

$$\phi = a\exp(i\beta), \tag{3.53}$$

where $\beta \in [0, 2\pi)$. It can be regarded as a circle of radius a. There is no classical ground state which would be invariant under the $U(1)$ transformations (3.44)—these transformations move the classical ground states along the vacuum manifold. Thus, we see that the Goldstone model exhibits spontaneous breaking of the $U(1)$ symmetry.

Let us compare the present example of SSB with the one discussed in the previous section. The main difference is that $U(1)$ is a continuous group, while the symmetry S together with the identity $I : \phi(x) \to \phi(x)$ form the two element discrete group Z_2, $Z_2 = \{S, I\}$. This difference has profound physical consequences. In particular, it turns out that the real fields ϕ_1 and ϕ_2 are not well suited to describe the physical contents of the model, analogously as, for instance, Cartesian coordinates are not the best choice when considering a problem which has only axial symmetry. A much better parametrization of the complex field ϕ of the Goldstone model is provided by two real fields $\chi(x)$ and $\Theta(x)$ introduced as follows:

$$\phi(x) = (a + \chi(x))\, e^{i\Theta(x)}, \tag{3.54}$$

where a is given by formula (3.52) and $\Theta \in [0, 2\pi)$. Thus, $\chi = 0$ and $\Theta = \text{const} = \beta$ corresponds to the classical ground state $a\exp(i\beta)$. This parametrization of the field space by χ and Θ is mathematically correct provided that $\chi \neq -a$. Let us assume for now that this is the case. In order to derive the equations of motion for χ and Θ one could use the Euler–Lagrange equation for the original field ϕ and formula (3.54). However, more enlightening is another way: we use the field transformation (3.54) directly in Lagrangian (3.48). This gives

$$\mathcal{L}(\chi, \Theta) = \partial_\mu \chi \, \partial^\mu \chi + (a+\chi)^2 \, \partial_\mu \Theta \, \partial^\mu \Theta - \frac{\lambda}{4!} \chi^2 (2a+\chi)^2. \tag{3.55}$$

This Lagrangian is used to generate the Euler–Lagrange equations for the fields χ and Θ. The resulting equations are equivalent to the ones obtained by substituting formula (3.54) into the Euler–Lagrange equation for ϕ. This follows from a general property of the stationary action principle, namely that nonsingular transformations of the fields in the action functional lead to equivalent Euler–Lagrange equations.

Let us prove this property. The transformation of the fields has the form $u_a = F_a(v_b)$, where (v_b) is the set of new fields. By definition, the action functional for the new fields has the form

$$\tilde{S}[v_b] = S[u_a]|_{u_a = F_a(v_b)} \, .$$

The functional derivatives of $\tilde{S}$ and S are related by the following formula

$$\frac{\delta \tilde{S}[v]}{\delta v_a(x)} = \int d^4y \, \frac{\delta S[u]}{\delta u_b(y)} K_{ba}(y, x), \tag{3.56}$$

where

$$K_{ba}(y, x) = \frac{\delta F_b(v_c(y))}{\delta v_a(x)}.$$

The assumption that the field transformation is nonsingular means that there exists $(K^{-1})_{ac}(x, z)$ such that

$$\int d^4x \, K_{ba}(y, x) \, (K^{-1})_{ac}(x, z) = \delta_{bc} \delta(y - z).$$

Therefore,

$$\int d^4x \, \frac{\delta \tilde{S}[v]}{\delta v_a(x)} (K^{-1})_{ac}(x, z) = \frac{\delta S[u]}{\delta u_c(z)}. \tag{3.57}$$

Relations (3.56) and (3.57) imply the equivalence of the Euler–Lagrange equations obtained from S and $\tilde{S}$,

$$\frac{\delta S[u]}{\delta u_a(x)} = 0 \Leftrightarrow \frac{\delta \tilde{S}[v]}{\delta v_a(x)} = 0.$$

In the case of a singular transformation, it could happen that the r.h.s. of formula (3.56) vanishes, and then $\delta\tilde{S}/\delta v_a = 0$, even if $\delta S[u]/\delta u_b(x) \neq 0$.

Lagrangian (3.55) does not contain any potential for the Θ field. It is invariant with respect to translations of values of this field of the form $\Theta(x) \to \Theta(x) + \Theta_0$, where Θ_0 is an arbitrary real constant. The Θ field is called the Goldstone field or Goldstone mode. The Euler–Lagrange equation for it has the form

$$\partial_\mu \left((a+\chi)^2 \, \partial^\mu \Theta(x) \right) = 0. \tag{3.58}$$

If χ is close to its ground state value 0 we may neglect χ in Lagrangian (3.55) and in (3.58). This is the so called London approximation, named after F. London and H. London who used an analogous approximation in their theory of superconductors. In this case (3.58) acquires the form of wave equation: the Klein–Gordon equation with $m^2 = 0$. Note also that when $\chi = 0$, i.e. when only the Goldstone field is present, the field $\phi = a \exp(i\Theta)$ does not leave the vacuum manifold $\mathcal{V}$. The energy density (3.50) is reduced to

$$T^0_{\ 0}(\Theta) = a^2 \, (\partial_0\Theta \, \partial_0\Theta + \partial_i\Theta \, \partial_i\Theta).$$

Note that it only contains terms with derivatives—this is a characteristic feature of Goldstone fields, seen also in other models.

Now let us have a look at the χ field. Assuming that its values are small in comparison with a, and keeping in Lagrangian (3.55) only the terms quadratic in χ and Θ, we obtain the so called free part of the Lagrangian,

$$\mathcal{L}_0 = \partial_\mu\chi\partial^\mu\chi - 2|m^2|\chi^2 + a^2\partial_\mu\Theta\partial^\mu\Theta. \tag{3.59}$$

It is clear that the Euler–Lagrange equations generated from $\mathcal{L}_0$ have the form of separate Klein–Gordon equations for Θ and χ, with mass coefficients equal to 0 and $2|m^2|$, respectively. Note another peculiarity of the Goldstone field Θ: Lagrangian (3.55) is already quadratic in Θ, hence we do not need any assumption that Θ is small.

We have seen in the previous section that the presence of a nontrivial vacuum manifold results in the presence of the domain walls, which are surface-like extended objects. Non triviality of the vacuum manifold in the Goldstone model suggests the existence of extended objects which are line-like. They are called vortices.

In the real scalar field model (3.27) the vacuum manifold consists of two points $\pm\phi_+$. Let us assume that there are two points $\vec{x}_1$ and $\vec{x}_2$ in the space, such that $\phi(t_0, \vec{x}_1) = \phi_+$ and $\phi(t_0, \vec{x}_2) = -\phi_+$ at a certain time t_0. Let us try to extend the field to the entire three-dimensional space. We take a certain small vicinity of $\vec{x}_1$ and assume that $\phi(t_0, \vec{x}) = \phi_+$ for all $\vec{x}$ within that vicinity. Similarly, we take a certain small vicinity of $\vec{x}_2$ and assume that $\phi(t_0, \vec{x}) = -\phi_+$ in it. Gradually increasing the two regions, we finally arrive at the stage where they fill the whole space and touch each other at a certain surface. The field ϕ is not continuous at that surface. In order to remove the discontinuity, let us replace this surface by a layer of finite thickness, and choose $\phi(t_0, \vec{x})$ such that it smoothly interpolates between $\pm\phi_+$ across the layer. The resulting field configuration $\phi(t_0, \vec{x})$, now defined on the whole space, is taken as initial data for the field equation (3.25). We also have to specify the time derivative of ϕ in order to have the complete set of initial data. We do not impose any special restrictions on this part of the initial data, we may take, for example, $\partial_t\phi(t, \vec{x})|_{t=t_0} = 0$. In this manner we have constructed a domain wall, which in general is curved. The solution of the field equation corresponding to such initial data has a nontrivial dependence on time. Note that the function $\phi(t_0, \vec{x})$ has to vanish on a surface lying somewhere inside the border layer, because it changes sign across it. On that surface the potential energy $U(\phi)$ has a local maximum. The planar domain wall discussed in the previous section is distinguished by the fact that it is static. The solution presented there shows that the field ϕ of the static domain wall reaches the ground state values asymptotically, and only at the spatial infinity in the directions perpendicular to the wall. Finally, let us stress that the existence of the domain wall is the consequence of the choice of the values of the function $\phi(t_0, \vec{x})$ at the points $\vec{x}_1, \vec{x}_2$, and of its continuity, irrespectively of the form of the Euler–Lagrange equation.

In the case of the Goldstone model the vacuum manifold is the circle given by formula (3.53). Therefore, in analogy with the case of domain walls, we choose a circle C of radius R_0 in the space, parameterized by the angle $\theta \in [0, 2\pi)$, and we assume that at different points of this circle the scalar field takes different values from the vacuum manifold. The simplest choice is $\phi(t_0, \vec{y}) = a\exp(i\theta)$ for $\vec{y} \in C$. Actually, there also exist other possibilities which we will discuss later. Now, let us try to define a smooth field $\phi_v(t_0, \vec{x})$ on the whole space. It is a more complicated task than in the case of the domain wall because ϕ is not constant on the circle. First, let us extend the circle C to an infinite cylinder $C \times R^1$ by adding at each point $\vec{y} \in C$ a straight-line perpendicular to the plane of the circle. On each such straight-line ϕ_v is a constant, equal to $\phi(t_0, \vec{y})$. Next, we expand the cylinder to the whole space. In this step, each point of the cylinder is translated along half of the straight-line perpendicular to the cylinder. We assume that $\phi_v(t_0, \vec{x})$ is constant on each such half-line. In this manner we have uniquely assigned a value to ϕ_v at each point of the space, except for the symmetry axis of the cylinder $C \times R^1$, where ϕ_v is not continuous—approaching this axis from various directions perpendicular to it, we obtain different values of ϕ_v. To remove this discontinuity we choose, inside the initial cylinder $C \times R^1$, a cylindrical volume U_ϵ around the symmetry axis, and allow the field ϕ_v to depart from the vacuum manifold within it.

Mathematical arguments based on the homotopy theory show that a smooth function $\phi_v(t_0, \vec{x})$ can be obtained only if ϕ_v vanishes somewhere in U_ϵ. To see this, suppose to the contrary, that ϕ_v does not vanish inside the cylinder $C \times R^1$. Then, the modulus of ϕ_v also does not vanish, and the phase factor $\phi_v/|\phi_v|$ is well-defined. It is a continuous function of $\vec{x}$ because by assumption ϕ_v is a continuous function. Phase factors can be regarded as points of the unit circle $S^1 = \{z : |z| = 1\}$ in the complex plane. Now, consider

$$f(\epsilon, \theta) = \phi_v(t_0, \vec{x})/|\phi_v(t_0, \vec{x})|$$

with points $\vec{x}$ restricted to a circle C_ϵ, which is co-planar and concentric with C and has the radius $\epsilon < R_0$. Each circle C_ϵ is parameterized by the same angle θ which parameterizes the circle C. Because $|f(\epsilon, \theta)| = 1$, it is clear that $f(\epsilon, \theta)$ with fixed ϵ is a continuous mapping from C_ϵ to the circle S^1. Moreover, the definition of $f(\epsilon, \theta)$ implies that it is a continuous function of ϵ, too. For $\epsilon = 0$ $f(\epsilon, \theta)$ is constant because the circle $C_{\epsilon=0}$ is just a point, the center of the circle C. For $\epsilon = R_0$ we obtain the initial circle C, and $f(R_0, \theta) = \exp(i\theta)$. Thus, we have constructed a continuous deformation of the constant mapping $f(0, \theta)$ into $f(R_0, \theta)$. But this contradicts a theorem from the homotopy theory which says that such deformations do not exist. Therefore, the assumption that $\phi_v \neq 0$ must be false. Because the presence of the zeros of ϕ is implied by homotopy theory, a branch of algebraic topology, they are called the topological zeros.

The nonexistence of continuous deformations of the constant mapping into the exponential mapping $\exp(i\theta)$ can be described in terms of the so called winding number $W[f]$, which characterizes any smooth mapping $f(\theta)$ from C to S^1. The winding number is defined as follows

$$W[f] = \frac{1}{2\pi i} \int_0^{2\pi} d\theta \, \frac{1}{f} \frac{df}{d\theta}, \tag{3.60}$$

where $|f(\theta)| = 1$. Let $\Theta_f(\theta)$ denote the phase of f, $f(\theta) = \exp(i\Theta_f(\theta))$. Formula (3.60) can be written in the form

$$W[f] = \frac{1}{2\pi} \int_0^{2\pi} d\theta \, \frac{d\Theta_f}{d\theta} = \frac{\Delta\Theta_f}{2\pi},$$

where $\Delta\Theta_f$ is the total change of the phase Θ_f during one pass along the circle C, $\Delta\Theta_f = \Theta_f(2\pi) - \Theta_f(0)$. Here by definition

$$\Theta_f(2\pi) = \lim_{\theta \to 2\pi_-} \Theta_f(\theta).$$

Because $f(\theta)$ is continuous on the circle C, we have $f(2\pi) = f(0)$ and $\Theta_f(2\pi) = \Theta(0) + 2\pi n$, where n is an integer. Therefore, $W[f] = n$ is an integer.

The winding number is constant under continuous deformations of the mapping f. In general, such a deformation $f \to g$ is represented by a function $h(\sigma, \theta)$ which is continuous in σ, differentiable in θ, and such that $h(0, \theta) = f(\theta)$ and $h(1, \theta) = g(\theta)$. Here $\sigma \in [0, 1]$ and θ parameterizes the circle C as before. Moreover, we demand that h has values in the unit circle S^1, i.e., that $|h(\sigma, \theta)| = 1$ for all σ and θ. Let us consider $W[h]$ obtained by inserting h on the r.h.s. of the formula (3.60). It is clear that the integral gives a continuous function of σ with integer values. Such a function has to be constant, hence $W[f] = W[g]$. For the constant mapping $f(0, \theta)$ the winding number is equal to zero. On the other hand, for $f(R_0, \theta) = \exp(i\theta)$ formula (3.60) gives $W[f(R_0, \theta)] = +1$. Therefore, these two mappings cannot be continuously deformed into each other.

From the mathematical arguments presented above we know that ϕ_v has to vanish at least at one point in each transverse cross section of the infinite cylinder $C \times R^1$. Note that at such points the potential energy $U(\phi_v)$ has a local maximum. Therefore, one may expect that the presence of several zeros of ϕ_v in these cross sections would increase the energy of the field. For this reason we assume that there is just one zero of ϕ_v in each transverse section of the cylinder. Let us choose one such cross section, e.g., the one with the circle C. The zero of ϕ_v in this cross section is enclosed by the circles C_ϵ of the arbitrarily small radius ϵ. Let us pick one such circle C_ϵ and shift it continuously through all planes parallel to the circle C in such a way that it does not pass through the zeros of ϕ_v. The winding number is constant during such translation. Because ϵ can be arbitrarily small, we see that the zeros have to form a continuous infinite line in the space.

Already at this point it is clear that such a field configuration has infinite total energy. The contribution to the potential energy from each finite segment of that line is proportional to its length. Analogously, as in the case of domain walls, this does not diminish the physical relevance of the vortices. Vortices akin to the ones discussed here are experimentally observed in superfluid ^4He.

Let us summarize our considerations. The field ϕ_v has the values $a \exp(i\theta)$ on the cylinder $C \times R^1$ and outside of it. Inside the cylinder, the field smoothly reaches the value zero on a continuous line extending to infinity in both directions. Such a field ϕ_v is taken as a part of the initial data for the field equation

$$\partial_\mu \partial^\mu \phi + \frac{\lambda}{12}\left(|\phi|^2 - \frac{12|m^2|}{\lambda}\right)\phi = 0, \tag{3.61}$$

which follows from Lagrangian (3.48). The remaining part of the initial data fixes $\partial_t \phi$ at the initial time t_0. There are no special restrictions on its choice. Such field $\phi(t_0, \vec{x})$ characterized by the unit winding number is called the infinite vortex. The field equation determines its time evolution. One may also construct initial data for which the line of zeros is closed. Such closed vortices have finite length, and finite total energy. Their time evolution can be rather nontrivial.

As in the case of domain walls, one may ask about a static vortex. To find it, we proceed analogously as in the case of the domain walls. We assume that the field has a special form ϕ_{sv}, frequently called the static vortex *Ansatz*, which is characterized by a high symmetry as explained below, and has a winding number equal to +1. In cylindrical coordinates (θ, ρ, z) on the R^3 space

$$\phi_{sv} = aF(\rho)e^{i\theta}, \tag{3.62}$$

where F is an unknown function of the cylindrical radius ρ. The presence of the topological zero is ensured by the assumption

$$F(0) = 0. \tag{3.63}$$

Thus, the z-axis coincides with the line of the topological zeros. The field ϕ_{sv} should approach the vacuum manifold at least when $\rho \to \infty$. Therefore, we also assume that

$$\lim_{\rho \to \infty} F(\rho) = 1 \tag{3.64}$$

(this does not exclude the possibility that $F(\rho) = 1$ for finite ρ). Formula (3.62) implies that ϕ_{sv} is homogeneous along the z-axis. Moreover, ϕ_{sv} is axially symmetric in the generalized sense that the effect of rotation around the z-axis can be compensated by a global $U(1)$ transformation. Indeed, after a rotation by θ_0 we have $\phi'_{sv}(\theta, \rho) = \exp(-i\theta_0)\phi_{sv}(\theta, \rho)$, and subsequent $U(1)$ symmetry transformation $\phi'_{sv}(\theta, \rho) \to \exp(i\theta_0)\phi'_{sv}(\theta, \rho)$ restores the initial field $\phi_{sv}(\theta, \rho)$.

For a field of the form (3.62) equation (3.61) is reduced to

$$\tilde{F}'' + \frac{\tilde{F}'}{s} - \frac{\tilde{F}}{s^2} + \tilde{F} - \tilde{F}^3 = 0, \tag{3.65}$$

where $s = \sqrt{|m^2|}\rho$ is the dimensionless variable replacing ρ, $\tilde{F}(s) = F(\rho)$, and $'$ denotes d/ds. Of course, $\tilde{F}$ also obeys conditions (3.63) and (3.64). Unfortunately, an exact analytic form of the solution F of (3.65) is not known. Assuming that $\tilde{F}(s)$ can be expanded in powers of s for small s, and solving (3.65) order by order in s we find that

$$\tilde{F}(s) \cong c_1 s + c_3 s^3 + \cdots,$$

where $c_3 = -c_1/8$. For large s, more natural is an expansion in powers of $1/s$, which gives

$$\tilde{F}(s) \cong 1 - \frac{1}{2s^2} + \cdots.$$

An approximate solution of (3.65) obeying conditions (3.63) and (3.64) can easily be found with the help of numerical methods. It has the form shown in Fig. 3.5. In particular, we find that $c_1 \approx 0.583$.

Note that for small values of ρ

$$\phi_{sv}(\theta, \rho) = c_1 a|m|(x + iy) + \ldots,$$

where x and y are the Cartesian coordinates in the plane perpendicular to the line of the topological zeros of ϕ_{sv} (the z axis). The dots denote terms of cubic and higher order in x and y, and i is the imaginary unit. Thus, first order derivatives of ϕ_{sv} with respect to x or y taken at $x = y = 0$ do not vanish. In this sense, the topological zero of ϕ_{sv} is of the first order.

We already know that the infinitely long vortex has infinite total energy. It also turns out that the energy per unit length is infinite. This energy is given by the integral

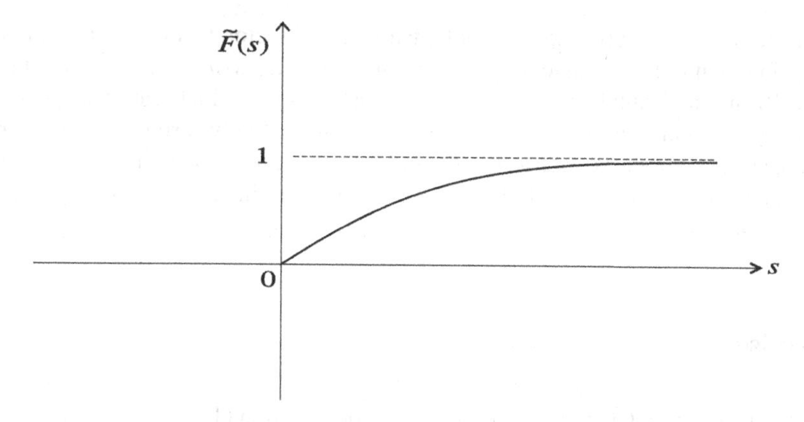

Fig. 3.5 The plot of the function $\tilde{F}(s)$

$$\int dxdy\, T^0_{\ 0} = 2\pi \int_0^\infty d\rho\, \rho \left[a^2 (\partial_\rho F)^2 + a^2 \frac{F^2}{\rho^2} + V(F^2) \right].$$

The factor 2π comes from integration over the angle θ. Because $F \cong 1$ for large s, the term F^2/ρ^2 gives a divergent contribution to the integral over ρ. Note that this term comes from the gradient energy $\partial_i \phi^* \partial_i \phi$. The energy density $T^0_{\ 0}$ of the rectilinear vortex described above has its maximal value on the line of topological zeros. In physical situations, the vortex is created within a vessel of a finite size. It ends on the walls of the vessel, or forms a loop inside of it. In all such cases the total energy is finite.

So far we have considered the simplest vortex which has a winding number equal to $+1$. Taking the exponential $\exp(-i\theta)$ instead of $\exp(i\theta)$, we obtain the so called anti-vortex which has a winding number equal to -1 and the same function $F(\rho)$ as the vortex. Furthermore, one can also take $\exp(in\theta)$ with integer n, $|n| > 1$. Such fields have winding numbers equal to n. The field equation (3.61) does not have static solutions of this type, except for very special cases. Physically, the reason is that the vortices, in general, interact with each other. Static multi-vortex solutions exist when (3.61) is modified by adding new terms corresponding to certain external forces acting on the vortices.

The presence of a vortex in the initial data has significant consequences for the time evolution of the complex scalar field. The total winding number is constant in time, because the field is a continuous function of time. For this reason, the winding number is called the topological charge. It is an integral of motion of a non-Noether type, because its existence is not related to some continuous global symmetry. The space of all fields ϕ is divided into so called topological sectors—each sector contains all fields which have the same winding number. Note that in the sectors with non vanishing winding number, decomposition (3.54) of the field ϕ into the Goldstone field Θ and the massive real field χ is not correct because of the presence of topological zeros.

Vortices can also appear in the topologically trivial sector, that is the one with the total winding number equal to 0. Here one can have vortices ($n > 0$) and antivortices ($n < 0$) in equal numbers, or finite (anti-)vortex loops, which can disappear by shrinking to a point. In this process the line of zeros gradually shrinks to a point and disappears. Nevertheless, such a closed vortex can live quite a long time. Vortices can annihilate with antivortices. All these processes are very interesting from a physical point of view.

Exercises

3.1 We know from Chap. 1 that the sine-Gordon equation (1.7)

$$\left(\partial_\tau^2 - \partial_\xi^2\right) \Phi(\xi, \tau) + \sin \Phi(\xi, \tau) = 0 \tag{3.66}$$

possesses a static solution of the form

$$\Phi_+(\xi) = 4 \arctan \mathrm{e}^\xi.$$

(a) Let $\Phi(\xi, \tau) = \Phi_+(\xi) + \varepsilon\chi(\xi, \tau)$. By inserting this *Ansatz* into (3.66), and keeping only the terms constant and linear in ε, find the approximated equation satisfied by $\chi(\xi, \tau)$.
(b) We shall look for the solution of the equation obtained in point (a) with the form

$$\chi(\xi, \tau) = \mathrm{e}^{i\omega\tau}\psi(\xi).$$

Prove that

$$\left(-\frac{d^2}{d\xi^2} + U_1(\xi)\right)\psi(\xi) = \omega^2\psi(\xi). \tag{3.67}$$

with

$$U_1(\xi) = \cos \Phi_+(\xi) = 1 - \frac{2}{\cosh^2 \xi}.$$

(c) Show that

$$\psi_0(\xi) = \partial_\xi \Phi_+(\xi) = \frac{2}{\cosh \xi}$$

is a solution of (3.67) with $\omega = 0$. How is this result—the existence of a zero mode of (3.67)—related to the fact that $\Phi_+(\xi - \xi_0)$ is a solution of the sine-Gordon equation for any constant ξ_0? Calculate (up to the terms of the order ε^2) the energy of the field $\Phi_+(\xi) + \varepsilon\phi_0(\xi)$. Is the result surprising?

(d) A solution of (3.67), which does not diverge for $|\xi| \to \infty$, can be either worked out by transforming this equation into the hypergeometric equation, or can be found in textbooks on quantum mechanics. It is of the form

$$\psi_k(\xi) = A_k f_1(k, \xi) + B_k f_2(k, \xi), \qquad k \in R_+,$$

where A_k and B_k are constants, $\omega = \sqrt{k^2 + 1}$, and

$$f_1(k, \xi) = N_1 (\tanh \xi \cos k\xi - k \sin k\xi),$$
$$f_2(k, \xi) = N_2 (\tanh \xi \sin k\xi + k \cos k\xi).$$

The normalization constants $N_\alpha,\ \alpha = 1, 2$, are chosen such that

$$\int_{-\infty}^{\infty} d\xi\ f_\alpha(k, \xi) f_\beta(k', \xi) = \delta_{\alpha\beta}\delta(k - k').$$

Using these functions, write down the general form of the perturbation $\chi(\xi, \tau)$, and show by a direct calculation that its contribution to the energy is (apart from the contribution from the zero mode $\psi_0(\xi)$) strictly positive. What does this mean for the stability of the perturbed sine-Gordon soliton?

3.2 Let $\hat{u}(x)$ be a smooth, matrix valued field in Minkowski space-time M, with values from the $SU(N)$ group. Thus, $\hat{u} = \hat{u}(x) \in \mathrm{SU}(N), x \in M$. Prove that for the current

$$j^\mu = \frac{1}{24\pi^2}\epsilon^{\mu\nu\rho\lambda}\mathrm{tr}\ \left(\hat{u}^\dagger\partial_\nu\hat{u}\,\hat{u}^\dagger\partial_\rho\hat{u}\,\hat{u}^\dagger\partial_\lambda\hat{u}\right),$$

the continuity equation $\partial_\mu j^\mu = 0$ holds. Next, check that the 'charge'

$$B = \frac{1}{24\pi^2}\int_{R^3} d^3x\ \varepsilon_{ijk}\mathrm{tr}\ \left(\hat{u}^\dagger\partial_i\hat{u}\,\hat{u}^\dagger\partial_j\hat{u}\,\hat{u}^\dagger\partial_k\hat{u}\right),$$

is conserved.

Note that here we do not assume that the field $\hat{u}(x)$ obeys any Euler–Lagrange equation, and that we do not invoke Noether's theorem. Such conservation laws are called topological ones.

3.3 An effective Lagrangian, describing interacting pion fields $\pi^a(t, \vec{x}),\ a = 1, 2, 3$, can be written with the help of the matrix field $\hat{u}(x)$, with values from the group SU(2), which we shall parameterize as

$$\hat{u}(x) = \exp\left\{-\frac{i}{F_\pi}\pi^a\sigma_a\right\}, \qquad \text{(the sum over } a \text{ is understood)},$$

where $\sigma_1 = \begin{pmatrix} 0 & 1 \\ 1 & 0 \end{pmatrix}$, $\sigma_2 = \begin{pmatrix} 0 & -i \\ i & 0 \end{pmatrix}$, $\sigma_3 = \begin{pmatrix} 1 & 0 \\ 0 & -1 \end{pmatrix}$, are the Pauli matrices, and F_π is a constant (called the pion decay constant). The effective Lagrangian has the form

$$\mathcal{L} = \frac{F_\pi^2}{4} \mathrm{tr}\left(\partial_\mu \hat{u}^\dagger \partial^\mu \hat{u}\right) + \frac{1}{4} m^2 F_\pi^2 \mathrm{tr}\left(\hat{u}^\dagger + \hat{u} - 2\right),$$

where m denotes the pion rest mass. Derive the Euler–Lagrange equations for the fields $\pi^a(x)$, and the formula for the energy which follows from $\mathcal{L}$.

3.4 Let us choose a specific form of the pion fields,

$$\pi^a(x) = F_\pi n^a P(r), \tag{3.68}$$

where $r = \sqrt{x^i x^i}$ is the radial coordinate in the spherical coordinate system, and n^a denotes the radial unit vector.
(a) Calculate, for the fields given by (3.68), the form of the charge B defined in Exercise 3.2.
(b) By inserting the *Ansatz* (3.68) into the Euler–Lagrange equations derived in the Problem 3.3 find the equation satisfied by $P(r)$.

3.5 Let the theory under consideration be specified by the Lagrangian

$$\mathcal{L} = \frac{1}{2} \partial_\mu \vec{\phi} \partial^\mu \vec{\phi} - U\left(\vec{\phi}(t, \vec{x})\right),$$

where $\vec{\phi}(t, \vec{x}) = \{\phi_i(t, \vec{x}), i = 1, \ldots, N\}$ is a set of N scalar fields in the $D + 1$ dimensional space-time (with D spatial dimensions).

Prove Derrick's theorem which states that for $D > 1$ there are no static, finite energy solutions to the Euler–Lagrange equations that follow from $\mathcal{L}$.

Hints: 1. Show that the equations of motion satisfied by a static configuration can be derived by minimizing the energy.
2. Write the total energy as a sum of kinetic and potential energy, and analyze how they behave under the variation of $\vec{\phi}(\vec{x})$ induced by the scaling of the spatial coordinates $\vec{x} \to \lambda \vec{x}$, i.e. for $\delta\vec{\phi}(\vec{x}) = \vec{\phi}\left((1 + \delta\lambda)\vec{x}\right) - \vec{\phi}(\vec{x})$ with arbitrary infinitesimal $\delta\lambda$. Show that for $D > 1$ the energy has no stationary points under this specific variation of the fields, and thus no finite energy, static solution exists.

Chapter 4
Vector Fields

Abstract The $U(1)$ gauge group. The parallel transport and gauge covariant derivatives. The Abelian gauge field and the minimal coupling prescription. The $SU(N)$ gauge group. The non-Abelian gauge field. The Yang–Mills equation. The gauge invariant energy-momentum tensor. The Higgs mechanism. The massive vector field (the Proca field).

A real or complex relativistic vector field $W_\mu(x)$ has, by definition, the following transformation law under Poincaré transformations[1]

$$W'_\mu(x') = L_\mu^{\ \nu}\, W_\nu(x), \tag{4.1}$$

where $x' = \hat{L}x + a, \ \hat{L} \in L_+^\uparrow$. The most important classes of vector fields are related to gauge transformations and gauge invariance (known also as local symmetry groups or gauge symmetries). The set of electromagnetic potentials A_μ introduced in Chap. 1 is the simplest example of a vector field of this kind. In this case the gauge transformations are related to the $U(1)$ group introduced in Chap. 3, formula (3.44). This group is Abelian and the corresponding vector field is generally called the Abelian gauge field. Another example is the non-Abelian gauge field, which is a matrix-valued vector field related to the $SU(N)$, $N \geq 2$, group. Yet another kind of vector field—the Proca field—appears when a continuous global symmetry is spontaneously broken in the presence of a gauge field. When introducing all of these fields we will pay attention to the related mathematical aspects, but only to the minimal level needed for a clear formulation of the theory.

In this and subsequent chapters we use so called natural units. They are obtained by attaching the constants c and $\hbar$ to the fields, or parameters or variables, in such a way that they disappear from all formulas in which they are present as coefficients. Often one says that in these units $c = \hbar = 1$, but this could be misleading—we remove these constants of Nature from the formulas by an appropriate redefinition of the fields and other quantities present in these formulas.

---

[1] There can be a caveat to this transformation law, see formula (4.43) and the remark preceding it.

H. Arodź and L. Hadasz, *Lectures on Classical and Quantum Theory of Fields*,
Graduate Texts in Physics, DOI 10.1007/978-3-319-55619-2_4

4.1 The Abelian Gauge Field

In this section we explain how the postulate of invariance under local $U(1)$ transformations can be satisfied by the introduction of a vector field: the Abelian gauge field. In Chap. 3 we introduced the Lagrangian

$$\mathcal{L}_0 = \partial_\mu \phi^* \partial^\mu \phi - m^2 \phi^* \phi \tag{4.2}$$

for the free complex scalar field ϕ. As we already know, this Lagrangian is invariant under the following linear transformations

$$\phi'(x) = e^{iq\alpha} \phi(x), \tag{4.3}$$

where q is a fixed integer (different from 0 in order to avoid the trivial case), and $\alpha \in [0, 2\pi)$. Because the factor $e^{iq\alpha}$ does not depend on x, these transformations are called global.

The $U(1)$ group is the set of all phase factors $z = \exp(i\alpha), \alpha \in [0, 2\pi)$ with the group multiplication given by the ordinary multiplication of complex numbers. This group is Abelian. Formula (4.3), which involves the phase factors $z^q = \exp(iq\alpha), \alpha \in [0, 2\pi)$, says that the field ϕ transforms under a representation of the $U(1)$ group. By definition, this means that the mapping

$$R : z \to z^q$$

has the following properties:

(a) it is continuous with respect to $z \in U(1)$,
(b) it preserves the product, that is

$$R(z_1 z_2) = R(z_1) R(z_2)$$

for all $z_1, z_2 \in U(1)$,

(c) $R(1) = 1$.

Condition (c) is not trivial—recall that 1^q is multi-valued for non-integer q. Let us show that conditions (a–c) imply that q has to be an integer. Applying R to both sides of the identity

$$1 = \lim_{\epsilon \to 0+} e^{i(2\pi - \epsilon)},$$

and using the condition of continuity we have

$$R(1) = R\left(\lim_{\epsilon \to 0+} e^{i(2\pi-\epsilon)}\right) = \lim_{\epsilon \to 0+} R\left(e^{i(2\pi-\epsilon)}\right) = \lim_{\epsilon \to 0+} e^{i2\pi q} e^{-i\epsilon q} = e^{i2\pi q}.$$

Because $R(1) = 1$, we obtain the condition $\exp(i2\pi q) = 1$ which is satisfied only by integer q.

The theory of gauge fields is based on gauge transformations and gauge invariance. The gauge transformations form an infinite dimensional group, generally called the gauge group. In the present case, the pertinent gauge group is a subgroup of the continuous direct product of copies of the $U(1)$ groups, and it is called the local $U(1)$ group. Let us recall that the direct product $G_1 \times G_2$ of two groups G_1 and G_2 is the set of all pairs (g_1, g_2), where $g_1 \in G_1$ and $g_2 \in G_2$, with the group multiplication defined by

$$(g_1, g_2)(g_1', g_2') = (g_1 g_1', g_2 g_2').$$

The pair (g_1, g_2) can be regarded as a mapping F defined on the two-element set $\{1, 2\}$, and such that $F(i) \in G_i$ for $i = 1, 2$. The set of all such mappings can be identified with the set $G_1 \times G_2$. Now, let us take a continuous index x with values in Minkowski space-time M. By definition, the continuous direct product $\prod_{x \in M} U(1)$ is the set of mappings $z(x)$ defined on M and such that $z(x) \in U(1)$ for each $x \in M$. The set of all such mappings is very large. It turns out that in field theoretic applications, it is sufficient to consider the subgroup of $\prod_{x \in M} U(1)$ consisting of all mappings $z(x)$ which are smooth functions of x and such that

$$z(x) \to 1 \quad \text{when} \quad |\vec{x}| \to \infty. \tag{4.4}$$

Only this subgroup, denoted as $U(1)_{loc}$, is called the local $U(1)$ group. The elements of $U(1)_{loc}$ can be written in the exponential form

$$z(x) = \exp(i\chi(x)), \tag{4.5}$$

where $\chi(x)$ is a smooth function of x. Moreover, we demand that χ and all its derivatives with respect to x^μ vanish at the spatial infinity, i.e., when $|\vec{x}| \to \infty$. One reason for the restriction to $U(1)_{loc}$ is that we want to exclude those transformations, which can change the asymptotic (that is at $|\vec{x}| \to \infty$) behavior of the derivatives of the field, and in consequence transform the scalar field configurations having finite total energy and momentum into ones with infinite energy or momentum. Another justification for the assumption that χ vanishes at the spatial infinity comes from quantum theory. It turns out that for a particular subset of such transformations, namely those with χ constant in time, there exists a simple implementation in the quantum theory of the field ϕ. A related example is given in Sect. 14.1, see formula (14.26).

By assumption, the local $U(1)$ transformations of the field ϕ have the form

$$\phi'(x) = e^{iq\chi(x)}\phi(x), \tag{4.6}$$

where $\chi(x)$ is the function introduced in formula (4.5). The non vanishing integer q has the same value as in formula (4.3)—in this sense both (4.3) and (4.6) involve the same representation of the $U(1)$ group.

Note that $\chi(x) = \text{constant} \neq 0$ is not allowed by the definition of the gauge group. Therefore, the global $U(1)$ group is not a subgroup of the local one. However, the $U(1)_{loc}$ contains elements, which in a sense approximate the elements of the global $U(1)$. Such elements have the function $\chi(x)$ constant in a compact region Ω in M, $\chi(x) = \alpha$ for $x \in \Omega$. By enlarging that region, we can have a function χ, which is constant on an arbitrarily large compact subset of M. Of course, we may combine the local and global $U(1)$ transformations—in this way we obtain transformations of the form

$$\phi'(x) = e^{iq(\alpha+\chi(x))}\phi(x)$$

with arbitrary constant $\alpha \in [0, 2\pi)$ and χ vanishing when $|\vec{x}| \to \infty$.

Lagrangian (4.2) is not invariant under the $U(1)_{loc}$ group because the gauge transformations change the form of the term with derivatives:

$$\partial_\mu\phi'^*(x)\partial^\mu\phi'(x) = \left(\partial_\mu\phi^*(x) - iq\partial_\mu\chi(x)\,\phi^*(x)\right)\left(\partial^\mu\phi(x) + iq\partial^\mu\chi(x)\,\phi(x)\right).$$

In order to make the Lagrangian invariant we first define a covariant derivative. The reason is that in the case of gauge transformations, the notion of derivative is not well represented by the ordinary partial derivative, which just compares values of the field at neighboring points,

$$\partial_\mu\phi(x) = \lim_{\epsilon\to 0}\frac{\phi(x^\nu + \delta^\nu_\mu\epsilon) - \phi(x^\nu)}{\epsilon}.$$

The problem lies in the difference present in the numerator on the r.h.s. of this formula: a meaningful difference should commute with the gauge transformations (4.6), while the one present in the numerator does not. The solution to this problem is well-known in mathematics: one should introduce a connection and the related covariant derivative. In the case of the local $U(1)$ group, the connection is represented by a vector field $A_\mu(x)$ which has the following transformation law under the local $U(1)$ transformations

$$A'_\mu(x) = A_\mu(x) - \partial_\mu\chi(x). \tag{4.7}$$

The covariant derivatives with respect to x^μ have the form

$$\begin{aligned} D_\mu(A)\phi(x) &= \partial_\mu\phi(x) + iqA_\mu(x)\phi(x), \\ D_\mu(A)\phi^*(x) &= \partial_\mu\phi^*(x) - iqA_\mu(x)\phi^*(x). \end{aligned} \tag{4.8}$$

They commute with the gauge transformations, for example,

$$D_\mu(A')\phi'(x) = \exp(iq\chi(x))\,D_\mu(A)\phi(x).$$

In physical literature, the connection A_μ is called the Abelian gauge field.

The connection can be used to define the parallel transport of the field ϕ along a directed path C in Minkowski space-time, M. Let x_0 be the starting point and y_0 the end point of the path C, $\phi_0 = \phi(x_0)$ is the value of the field ϕ at the point x_0. By definition, the parallel transport of ϕ_0 to the point y_0 along the path C yields the complex number $W[y_0, x_0; C; A]\phi_0$, where

$$W[y_0, x_0; C; A] = \exp\left(-iq \int_C dx^\mu A_\mu\right). \tag{4.9}$$

Note that $|W[y_0, x_0; C; A]| = 1$, hence $W[y_0, x_0; C; A] \in U(1)$. When the line C is smoothly parameterized by $\sigma \in [0, 1]$ with $x(0) = x$ and $x(1) = y$, the line integral can be written as the integral over σ,

$$\int_C dx^\mu A_\mu = \int_0^1 d\sigma \frac{dx^\mu}{d\sigma} A_\mu(x(\sigma)).$$

The parallel transport commutes with the gauge transformations in the following sense

$$W[y_0, x_0; C; A']\, \phi'(x_0) = \exp(iq\chi(y_0))\, W[y_0, x_0; C; A]\, \phi(x_0), \tag{4.10}$$

where ϕ', A'_μ are given by formulas (4.6), (4.7). On the l.h.s. of formula (4.10), we first perform the gauge transformation and next the parallel transport, while on the r.h.s. the order of these operations is reversed. As the meaningful difference of values of ϕ at different points x and y one can take, for instance,

$$\phi(y) - W[y, x; C; A]\phi(x).$$

According to this formula, we first parallel transport $\phi(x)$ to the point y, and then compare it with $\phi(y)$. Note that such a difference depends on the directed path C connecting x with y. Let us take $y^\mu = x^\mu + \epsilon\delta^\mu_\nu$, and the rectilinear segment connecting y with x (directed from y to x) as the path C. Then

$$D_\nu(A)\phi(x) = \lim_{\epsilon\to 0} \frac{W[x, y; C; A]\phi(y) - \phi(x)}{\epsilon} \tag{4.11}$$

(Exercise 4.1).

In order to obtain a Lagrangian which is invariant with respect to the gauge transformations, it suffices to replace the ordinary partial derivatives in Lagrangian $\mathcal{L}_0$ by the covariant ones,

$$\mathcal{L}_1 = D_\mu(A)\phi^* D^\mu(A)\phi - m^2\phi^*\phi. \tag{4.12}$$

This simple recipe is called 'the minimal coupling prescription'. Lagrangian $\mathcal{L}_1$ contains two fields: A_μ and ϕ.

Because $\mathcal{L}_1$ does not contain derivatives of A_μ, the Euler–Lagrange equation for A_μ has the form

$$0 = \frac{\partial \mathcal{L}_1}{\partial A_\mu} = iq\phi\, \partial^\mu \phi^* - iq\phi^*\, \partial^\mu \phi + 2q^2 \phi^* \phi A^\mu.$$

This equation implies that $A_\mu(x)$ remains undetermined at points x such that $\phi(x) = 0$, and

$$A_\mu = \frac{i}{2q}\left(\frac{\partial_\mu \phi}{\phi} - \frac{\partial_\mu \phi^*}{\phi^*}\right)$$

if $\phi(x) \neq 0$. Thus, the model defined by Lagrangian $\mathcal{L}_1$ is acceptable only if we add the assumption that $\phi(x) \neq 0$ on the whole Minkowski space-time. Then, the gauge field is expressed by the scalar field. This is an example of the so called composite gauge field: it has the right behavior with respect to Poincaré and $U(1)_{loc}$ gauge transformations, but it is not an independent field when the Euler–Lagrange equations are taken into account. Quite interesting models of this kind are obtained if the single complex scalar field ϕ is replaced by a multiplet $\vec{\phi}$ of $n > 1$ complex scalar fields $\phi_1, \phi_2, \ldots, \phi_n$ which belong to the same representation of the $U(1)$ group and obey the condition $\vec{\phi}^* \vec{\phi} = 1$, which excludes $\vec{\phi} = 0$. Then, the Lagrangian has the form (4.12) with ϕ and ϕ^* replaced by $\vec{\phi}$ and $\vec{\phi}^*$. Note that due to the condition $\vec{\phi}^* \vec{\phi} = 1$, one field out of the $2n$ real scalar fields Re ϕ_1, Im ϕ_1, Re ϕ_2, Im $\phi_2, \ldots$ can be expressed by the remaining ones, so we have $2n - 1$ independent real scalar fields. These models are called the CP^{n-1} models.

Another gauge invariant model is obtained by adding to the Lagrangian $\mathcal{L}_1$ a certain Lagrangian $\mathcal{L}_A(A_\mu, \partial_\nu A_\mu)$ for the A_μ field. Of course, $\mathcal{L}_A$ should be invariant under the local $U(1)$ transformations. We also assume that the Lagrangian $\mathcal{L}_A$ is local. Let us take a gauge transformation (4.7) with $\chi(x)$ of the form

$$\chi(x) = -a_\mu x^\mu g(x),$$

where $g(x)$ is a smooth function such that $g(x) = 1$ in a vicinity of certain point x_0 in M and $g(x) = 0$ far away from it, a_μ are arbitrary real constants. Formula (4.7) gives $A'_\mu(x_0) = A_\mu(x_0) + a_\mu$. Because x_0 can be any point in M, we see that the gauge invariance of $\mathcal{L}_A$ is possible only if this Lagrangian does not depend on A_μ. Moreover, the dependence on the derivatives has to be restricted. Let us consider the symmetric part of the tensor $\partial_\nu A_\mu$, that is $(\partial_\nu A_\mu + \partial_\mu A_\nu)/2$. The gauge transformations with $\chi(x) = -a_{\mu\nu} x^\mu x^\nu g(x)/4$, where $a_{\mu\nu}$ are arbitrary real constants such that $a_{\mu\nu} = a_{\nu\mu}$, change the symmetric part at the point x_0 by $a_{\mu\nu}$. On the other hand, the antisymmetric part of the tensor $\partial_\nu A_\mu$, or equivalently,

$$F_{\mu\nu} = \partial_\mu A_\nu - \partial_\nu A_\mu, \tag{4.13}$$

is invariant under all gauge transformations. Therefore, the requirement of gauge invariance implies that $\mathcal{L}_A$ can be a function of $F_{\mu\nu}$ only.

Because $\mathcal{L}_A$ should also be a Lorentz invariant, it has to be a function of the invariants

$$I_1 = F_{\mu\nu}F^{\mu\nu}, \quad I_2 = \epsilon_{\mu\nu\lambda\rho}F^{\mu\nu}F^{\lambda\rho},$$

where $\epsilon_{\mu\nu\lambda\rho}$ is the totally antisymmetric symbol, $\epsilon_{0123} = +1$. I_2 is not invariant under the reflections T and P introduced in Sect. 3.1. Therefore, if we add I_2 to the Lagrangian, the resulting model will not be invariant with respect to these reflections. It turns out that in physical applications of the four-dimensional Abelian gauge field, the I_2 term is not needed.

The Lagrangian which gives the Euler–Lagrange equations for A_μ of the Klein–Gordon type has the form

$$\mathcal{L}_A = -\frac{1}{4e^2}F_{\mu\nu}F^{\mu\nu}, \tag{4.14}$$

where e^2 is an arbitrary positive constant. In natural units ('$\hbar = 1 = c$') the action functional is dimensionless, $A_\mu(x)$ has the dimension cm^{-1} as implied by formulas (4.7), (4.8), therefore e^2 is dimensionless.[2] The minus sign in formula (4.14) is present, because then the corresponding energy density is non-negative, see formula (4.49) below.

To summarize, the requirement of invariance with respect to local $U(1)$ symmetry is satisfied when the initial Lagrangian (4.2) is extended by including the Abelian gauge field $A_\mu(x)$. When this field is a dynamical field which is independent of ϕ, the simplest gauge invariant Lagrangian has the form

$$\mathcal{L} = D_\mu(A)\phi^*\, D^\mu(A)\phi - m^2\phi^*\phi - \frac{1}{4e^2}F_{\mu\nu}F^{\mu\nu}. \tag{4.15}$$

Instead of the A_μ field, we could use the equivalent field $B_\mu(x) = A_\mu(x)/e$. After rewriting the Lagrangian $\mathcal{L}$ with the use of the field B_μ, the constant e would appear only in the covariant derivatives,

$$D_\mu(B)\phi = \partial_\mu\phi + ieqB_\mu(x)\phi(x).$$

From Lagrangian (4.15) (with the field A_μ) we obtain the following Euler–Lagrange equations:

$$\partial_\nu F^{\nu\mu} = j^\mu, \tag{4.16}$$

where

$$j^\mu = qe^2\left(i\phi^*\partial^\mu\phi - i\phi\partial^\mu\phi^* - 2qA^\mu\phi^*\phi\right), \tag{4.17}$$

---

[2]This is true only in the case of four-dimensional space-time. In D-dimensional space-time the volume element d^Dx in the action functional does not cancel the dimension of I_1, and in consequence e^2 has the dimension equal to cm^{D-4}.

and

$$D_\mu(A)D^\mu(A)\,\phi + m^2\phi = 0. \tag{4.18}$$

Equation (4.16) has the form of the Maxwell equation with current density j^μ. Therefore, the electromagnetic field can be regarded as an example of the $U(1)$ gauge field. The model with Lagrangian (4.15) is known by the name 'scalar electrodynamics', which emphasizes the fact that the current j^μ is constructed from the scalar field.

The presence of gauge invariance is a signal that the model is formulated in terms of fields, some components of which are redundant. The redundant components are not needed to describe physical phenomena—their only role is to simplify the mathematical formulation of the model. Observables, that is quantities which are at least in principle measurable, do not depend on them. Therefore, as far as the observables are concerned, the redundant components can have arbitrary values. The gauge transformations change the redundant components, and do not change the physically relevant ones.

In the case of the $U(1)$ gauge field $A_\mu(x)$, the redundant component can be found explicitly. We assume that each function $A_\mu(x)$ and its derivatives vanish sufficiently quickly in the limit $|\vec{x}| \to \infty$. The redundant component is related to the longitudinal part, $\vec{A}_L$, of the vector potential $\vec{A}$, defined as follows

$$\vec{A}_L = \nabla\psi,$$

where

$$\psi = \Delta^{-1}(\mathrm{div}\vec{A}(x)).$$

Here Δ^{-1} denotes a Green's function of the 3-dimensional Laplace operator Δ, see formula (1.28). Gauge transformations (4.7) imply that $\vec{A}' = \vec{A} + \nabla\chi$ (because $\vec{A} = (A^i) = (-A_i)$), hence

$$\psi'(x) = \Delta^{-1}(\mathrm{div}\vec{A}'(x)) = \psi(x) + \chi(x).$$

Therefore, the redundant component is given precisely by ψ. Let us introduce the transverse part, $\vec{A}_T$, of the vector potential $\vec{A}$ and the longitudinal part, $\vec{E}_L$, of the electric field $\vec{E}$:

$$\vec{A}_T = \vec{A} - \vec{A}_L, \quad \vec{E}_L = -\nabla A_0 - \partial_0\vec{A}_L.$$

Both $\vec{A}_T$ and $\vec{E}_L$ are gauge invariant because

$$\vec{A}'_L = \vec{A}_L + \nabla\chi.$$

The gauge field A_μ can be decomposed into the gauge invariant (physical) part and the part containing only ψ (the so called gauge part):

$$\vec{A} = \vec{A}_T + \nabla\psi, \quad A_0 = -\Delta^{-1}\mathrm{div}\vec{E}_L - \partial_0\psi.$$

As we can see, explicit separation of the physical and the redundant components of the Abelian gauge field is possible. However, one should add that in most cases such separation only complicates calculations because the formula defining ψ is rather complicated and, moreover, it is not Lorentz covariant. In the case of non-Abelian gauge fields, discussed in the next section, such explicit extraction of the gauge component of the field is not possible.

The Abelian gauge field is an example of those constrained systems mentioned in Chap. 2. Equation (4.16) can be written in the following form

$$\begin{pmatrix} 0\,0\,0\,0 \\ 0\,1\,0\,0 \\ 0\,0\,1\,0 \\ 0\,0\,0\,1 \end{pmatrix} \partial_0^2 \begin{pmatrix} A^0 \\ \vec{A} \end{pmatrix} = \triangle \begin{pmatrix} A^0 \\ \vec{A} \end{pmatrix} + \begin{pmatrix} \partial_0 \nabla \vec{A} + j^0 \\ -\nabla(\partial_0 A^0 + \nabla \vec{A}) + \vec{j} \end{pmatrix}.$$

The matrix on the l.h.s. is singular. The corresponding constraint is obtained by multiplying both sides of this equation by the four-vector $(1, 0, 0, 0)$. It has the form

$$\triangle A^0 + \partial_0 \nabla \vec{A} + j^0 = 0.$$

Note that it coincides with the $\mu = 0$ component of (4.16) (the Gauss law of electrodynamics).

4.2 Non-Abelian Gauge Fields

Let us consider the following generalization of Lagrangian (4.2)

$$\mathcal{L}_0 = \partial_\mu \vec{\phi}^{\,\dagger} \partial^\mu \vec{\phi} - m^2 \vec{\phi}^{\,\dagger} \vec{\phi}, \tag{4.19}$$

where $\vec{\phi}$ is a multiplet of N complex scalar fields, † denotes Hermitian conjugation. This Lagrangian is invariant under global $U(N)$ transformations

$$\vec{\phi}'(x) = u\, \vec{\phi}(x),$$

where $u \in U(N)$, $U(N)$ denotes the group of all unitary N by N matrices. Thus, $u^\dagger u = I_N$ and $|\det u| = 1$, where I_N is the N by N unit matrix.

The $U(N)$ group contains a subgroup isomorphic to the $U(1)$ group. It consists of all matrices of the form $\exp(i\alpha) I_N$, $\alpha \in [0, 2\pi)$. Determinants of these matrices are equal to $\exp(iN\alpha)$. Because the $U(1)$ gauge group was already considered in connection with the Abelian gauge field, we would like to exclude this subgroup of $U(N)$. Therefore we consider the $SU(N)$ group, which is a subgroup of $U(N)$, formed by all unitary matrices which have determinant equal to $+1$. The $SU(N)$ group also contains matrices which belong to the $U(1)$ subgroup, namely matrices of the form $\exp(i2\pi k/N)I_N$, where $k = 0, 1, \ldots, N-1$. These matrices form a discrete subgroup of $U(1)$, denoted by Z_N and called the center of the $SU(N)$ group.

Let us note that there is no gauge field in Minkowski space-time associated with a local version of the Z_N group alone. The point is that the corresponding gauge transformations, and the functions $\chi(x)$ in (4.5), cannot be continuous functions of $x \in M$ unless the transformation is the trivial one (multiplication of $\vec{\phi}$ by 1). Therefore, formula (4.7) cannot be applied here—it contains derivatives of the discontinuous function $\chi(x)$. The Z_N gauge field is feasible if continuous space-time is replaced by a discrete set of points, e.g., an infinite lattice.

The $SU(N)$ gauge transformations of the multiplet of scalar fields have the form

$$\vec{\phi}'(x) = \omega(x)\vec{\phi}(x), \tag{4.20}$$

where $\omega(x) \in SU(N)$ for all x from M. In analogy to the case of the $U(1)_{loc}$ group, we require that the matrix elements of $\omega(x)$ be smooth functions on M, and that $\omega(x) \to I_N$ when $|\vec{x}| \to \infty$. All such mappings $\omega(x)$ form the local $SU(N)$ group, denoted by $SU(N)_{loc}$. Lagrangian (4.19) is not invariant under such local transformations, and the cure is the same as before—the ordinary derivatives should be replaced by covariant ones. According to the mathematical theory of connections, in the present case, the covariant derivative has the form

$$D_\mu(A)\vec{\phi}(x) = \partial_\mu\vec{\phi} + i\hat{A}_\mu(x)\vec{\phi}, \tag{4.21}$$

where the connection, or the non-Abelian gauge field, $\hat{A}_\mu(x)$ for all $\mu = 0, 1, 2, 3$ and $x \in M$ belongs to the Lie algebra of the $SU(N)$ group. This algebra consists of all N by N, Hermitian and traceless matrices:

$$\hat{A}^\dagger_\mu(x) = \hat{A}_\mu(x), \quad \mathrm{tr}\hat{A}_\mu(x) = 0. \tag{4.22}$$

Furthermore, the connection has the following transformation law under the $SU(N)$ gauge transformations

$$\hat{A}'_\mu(x) = \omega_x\, \hat{A}_\mu(x)\, \omega_x^{-1} + i\partial_\mu\omega_x\, \omega_x^{-1}, \tag{4.23}$$

where we have introduced the short notation

$$\omega_x \equiv \omega(x).$$

The form of transformation law (4.23) is such that

$$D_\mu(A')\vec{\phi}'(x) = \omega_x\, D_\mu(A)\vec{\phi}(x).$$

This formula justifies the name 'covariant derivative' for $D_\mu(A)$.

Formula (4.11), which relates the covariant derivative to parallel transport, holds also in the case of the non-Abelian covariant derivative (4.21) if the phase factor $W[x, y; C; A]$ is replaced by the unitary matrix

$$\hat{W}[x, y; C; \hat{A}] = P \exp\left(-i \int_C dx^\mu \, \hat{A}_\mu\right).$$

The symbol P means that the exponential is path-ordered. Such an exponential is a rather complicated object. In order to compute it, one first has to solve the differential equation

$$i \frac{d\hat{W}[\sigma; \hat{A}]}{d\sigma} = \dot{x}^\mu(\sigma) \, \hat{A}_\mu(x(\sigma)) \, \hat{W}[\sigma; \hat{A}],$$

with the initial condition

$$\hat{W}[0; \hat{A}] = I_N.$$

Here σ is the parameter along the path C, introduced below formula (4.9). The path ordered exponential is given by $\hat{W}[1; \hat{A}]$:

$$P \exp\left(-i \int_C dx^\mu \, \hat{A}_\mu\right) = \hat{W}[1; \hat{A}].$$

The calculations are nontrivial because the matrices $\dot{x}^\mu(\sigma) \, \hat{A}_\mu(x(\sigma))$ with different values of σ generally do not commute. In the Abelian case, this problem does not appear and the path ordered exponential coincides with the ordinary one.

Transformation law (4.23) preserves the Hermiticity and tracelessness of $\hat{A}_\mu(x)$. Hermiticity of the first term on the r.h.s. of formula (4.23) is obvious. The Hermitian conjugation of the second term gives $-i\omega_x \, \partial_\mu \omega_x^{-1}$. Using the formula

$$\partial_\mu(\omega_x^{-1}) = -\omega_x^{-1} \, \partial_\mu \omega_x \, \omega_x^{-1},$$

which follows from the identity

$$0 = \partial_\mu I_N = \partial_\mu \left(\omega_x \, \omega_x^{-1}\right) = \omega_x \, \partial_\mu(\omega_x^{-1}) + \partial_\mu \omega_x \, \omega_x^{-1},$$

we recover the $i\partial_\mu \omega_x \, \omega_x^{-1}$ term. Therefore, $\hat{A}'_\mu(x)$ is a Hermitian matrix too. Now let us compute $\mathrm{tr}\hat{A}'_\mu(x)$:

$$\mathrm{tr}\hat{A}'_\mu(x) = \mathrm{tr}\hat{A}_\mu(x) + i \, \mathrm{tr}\left(\partial_\mu \omega_x \, \omega_x^{-1}\right) = \mathrm{tr}\hat{A}_\mu(x),$$

because

$$\mathrm{tr}\left(\partial_\mu \omega_x \, \omega_x^{-1}\right) = 0. \tag{4.24}$$

This last formula follows from the fact that

$$1 = \det(\omega_{x+\epsilon})\det(\omega_x^{-1}) = \det\left(\omega_{x+\epsilon}\,\omega_x^{-1}\right) = \det\left(I + \epsilon^\mu \partial_\mu \omega_x\,\omega_x^{-1} + \ldots\right)$$
$$= 1 + \epsilon^\mu \operatorname{tr}\left(\partial_\mu \omega_x\,\omega_x^{-1}\right) + \ldots,$$

where the dots denote terms with second and higher powers of ϵ^μ. Differentiating with respect to ϵ^μ and substituting $\epsilon = 0$ gives formula (4.24). Thus, $\operatorname{tr}\hat{A}_\mu(x)$ is invariant under the gauge transformations (4.23).

The conditions (4.22) define an N^2-1 dimensional subset of the N by N complex matrices which is called the Lie algebra of the $SU(N)$ group. It is a linear space over real numbers (and not over complex numbers, because linear combinations with complex coefficients do not preserve Hermiticity of matrices). Let $(\hat{T}_a)$, $a = 1, 2, \ldots, N^2-1$, be a basis in this subspace. The matrices $\hat{T}_a$ are of course Hermitian and traceless. For simplicity, we use only an orthogonal basis, that is such that

$$\operatorname{tr}(\hat{T}_a \hat{T}_b) = \frac{1}{2}\delta_{ab}. \tag{4.25}$$

The matrix commutator $[\hat{T}_a, \hat{T}_b]$ is anti-Hermitian and traceless. Multiplying it by $-i$ we obtain an element of the Lie algebra, and therefore it can be written as a linear combination of the matrices $\hat{T}_a$ with real coefficients. Hence,

$$-i[\hat{T}_a, \hat{T}_b] = f_{abc}\hat{T}_c, \tag{4.26}$$

where f_{abc} are real numbers, called the structure constants of the Lie algebra in the chosen basis. The Jacobi identity for matrix commutators, $[[T_a, T_b], T_c] + [[T_c, T_a], T_b] + [[T_b, T_c], T_a] = 0$, implies the Jacobi identity for the structure constants,

$$f_{abd}f_{dce} + f_{cad}f_{dbe} + f_{bcd}f_{dae} = 0.$$

It turns out that condition (4.25) implies that the structure constants are antisymmetric in all three indices (Exercise 4.2).

The gauge field $\hat{A}_\mu(x)$ can be expanded in the basis $(\hat{T}_a)$,

$$\hat{A}_\mu(x) = \hat{T}_a A^a_\mu(x), \tag{4.27}$$

where the vector fields $A^a_\mu(x)$ have real values, and $a = 1, \ldots, N^2 - 1$. Thus, the number of these fields is equal to N^2-1. We may equivalently use the matrix notation $\hat{A}_\mu$, or the multiplet notation $A^a_\mu(x)$.

In physical applications such as in the theory of electro-weak interactions (the Glashow–Salam–Weinberg model, $N = 2$), or in the theory of strong interactions of quarks (quantum chromodynamics, $N = 3$), the non-Abelian gauge field appears as the dynamical field, not reducible to other fields. In the first step in the construction of the gauge invariant Lagrangian for this field, we find the non-Abelian counterpart of the field strength tensor $F_{\mu\nu}$. In the Abelian case it is given by formula (4.13), but $F_{\mu\nu}$ in that form, generalized by merely replacing the Abelian gauge field by $\hat{A}_\mu$, has a rather complicated transformation law under the non-Abelian gauge transformations (4.23). The correct non-Abelian field strength tensor $\hat{F}_{\mu\nu}$ with a simple transformation law is obtained by calculating the commutator of the covariant derivatives:

$$D_\mu(A)D_\nu(A)\vec{\phi}(x) - D_\nu(A)D_\mu(A)\vec{\phi}(x) = i\hat{F}_{\mu\nu}(A)(x)\vec{\phi}(x), \tag{4.28}$$

where

$$\hat{F}_{\mu\nu}(A)(x) = \partial_\mu \hat{A}_\nu(x) - \partial_\nu \hat{A}_\mu(x) + i[\hat{A}_\mu(x), \hat{A}_\nu(x)]. \tag{4.29}$$

The gauge transformation of $\hat{F}_{\mu\nu}$ follows directly from this definition,

$$\hat{F}_{\mu\nu}(A')(x) = \omega_x \hat{F}_{\mu\nu}(A)(x)\omega_x^{-1}, \tag{4.30}$$

where $\hat{A}'_\mu$ is given by formula (4.23). $\hat{F}_{\mu\nu}$ is antisymmetric in indices $\mu\,\nu$, and it has values in the Lie algebra of the $SU(N)$ group. Its expansion in the basis $\hat{T}_a$ has the form

$$\hat{F}_{\mu\nu}(x) = \hat{T}_a F^a_{\mu\nu}(x), \tag{4.31}$$

where

$$F^a_{\mu\nu}(x) = \partial_\mu A^a_\nu(x) - \partial_\nu A^a_\mu(x) - f_{abc} A^b_\mu(x) A^c_\nu(x). \tag{4.32}$$

This last formula is obtained by substituting formula (4.27) into definition (4.29), and using (4.26). By analogy with the Abelian case,

$$\hat{E}^i = \hat{F}_{0i}, \quad \hat{B}^k = -\frac{1}{2}\epsilon_{ijk}\hat{F}_{ij}$$

are called the non-Abelian electric and magnetic fields, respectively. Their physical significance is not so profound as in the Abelian case because they are not invariant with respect to gauge transformations, and only gauge-invariant quantities are accepted as observables.

As the Lagrangian for the non-Abelian gauge field $\hat{A}_\mu$ we take

$$\mathcal{L} = -\frac{1}{2g^2}\mathrm{tr}(\hat{F}_{\mu\nu}\hat{F}^{\mu\nu}) = -\frac{1}{4g^2}F^a_{\mu\nu}F^{a\mu\nu}, \tag{4.33}$$

where g is a dimensionless positive constant. This Lagrangian is invariant with respect to gauge transformations, Poincaré transformations, and P, T reflections.

As in the Abelian case, we may rescale the field

$$\hat{A}_\mu = g\hat{B}_\mu.$$

Then

$$\hat{B}'_\mu(x) = \omega_x\, \hat{B}_\mu(x)\, \omega_x^{-1} + \frac{i}{g}\partial_\mu \omega_x\, \omega_x^{-1},$$

and

$$\hat{F}_{\mu\nu}(A) = g\hat{F}_{\mu\nu}(B),$$

where

$$\hat{F}_{\mu\nu}(B) = \partial_\mu \hat{B}_\nu - \partial_\nu \hat{B}_\mu + ig[\hat{B}_\mu, \hat{B}_\nu].$$

Because such a rescaling is a nonsingular transformation of the field, the formulations using $\hat{A}_\mu$ or $\hat{B}_\mu$ are equivalent.

The Euler–Lagrange equation for the non-Abelian gauge field has the form

$$\frac{\partial \mathcal{L}}{\partial A^a_\mu} - \partial_\nu \left(\frac{\partial \mathcal{L}}{\partial (A^a_{\mu,\nu})} \right) = 0,$$

where $\mathcal{L}$ is given by formula (4.33), and $A^a_{\mu,\nu} = \partial_\nu A^a_\mu$. Because

$$\frac{\partial \mathcal{L}}{\partial A^a_\mu} = \frac{1}{g^2} f_{abc}\, A^b_\nu\, F^{c\mu\nu}, \qquad \frac{\partial \mathcal{L}}{\partial(\partial_\nu A^a_\mu)} = -\frac{1}{g^2} F^{a\nu\mu},$$

we obtain the following equation

$$\partial_\nu F^{a\nu\mu} - f_{abc} A^b_\nu F^{c\nu\mu} = 0, \tag{4.34}$$

which is known as the Yang–Mills equation. Comparing it with (4.16) for the Abelian gauge field (with $j^\mu = 0$), the main difference is the presence of several terms with the structure constants f_{abc}—all are nonlinear with respect to A^a_μ. If these terms were absent ($f_{abc} = 0$) we would obtain $N^2 - 1$ linear equations of the form (4.16) with $j^\mu = 0$, and A^a_μ could be regarded as a set of $N^2 - 1$ independent copies of the Abelian gauge field. Because of the presence of these nonlinear terms, the Yang-Mills equation is rather difficult to solve. Only very few explicit analytic solutions of it are known.

The Yang–Mills equation can be rewritten in the form

$$\partial_\nu \left(\partial^\nu A^{a\mu} - \partial^\mu A^{a\nu} \right) = j^{a\mu}_{YM}, \tag{4.35}$$

where

$$j^{a\mu}_{YM} = f_{abc} \left[\partial_\nu (A^{b\nu} A^{c\mu}) + A^b_\nu F^{c\nu\mu} \right].$$

Taking ∂_μ on both sides of (4.35) we obtain the continuity equation

$$\partial_\mu j^{a\mu}_{YM} = 0.$$

The l.h.s. of (4.35) has the same form as the l.h.s. of (4.16), but the conserved current $j^{a\mu}_{YM}$ is constructed only from the non-Abelian gauge field. Therefore, we may say that the non-Abelian gauge field is charged. The charge density is given by the $\mu = 0$ component of the conserved current $j^{a\mu}_{YM}$. Of course, this charge is the non-Abelian one, not related at all to the electric charge.

The Yang–Mills equation can also be written in matrix form. Multiplying both sides of (4.34) by $\hat{T}_a$, and using formula (4.26) in order to eliminate f_{abc}, we obtain

$$\partial_\nu \hat{F}^{\nu\mu} + i \left[\hat{A}_\nu, \hat{F}^{\nu\mu} \right] = 0. \tag{4.36}$$

Each term with a fixed value of ν on the l.h.s. of this equation is a particular case of a covariant derivative of the field-strength tensor, in general defined by the following formula

$$D_\rho(A) \hat{F}^{\mu\nu} = \partial_\rho \hat{F}^{\mu\nu} + i \left[\hat{A}_\rho, \hat{F}^{\mu\nu} \right]. \tag{4.37}$$

Simple calculation shows that

$$D_\rho(A') \hat{F}'^{\mu\nu}(x) = \omega_x \; D_\rho(A) \hat{F}^{\mu\nu}(x) \; \omega_x^{-1},$$

where $\hat{A}'_\mu$, $\hat{F}'^{\mu\nu}$ are given by formulas (4.23), (4.30), respectively. In the component notation,

$$(D_\rho(A) \hat{F}^{\mu\nu})^a = \partial_\rho F^{a\mu\nu} - f_{abc} A^b_\rho F^{c\mu\nu}.$$

Notice that so far we have introduced two covariant derivatives, see formulas (4.21) and (4.37). They have different forms because they act on objects which transform in different ways under the $SU(N)$ gauge transformations, cf. formulas (4.20), (4.30). For that matter, these transformation laws define two representations of the $SU(N)$ group, namely the fundamental representation in the case of (4.20), and the adjoint one in the case of (4.30). Instead of the multiplet of N scalar fields transforming under the fundamental representation according to formula (4.20), one may consider other multiplets $\vec{\Phi}$ which transform under an arbitrary (nontrivial) representation R of the $SU(N)$ group. Then, the transformation law and the covariant derivative have the form

$$\vec{\Phi}'(x) = \hat{R}(\omega(x)) \, \vec{\Phi}(x) \tag{4.38}$$

$$D_\mu(A) \vec{\Phi}(x) = \partial_\mu \vec{\Phi} + i A^a_\mu(x) \hat{R}_a \vec{\Phi}(x), \tag{4.39}$$

where $\hat{R}_a$ are counterparts of the matrices $\hat{T}_a$. They are called generators of the representation $\hat{R}$. Specifically, $\hat{R}_a$ can be obtained from the following formula

$$\hat{R}_a = -i \left. \frac{\partial \hat{R}(\exp(i\epsilon^a \hat{T}_a))}{\partial \epsilon^a} \right|_{\vec{\epsilon}=0}.$$

One can prove that the commutator of the generators $\hat{R}_a$ contains the same structure constants as are present in formula (4.26),

$$[\hat{R}_a,\ \hat{R}_b] = i f_{abc} \hat{R}_c. \tag{4.40}$$

Needless to say, the multiplet of real vector fields A^a_μ present in the covariant derivative (4.39) is the non-Abelian gauge field discussed earlier in this section.

The Yang–Mills equation (4.34) corresponds to (1.19a, b) of Chap. 1 (with $\rho = 0,\ \vec{j} = 0$). The remaining equations (1.19c, d) also have their non-Abelian counterpart, namely

$$D_\mu \hat{F}_{\nu\rho} + D_\rho \hat{F}_{\mu\nu} + D_\nu \hat{F}_{\rho\mu} = 0. \tag{4.41}$$

Inserting here the definition (4.29) of the field-strength tensor, and using the Jacobi identity for commutators of matrices, we find that (4.41) is just an identity. It is called the Bianchi identity. There is an interesting theorem which says that if some $\hat{H}_{\mu\nu}(x)$ has values in the Lie algebra of the $SU(N)$ group, is antisymmetric in μ, ν, and obeys the Bianchi identity with arbitrary $\hat{A}_\mu$, then it coincides with $\hat{F}_{\mu\nu}$ up to multiplication by a real constant.

The energy-momentum tensor for the non-Abelian gauge field follows from Noether's formula (2.32). As far as translations in space-time are concerned, the vector field behaves like a set of independent scalar fields, see formula (4.1) with $\hat{L} = I_4$. Therefore

$$\mathcal{D}_\alpha A^a_\mu(x) = -\partial_\alpha A^a_\mu(x).$$

Of course, $K^\nu_\alpha = 0$, $\xi^\nu_\alpha = \delta^\nu_\alpha$, and the fields u_a now coincide with A^a_μ. The general formula (2.32) gives

$$T_\nu{}^\mu = \frac{1}{g^2} \partial_\nu A^a_\rho\ F^{a\rho\mu} - \delta^\mu_\nu \mathcal{L}, \tag{4.42}$$

where $\mathcal{L}$ has the form (4.33). The first term on the r.h.s. of formula (4.42) can be written in the form $2\text{tr}\,(\partial_\nu \hat{A}_\rho \hat{F}^{\rho\mu})/g^2$ which shows that it is not invariant with respect to the $SU(N)$ gauge transformations. This means that $T^0_{\ 0}$ and $T^0_{\ i}$ computed from formula (4.42) cannot be accepted as energy and momentum densities, respectively, because such important characteristics of the gauge field should belong to the set of observables. There is a simple way to find an improved energy-momentum tensor $\overline{T}^\mu{}_\nu$ which is conserved and gauge invariant [4]. The trick is based on the observation, that thanks to gauge invariance, it is possible to modify the transformation law (4.1) by combining a certain gauge transformation with the Poincaré transformation. Then, formula (4.1) is replaced by a more general transformation law of the form

$$\hat{A}'_\mu(x') = L_\mu{}^\nu \left(\omega_x\ \hat{A}_\nu(x)\ \omega_x^{-1} + i \partial_\nu \omega_x\ \omega_x^{-1} \right). \tag{4.43}$$

Furthermore, ω_x is adjusted in order to give a suitably modified Lie derivative $\mathcal{D}_\alpha$.

For the present goal of computing the conserved gauge invariant energy-momentum tensor, it is sufficient to consider infinitesimal transformations. Then,

$$\omega_x = I_N + i\hat{X}(x) + \dots, \tag{4.44}$$

where $\hat{X}$ is Hermitian and traceless in order to ensure that $\omega_x \in SU(N)$. The dots stand for terms with higher powers of $\hat{X}$. Formula (4.44) follows from the exponential parametrization of the $SU(N)$ group in a vicinity of the unit matrix, $\omega_x = \exp(i\hat{X}(x))$. Inserting formula (4.44) on the r.h.s. of (4.23) we obtain the infinitesimal form of the gauge transformations of $\hat{A}_\mu$

$$\hat{A}'_\mu(x) = \hat{A}_\mu(x) + i[\hat{X}, \hat{A}_\mu] - \partial_\mu \hat{X} + \dots. \tag{4.45}$$

Now, let us consider infinitesimal translations in the α direction. The corresponding Killing vector is $\xi^\mu_\alpha = \delta^\mu_\alpha$. For these translations we choose $\hat{X}_\alpha(x) = \hat{A}_\alpha(x)$. Then the calculation of the Lie derivative in the case of transformation law (4.43) gives

$$\mathcal{D}_\alpha \hat{A}_\mu = \hat{F}_{\mu\alpha}(x). \tag{4.46}$$

Now Noether's formula (2.32) gives the improved energy-momentum tensor

$$\overline{T}^{\mu}{}_{\nu} = -\frac{1}{g^2} F^a_{\nu\rho}\, F^{a\mu\rho} - \delta^\mu_\nu \mathcal{L}, \tag{4.47}$$

or in the matrix form

$$\overline{T}^{\mu}{}_{\nu} = -\frac{2}{g^2} \mathrm{tr}(\hat{F}_{\nu\rho}\hat{F}_{\mu\rho}) - \delta^\mu_\nu \mathcal{L}. \tag{4.48}$$

From formula (4.48) we immediately see that indeed, $\overline{T}^{\mu}{}_{\nu}$ is gauge invariant. In particular, the gauge invariant energy density of the non-Abelian gauge field has the form

$$\overline{T}^{0}{}_{0} = \frac{1}{2g^2}\left(F^a_{0k}F^a_{0k} + \frac{1}{2}F^a_{ik}F^a_{ik}\right). \tag{4.49}$$

It is clear that $\overline{T}^{0}{}_{0}$ is non negative.

Formula (4.47) for the improved energy-momentum tensor can also be used for the Abelian gauge field: we just put $f_{abc} = 0$, and assume that the index a has only one value so that this index can be omitted. In this case the matrix notation for the field A_μ is of course superfluous, and the matrix ω_x should be replaced by the phase factor $z(x) = \exp(i\chi(x))$.

The theory of non-Abelian gauge fields is very intricate and beautiful. Combining rather elegant mathematical formalism with highly nontrivial physics, it belongs to the most interesting branches of modern theoretical physics. We shall return to it in Chap. 12.

4.3 The Higgs Mechanism and a Massive Vector Field

Lagrangians (4.14) and (4.33) do not contain any dimensional parameters (in the natural units). This fact is often rephrased as the statement that the gauge fields are massless, but this is not quite correct. First, there does not exist any notion of the mass of a field—one can meaningfully talk only about the mass of a particle obtained after quantization of the field in a Fock space. In the case of the Abelian gauge field, such particles have the physical properties of photons, in particular they are massless, that is their four-momentum is light-like, see Chap. 6. Thus, in this case the term 'massless' is, to some extent, justified. In the case of non-Abelian gauge fields quantization is rather nontrivial, and properties of the quantum version of these fields are still under investigation. Apparently, there exist several versions of the quantum theory of non-Abelian gauge fields, in the literature they are called 'phases'. The one which seems to describe the observed strong interactions of quarks inside hadrons, does not actually contain massless particles corresponding to the gauge field. Instead, it predicts the existence of massive particles called glueballs, which correspond to some composite fields built from the non-Abelian gauge field $\hat{A}_\mu$. Therefore, in this case the term 'massless' is not appropriate.

In the case of the so called massive vector field, the corresponding Lagrangian contains a parameter with the dimension of mass (cm^{-1} in natural units). It turns out that the quantum theory of these fields leads to particles with non vanishing rest mass.

Perhaps the most natural way to introduce the massive vector fields is through the so called Higgs mechanism. We shall only present the Abelian version of it within the framework of scalar electrodynamics with Lagrangian (4.15) modified in the scalar field sector: the mass term $-m^2\phi^*\phi$ is replaced by the potential (3.49). Thus, the total Lagrangian now has the form

$$\mathcal{L} = D_\mu\phi^* D^\mu\phi - \frac{\lambda}{4!}\left(\phi^*\phi - \frac{12|m^2|}{\lambda}\right)^2 - \frac{1}{4e^2}F_{\mu\nu}F^{\mu\nu}, \tag{4.50}$$

where the covariant derivatives are given by formulas (4.8). This version of scalar electrodynamics is known as the Abelian Higgs model, and the scalar field ϕ is called the Higgs field.

The vacuum manifold in the Abelian Higgs model is given by the conditions

$$|\phi| = \sqrt{\frac{12|m^2|}{\lambda}}, \quad F_{\mu\nu} = 0, \quad D_\mu\phi = 0, \tag{4.51}$$

which are obtained by minimizing the gauge-invariant energy density $T^0_{\ 0}$ obtained from Lagrangian (4.50) and Noether's formula, with the translational Lie derivative improved in the manner described at the end of the previous section. The general solution of (4.51) has the form

$$\phi_0 = a \exp(iq\beta(x)), \; A^0_\mu = -\partial_\mu \beta(x), \tag{4.52}$$

where $\beta(x)$ is an arbitrary smooth function of x and $a = \sqrt{12|m^2|/\lambda}$. It is clear that $\beta(x)$ and $\beta(x) + 2\pi k/q$, where k is an integer, give the same vacuum fields. Solution (4.52) contains fields which are gauge equivalent and therefore describe the same physical situation. Nevertheless, gauge transformations (4.6) are not sufficient to completely remove the phase factor $\exp(iq\beta(x))$ because the elements of the local $U(1)$ group have to be equal to 1 in the limit $|\vec{x}| \to \infty$. In particular, $\beta =$ constant $\neq 0$ cannot be removed by the gauge transformations. Thus, in spite of the gauge symmetry we have an infinite number of classical ground states of the form (4.52), including the ones with constant phases $\beta \in [0, 2\pi/q)$ and $A^0_\mu = 0$. Nontrivial global $U(1)$ symmetry transformations change one such ground state into another. Thus, the Abelian Higgs model exhibits spontaneous breaking of the global $U(1)$ symmetry. In this respect, it is similar to the Goldstone model of Chap. 3.

The reasoning which lead to vortices in the Goldstone model, Sect. 3.3, can also be repeated in the present case—vortices also exist in the Abelian Higgs model. A single static vortex can be described as a narrow, rectilinear flux of magnetic field surrounded by an axially symmetric current of the $U(1)$ charge carried by the scalar field. It turns out that such vortices have finite energy per unit length, in contradistinction to the vortices of the Goldstone model.

The Higgs mechanism works in the sector of the configuration space of the scalar field which is defined by the condition $\phi(x) \neq 0$ for all $x \in M$. The vacuum manifold belongs to this sector, while the vortices do not because of 'topological zeros' which are of the same origin as in the case of the vortices discussed in Chap. 3. Thus, we now consider only the functions $\phi(x)$ which do not vanish on the whole Minkowski space-time M. Such $\phi(x)$ can be uniquely decomposed into modulus and phase,

$$\phi(x) = (a + H(x))\, e^{i\Theta(x)}, \tag{4.53}$$

cf. formula (3.54). Here $H(x)$ is a real scalar field such that $H > -a$. The field transformation $(\mathrm{Re}\phi, \mathrm{Im}\phi) \to (H, \Theta)$ is nonsingular in the considered sector of the configuration space of the scalar field. Let us insert parametrization (4.53) into Lagrangian (4.50). We obtain

$$\begin{aligned}\mathcal{L} = {} & \partial_\mu H \partial^\mu H + q^2 (a + H)^2 W_\mu W^\mu \\ & - \frac{\lambda}{4!}(2a + H)^2 H^2 - \frac{1}{4e^2} Z_{\mu\nu} Z^{\mu\nu},\end{aligned} \tag{4.54}$$

where

$$W_\mu = A_\mu + \frac{1}{q}\partial_\mu \Theta, \quad Z_{\mu\nu} = \partial_\mu W_\nu - \partial_\nu W_\mu. \tag{4.55}$$

We see that the Θ field has completely disappeared from Lagrangian (4.54). In fact, the new form (4.54) of Lagrangian (4.50) is more transparent where the physical contents of the Abelian Higgs model in the sector without vortices is concerned. The point is that the Euler–Lagrange equations, as well as observables like the energy-momentum tensor, explicitly contain the H and W_μ fields which are gauge invariant, while Θ and A_μ are hidden inside W_μ.

The Euler–Lagrange equations derived from Lagrangian (4.54) have the form

$$\partial_\mu \partial^\mu H + 2|m^2| H = -\frac{\lambda a}{4} H^2 - \frac{\lambda}{12} H^3 + q^2(a + H)\, W_\mu W^\mu, \tag{4.56}$$

$$\partial_\mu Z^{\mu\nu} + 2q^2 e^2 a^2 W^\nu = -2q^2 e^2\, (2a + H)\, H\, W^\nu. \tag{4.57}$$

In the limit of weak fields, we may neglect all terms on the r.h.s.'s of these equations. Then the field H obeys the Klein–Gordon equation with positive mass $m_H^2 = 2|m^2|$, while (4.57) is reduced to the so called Proca equation:

$$\partial_\mu Z^{\mu\nu} + m_W^2 W^\nu = 0, \tag{4.58}$$

where $m_W^2 = 2q^2 e^2 a^2 > 0$. Note that acting with ∂_ν on both sides of (4.58) we obtain the constraint

$$\partial_\nu W^\nu = 0. \tag{4.59}$$

The vector field which obeys the Proca equation with $m_W^2 > 0$ is called the Proca field. The quantum theory of this field leads to particles which have positive mass m_W and spin equal to one.

The name 'Higgs mechanism' refers to the previously described procedure of hiding the original $U(1)$ gauge field $A_\mu(x)$ and the would-be Goldstone field $\Theta(x)$, and forming the physically relevant massive vector field W_μ. The presence of covariant derivatives in the initial Lagrangian (4.50) is one of the prerequisites for this mechanism to work. The non-Abelian version of the Higgs mechanism is a key ingredient of the Glashow–Salam–Weinberg model of electroweak interactions.

Exercises

4.1 Check formula (4.11).

Hint: Parameterize the segment C as follows

$$x(\sigma) = x + (1 - \sigma)\, \epsilon\, e_{(\nu)},$$

where $e_{(\nu)}$ is the unit 4-vector in the direction ν, $e^\mu_{(\nu)} = \delta^\mu_\nu$, and write the numerator in formula (4.11) in the form

$$W[x, y; C; A]\, \phi(y) - \phi(x) = iq\epsilon A_\nu(x) + \epsilon \partial_\nu \phi(x) + \mathcal{O}(\epsilon^2).$$

4.2 Prove that condition (4.25) implies that the structure constants f_{abc} are anti-symmetric in all indices.

Hints: Definition (4.26) implies that

$$f_{abc} = -f_{bac}.$$

In order to show that also $f_{abc} = -f_{acb}$ multiply both sides of (4.26) by $\hat{T}_c$ and take the trace in order to obtain the formula

$$\mathrm{tr}\left([\hat{T}_a, \hat{T}_b]\hat{T}_c\right) = \frac{i}{2} f_{abc}.$$

Next show that the l.h.s. of this formula is equal to $\mathrm{tr}\left(\hat{T}_c\hat{T}_a\hat{T}_b - \hat{T}_a\hat{T}_c\hat{T}_b\right)$, and therefore to $-\frac{i}{2} f_{acb}$.

4.3 Prove that

$$\hat{W}[x, y; C; \hat{A}'] = \omega(y)\, \hat{W}[x, y; C; \hat{A}]\, \omega^{-1}(x),$$

where $\hat{A}'$ is given by formula (4.23).

Hint: Find the relation between $\hat{W}[\sigma; \hat{A}]$ and the solution of the equation

$$i\frac{d\hat{W}[\sigma; \hat{A}']}{d\sigma} = \dot{x}^{\mu}(\sigma)\, \hat{A}'_{\mu}(x(\sigma))\, \hat{W}[\sigma; \hat{A}'],$$

with the initial condition

$$\hat{W}[\sigma = 0; \hat{A}'] = I_N.$$

4.4 Assuming the transformation law (4.23) for $\hat{A}_{\mu}$ prove that

$$D_{\mu}(A)\vec{\Phi}'(x) = \hat{R}(\omega(x))\, D_{\mu}(A)\vec{\Phi}(x).$$

Hints: 1. In the case of representation R the non-Abelian version of formula (4.11) has the form

$$D_{\nu}(A)\vec{\Phi}(x) = \lim_{\epsilon\to 0} \frac{\hat{R}(\hat{W}[x, y; C; \hat{A}])\, \vec{\Phi}(y) - \vec{\Phi}(x)}{\epsilon}.$$

2. Use the formula proved in Exercise 4.3.

4.5 ${}^*\hat{F}^{\mu\nu} = \frac{1}{2}\varepsilon^{\mu\nu\alpha\beta}\hat{F}_{\alpha\beta}$ is called the dual tensor of the non-Abelian field strength tensor $\hat{F}^{\mu\nu}$. Prove that

$$\mathrm{tr}\left({}^*\hat{F}^{\mu\nu}\hat{F}_{\mu\nu}\right) = \partial_\mu K^\mu,$$

where

$$K^\mu = \varepsilon^{\mu\nu\alpha\beta}\,\mathrm{tr}\left(\hat{F}_{\nu\alpha}\hat{A}_\beta - \frac{2i}{3}\hat{A}_\nu\hat{A}_\alpha\hat{A}_\beta\right).$$

Hint: In order to facilitate the calculations, consider separately the terms with two, three and four $\hat{A}'s$.

4.5 The Georgi–Glashow model describes a three component real scalar field ϕ^a, $a = 1, 2, 3$, interacting with a non-Abelian gauge field A^a_μ of the $SU(2)$ type. It has the following Lagrangian

$$\mathcal{L} = -\frac{1}{4}F^a_{\mu\nu}F^{a\mu\nu} + \frac{1}{2}\left(D_\mu\phi\right)^a(D^\mu\phi)^a - \frac{\lambda}{4}\left(\phi^a\phi^a - \mu^2\right)^2,$$

where $\left(D_\mu\phi\right)^a = \partial_\mu\phi^a - \varepsilon_{abc}A^b_\mu\phi^c$, $F^a_{\mu\nu} = \partial_\mu A^a_\nu - \partial_\nu A^a_\mu - \varepsilon_{abc}A^b_\mu A^c_\nu$. Assume that

$$A^a_0 = 0, \quad A^a_i = \varepsilon_{aik}n^k\frac{P(r)-1}{r}, \quad \phi^a = -n^a\frac{H(r)}{r},$$

where the indices $a, i,$ and k take values 1, 2, and 3, $r = \sqrt{x^k x^k}$ is the radial coordinate, and $n^a = \frac{x^a}{r}$ is the radial unit vector. Find the equations for $P(r)$ and $H(r)$ which follow from the Euler–Lagrange equations.
Answer:

$$r^2H'' + \frac{\lambda}{g^2}\left(H^2 - \mu^2g^2r^2\right)H = 2HP^2,$$

$$r^2P'' + PH^2 = P(P^2-1),$$

where $'$ denotes the derivative d/dr.

4.6 The Lagrangian for a complex Proca field has the form

$$\mathcal{L} = -\frac{1}{2}Z^*_{\mu\nu}Z^{\mu\nu} - m^2W^*_\mu W^\mu,$$

where $Z_{\mu\nu} = \partial_\mu W_\nu - \partial_\nu W_\mu$, and $m^2 > 0$. Find the formula for the energy density T_{00} of this field. Prove that the total energy $E = \int d^3x \ T_{00}$ is non negative if W_μ obeys the corresponding Euler–Lagrange equation.

Hints: The field W_ν has the transformation law (4.1). The energy density is obtained with the help of the formalism of Chap. 2:

$$T_{00} = \partial_0 W_i^* \partial_0 W_i + \frac{1}{2} Z_{ik}^* Z_{ik} + m^2 W_i^* W_i - \partial_i W_0^* \partial_i W_0 - m^2 W_0^* W_0.$$

The problem lies in the negative contribution of the W_0 component. The Euler–Lagrange equation has the Proca form (4.58), hence $\partial_\mu W^\mu = 0$ and in consequence

$$\partial_\mu \partial^\mu W_\nu + m^2 W_\nu = 0.$$

In order to prove that $E \geq 0$ first show that

$$E = \int d^3x \left[\frac{1}{2} Z_{ik}^* Z_{ik} + m^2 W_i^* W_i + Z_{0i}^* Z_{0i} - \partial_i W_0^* \partial_i W_0 \right.$$

$$\left. + \partial_0 W_i^* \partial_i W_0 + \partial_i W_0^* \partial_0 W_i - \partial_i W_0^* \partial_i W_0 - m^2 W_0^* W_0 \right].$$

Next, by applying integration by parts (assume that all components of the field vanish sufficiently fast when $\vec{x} \to \infty$), the Proca equation, and the condition $\partial_\mu W^\mu = 0$, prove that

$$E = \int d^3x \left[\frac{1}{2} Z_{ik}^* Z_{ik} + m^2 W_i^* W_i + Z_{0i}^* Z_{0i} + m^2 W_0^* W_0 \right] \geq 0.$$

Chapter 5
Relativistic Spinor Fields

Abstract The Dirac equation. The transformation law of a relativistic bispinor. The $SL(2, C)$ and $Spin(4)$ groups. The free classical Dirac field. The Weyl spinor fields. The $U(1) \times U(1)$ symmetry of the massless Dirac field. The Majorana field. The Grassmann versions of the classical (bi-)spinor fields.

5.1 The Dirac Equation, $Spin(4)$ and $SL(2, C)$ Groups

Discovery of the relativistic wave equation for spin 1/2 particles (electrons) by P.A.M. Dirac in 1928 is regarded as one of the most outstanding achievements of theoretical physics in the 20th century. Apart from the well-known physical consequences, it has revealed a new class of relativistic wave equations, and subsequently, new relativistic fields with intricate mathematical properties. In this section we recall the main facts about the Dirac equation. The classical Dirac field, as well as certain related fields, are introduced in the next sections.

The Dirac equation is the basic equation of relativistic quantum mechanics for a single spin one-half particle. It governs the time evolution of the wave function of such a particle,[1] replacing in this role the non-relativistic, time-dependent Schroedinger equation. Therefore, in this section we consider the quantum mechanics of a single particle, which has a finite number of degrees of freedom, and not a field theory.

The wave function of a single Dirac particle has the form of a column of four complex numbers, $\psi(x) = (\psi^{\alpha}(x))$, $\alpha = 1, 2, 3, 4$. It is called the Dirac bispinor. In the absence of interactions with the Dirac particle, it obeys the Dirac equation

$$i\gamma^{\mu}\partial_{\mu}\psi - m\psi = 0. \tag{5.1}$$

The 4 by 4 matrices γ^{μ} satisfy the Dirac relations

$$\{\gamma^{\mu}, \gamma^{\nu}\} = 2\eta^{\mu\nu} I_4, \tag{5.2}$$

[1] Often called the Dirac particle.

H. Arodź and L. Hadasz, *Lectures on Classical and Quantum Theory of Fields*, Graduate Texts in Physics, DOI 10.1007/978-3-319-55619-2_5

where $\{A, B\} = AB + BA$ (the anticommutator of matrices), $\eta^{\mu\nu}$ is the metric in Minkowski space-time in Cartesian coordinates, and I_4 denotes the 4 by 4 unit matrix. The first question about the Dirac relations is whether there exist matrices which obey it. Dirac showed that one may take, for example,

$$\gamma^0 = \gamma^0_D \equiv \begin{pmatrix} \sigma_0 & 0 \\ 0 & -\sigma_0 \end{pmatrix}, \quad \gamma^i = \gamma^i_D = \begin{pmatrix} 0 & \sigma_i \\ -\sigma_i & 0 \end{pmatrix}, \tag{5.3}$$

where σ_0 is the 2 by 2 unit matrix, and σ_i, $i = 1, 2, 3$, are the Pauli matrices. Matrices (5.3) are called the Dirac representation of the γ^μ matrices. There exists a mathematical theorem which says that all γ^μ matrices can be obtained from an arbitrary particular representation by a similarity transformation. Therefore, any other set of Dirac matrices γ^μ can be obtained from γ^μ_D,

$$\gamma^\mu = A^{-1}\gamma^\mu_D A, \tag{5.4}$$

where A is a nonsingular 4 by 4 matrix ($\det A \neq 0$) [5]. Solutions of Dirac equation (5.1) with the various choices of γ^μ matrices are of course related, namely

$$\psi(x) = A^{-1}\psi_D(x), \tag{5.5}$$

where ψ_D is a solution of (5.1) with γ^μ matrices in the Dirac representation.

Before we conclude that the quantum mechanical models with the various choices of γ^μ matrices are equivalent, we also have to check whether the scalar product of Dirac bispinors is independent of the choice of representation. Such a scalar product has the form

$$\langle\psi_1|\psi_2\rangle = \int d^3x\, \overline{\psi}_1\gamma^0\psi_2, \tag{5.6}$$

where ψ_1, ψ_2 are Dirac bispinors, and

$$\overline{\psi}(x) = \psi^\dagger(x)A^\dagger A\gamma^0. \tag{5.7}$$

Here † denotes the matrix Hermitian conjugation of the bispinor regarded as a one-column, complex matrix. The scalar product (5.6) can be written in the form

$$\langle\psi_1|\psi_2\rangle = \int d^3x\, \psi_1^\dagger A^\dagger A\psi_2 = \int d^3x\, \psi_{1D}^\dagger\psi_{2D} \tag{5.8}$$

from which we see that its value does not depend on the choice of representation of γ^μ matrices. Therefore, the quantum mechanical models based on the Dirac equation (5.1) and scalar product (5.6) are indeed equivalent. Note that formula (5.8) also shows that the scalar product is positive definite.

The Dirac equation (5.1) is invariant under Poincaré transformations $x' = \hat{L}x + a$. The transformation law of the Dirac bispinor under such transformations has the form

$$\psi'(x') = S(\hat{L})\psi(x), \tag{5.9}$$

or equivalently

$$\psi'(x) = S(\hat{L})\psi(\hat{L}^{-1}(x - a)), \tag{5.10}$$

where $S(\hat{L})$ is a nonsingular 4 by 4 matrix which obeys the following condition

$$S^{-1}(\hat{L})\gamma^\mu S(\hat{L}) = L^\mu{}_\nu \gamma^\nu. \tag{5.11}$$

By definition, the invariance of (5.1) means that if $\psi(x)$ obeys that equation, then so does $\psi'(x)$. We check that indeed this is the case by inserting $\psi'(x)$ in (5.1) and using condition (5.11) together with the relation

$$\frac{\partial\psi'(x)}{\partial x^\mu} = S(\hat{L})(\hat{L}^{-1})^\lambda{}_\mu \frac{\partial\psi(y)}{\partial y^\lambda},$$

where $y^\lambda = (\hat{L}^{-1})^\lambda{}_\nu (x^\nu - a^\nu)$.

The existence of the matrix $S(\hat{L})$ for an arbitrary Lorentz transformation follows from the quoted theorem about the equivalence of all representations of γ^μ matrices. Let us denote the r.h.s. of condition (5.11) by γ'^μ. Because

$$\{\gamma'^\mu, \gamma'^\nu\} = L^\mu{}_\rho L^\nu{}_\sigma \{\gamma^\rho, \gamma^\sigma\} = 2L^\mu{}_\rho L^\nu{}_\sigma \eta^{\rho\sigma} I_4 = 2\eta^{\mu\nu} I_4,$$

the matrices γ'^μ obey the Dirac relations (5.2). Here we have used the relation

$$\eta^{-1} = \hat{L}\eta^{-1}\hat{L}^T,$$

which is obtained by taking the matrix inverse of both sides of condition (3.5) in which $\hat{L}$ is replaced by $\hat{L}^{-1}$ (which is a Lorentz matrix as well). Therefore, γ'^μ are related to the γ^μ matrices by a similarity transformation of the form (5.4) with $S(\hat{L})$ playing the role of the matrix A.

Condition (5.11) determines the matrix $S(\hat{L})$ up to multiplication by a number which can depend on $\hat{L}$. In order to prove this assertion, let us suppose that two matrices $S_1(\hat{L})$ and $S_2(\hat{L})$ obey condition (5.11) with the same Lorentz matrix $\hat{L}$. Then

$$S_1^{-1}(\hat{L})\gamma^\mu S_1(\hat{L}) = S_2^{-1}(\hat{L})\gamma^\mu S_2(\hat{L})$$

and

$$S_1(\hat{L})S_2^{-1}(\hat{L})\gamma^\mu = \gamma^\mu S_1(\hat{L})S_2^{-1}(\hat{L})$$

for $\mu = 0, 1, 2, 3$. Next, we use a lemma which says that any nonsingular matrix which commutes with all γ^μ matrices has the form $c_0 I_4$, where c_0 is a complex number different from 0. Therefore,

$$S_1(\hat{L})\, S_2^{-1}(\hat{L}) = c_0(\hat{L}) I_4, \quad S_1(\hat{L}) = c_0(\hat{L})\, S_2(\hat{L}).$$

The lemma used above can be proved first in the Dirac representation simply by explicit calculation, that is by writing down the four commutativity conditions and solving them for the matrix elements of $S_1(\hat{L})_D\, S_2^{-1}(\hat{L})_D$. Next we transform the $S(\hat{L})_D$ matrices to the original representation with the help of formula

$$S(\hat{L})_D = A S(\hat{L}) A^{-1}, \tag{5.12}$$

which follows from (5.5) and (5.9).

The arbitrary multiplicative constant c_0 in each matrix $S(\hat{L})$ can be used to adjust the determinant of this matrix. We choose it in such a way that the matrix $S(\hat{L})$ has unit determinant,

$$\det S(\hat{L}) = 1. \tag{5.13}$$

This condition still leaves the freedom of multiplying $S(\hat{L})$ by -1, or $\pm i$ because $(-1)^4 = (\pm i)^4 = 1$. In the next paragraph we eliminate $\pm i$ from this list.

In the considerations presented above we have not yet used the fact that $L^{\mu}{}_{\nu}$ in condition (5.11) are real. In order to derive the consequences of this for $S(\hat{L})$, we use the so called Majorana representation of Dirac matrices, in which all of the Dirac matrices have imaginary elements. For example, we may take

$$\gamma_M^0 = i \begin{pmatrix} 0 & -\sigma_1 \\ \sigma_1 & 0 \end{pmatrix}, \quad \gamma_M^1 = i \begin{pmatrix} \sigma_0 & 0 \\ 0 & -\sigma_0 \end{pmatrix}, \tag{5.14}$$

$$\gamma_M^2 = \begin{pmatrix} 0 & -\sigma_2 \\ \sigma_2 & 0 \end{pmatrix}, \quad \gamma_M^3 = i \begin{pmatrix} 0 & \sigma_0 \\ \sigma_0 & 0 \end{pmatrix}.$$

In this representation the r.h.s. of condition (5.11) is imaginary. Thus, the $S(\hat{L})_M$ matrices transform imaginary matrices γ_M^{μ} into imaginary matrices. It turns out that there exist matrices $S(\hat{L})_M$ which are real. They are crucial for the relativistic invariance of the theory of the Majorana field discussed in Sect. 5.4. Therefore, it is natural to add one more restriction on the matrices $S(\hat{L})$ in the original representation: they should become real when transformed to the Majorana representation. Due to this reality condition it is not possible to multiply $S(\hat{L})$ by $\pm i$.

To summarize, condition (5.11) strengthened by assumption (5.13) and the reality condition, determine the matrix $S(\hat{L})$ in the Majorana representation up to an overall sign factor. Next, we may pass to the other representations by a similarity transformation analogous to (5.12). It is easy to see from (5.11) that the similarity transformation of $S(\hat{L})$ can have exactly the same form as the transformation of the Dirac matrices γ^{μ}.

Condition (5.11) applied twice gives

$$S^{-1}(\hat{L}_1) S^{-1}(\hat{L}) \gamma^{\mu} S(\hat{L}) S(\hat{L}_1) = L^{\mu}{}_{\nu} S^{-1}(\hat{L}_1) \gamma^{\nu} S(\hat{L}_1) = (\hat{L}\hat{L}_1)^{\mu}{}_{\rho} \gamma^{\rho}.$$

This formula implies that

$$S(\hat{L})S(\hat{L}_1) = c_0\, S(\hat{L}\hat{L}_1), \tag{5.15}$$

where $c_0 = \pm 1$.

The notation $S(\hat{L})$ suggests that we consider a function of $\hat{L}$. Actually, there is a subtlety which should be discussed. The point is that for a given $\hat{L}$ there exist two matrices $S(\hat{L})$ satisfying condition (5.11). They differ only by their sign. It turns out that this ambiguity can be removed by a more restrictive definition of $S(\hat{L})$, only at the price that $S(\hat{L})$ would not be a continuous function of $\hat{L}$, but we shall not discuss this mathematical point in detail. Let us only mention that the sign ambiguity is related to the fact that the $L_+^\uparrow$ group, regarded as a topological space, is not simply connected (that is, there exist closed paths (loops) in it, which cannot be contracted to a point without leaving the group on some intermediate stages of the contraction). The situation is analogous to the problem of removing the ambiguity of sign in $\sqrt{z}$, where z is a complex number. We have the choice: either $\sqrt{z}$ is not continuous along a cut in the complex plane or it is double valued. We assume that $S(\hat{L})$ is a continuous function of $\hat{L}$, therefore it has to be double valued. We may write (5.15) in the form

$$S(\hat{L}_1)S(\hat{L}_2) = S(\hat{L}_1\hat{L}_2), \tag{5.16}$$

but it is understood that $S(\hat{L})$ is double valued. Also $S(I_4) = I_4$ in the same sense, that is actually $S(I_4) = \pm I_4$. Strictly speaking, the matrices $S(\hat{L})$ do not form a representation of the $L_+^\uparrow$ group, because for a representation the mapping $\hat{L} \to S(\hat{L})$ has to be single valued.

It remains to compute the matrices $S(\hat{L})$ obeying (5.11), (5.13) and (5.16). We first find an explicit formula for these matrices, which is valid in the vicinity of the trivial Lorentz matrix $\hat{L} = I_4$ in which the exponential parametrization (3.9) is defined. Because the $L_+^\uparrow$ group is connected, every element $\hat{L}$ of it can be obtained as a product of elements $\hat{L}_1, \hat{L}_2, \ldots, \hat{L}_n$ from that vicinity, $\hat{L} = \hat{L}_1\hat{L}_2\ldots\hat{L}_n$. Therefore, with the help of formula (5.16) we obtain $S(\hat{L}) = S(\hat{L}_1)S(\hat{L}_2)\ldots S(\hat{L}_n)$. In spite of the fact that $\hat{L}$ can be written as such a product in many ways, $S(\hat{L})$ is determined uniquely, except for the factor $c_0 = \pm 1$. The reason is that such an $S(\hat{L})$ obeys condition (5.11) with fixed $L^\mu{}_\nu$, it has a unit determinant and it is real in the Majorana representation. Therefore, the above reasoning which proves that $c_0 = \pm 1$ also applies to this matrix.

The formula for $S(\hat{L})$ has the form

$$S(\hat{L}) = \pm \exp\left(\frac{1}{8}\omega_{\mu\nu}[\gamma^\mu, \gamma^\nu]\right). \tag{5.17}$$

Here [,] denotes the commutator of the matrices, and $\omega_{\mu\nu} = \eta_{\mu\sigma}\omega^\sigma{}_\nu$. The matrix $\hat{\omega} = (\omega^\mu{}_\nu)$ is related to $\hat{L}$ through the exponential parametrization,

$$\hat{L} = \exp\hat{\omega}.$$

$S(\hat{L})$ given by formula (5.17) is real in the Majorana representation. In order to check its determinant, we use the formula $\det(\exp \hat{a}) = \exp(\mathrm{tr}\, \hat{a})$ which is valid for any matrix $\hat{a}$. The determinant is equal to 1, because the trace of the commutator of matrices vanishes. Condition (5.11) can be checked in the following way. First, we introduce auxiliary matrices

$$X(\tau) = \pm \exp\left(\frac{1}{8}\omega_{\mu\nu}[\gamma^\mu, \gamma^\nu]\tau\right),$$

$$\Gamma^\rho(\tau) = X(\tau)^{-1}\gamma^\rho X(\tau),$$

where τ is a real parameter. In particular,

$$X(1) = S(\hat{L}), \quad \Gamma^\rho(0) = \gamma^\rho. \tag{5.18}$$

Let us compute $d\Gamma^\rho(\tau)/d\tau$,

$$\frac{d\Gamma^\rho(\tau)}{d\tau} = -\frac{1}{4}\omega_{\mu\nu} X(\tau)^{-1}(\gamma^\mu\gamma^\nu\gamma^\rho - \gamma^\rho\gamma^\mu\gamma^\nu)X(\tau).$$

Applying the Dirac relation (5.2) to the r.h.s. of this formula we obtain

$$\frac{d\Gamma^\rho(\tau)}{d\tau} = \omega^\rho{}_\nu \Gamma^\nu(\tau)$$

(see Exercise 5.1). Solution of this equation consistent with the second condition (5.18) has the form

$$\Gamma^\rho(\tau) = \left(\exp(\tau\hat{\omega})\right)^\rho{}_\nu \gamma^\nu.$$

Putting $\tau = 1$ we obtain relation (5.11).

Matrices of the form (5.17) and their products form a group called $Spin(4)$. When constructing this group we have used the Dirac matrices in a fixed representation. However, the $Spin(4)$ groups obtained for various choices of representation are related by similarity transformations of the form (5.12), hence all these groups are isomorphic to each other. It turns out that the $Spin(4)$ group is isomorphic to the $SL(2, C)$ group, which consists of all 2×2 complex matrices with the unit determinant. This isomorphism is seen directly when we construct the $Spin(4)$ group in the so called spinor representation of the Dirac matrices, where

$$\gamma^0 = \begin{pmatrix} 0 & \sigma_0 \\ \sigma_0 & 0 \end{pmatrix}, \quad \gamma^k = \begin{pmatrix} 0 & -\sigma_k \\ \sigma_k & 0 \end{pmatrix}. \tag{5.19}$$

Simple calculation then gives

$$\frac{1}{8}\omega_{\mu\nu}[\gamma^\mu, \gamma^\nu] = \begin{pmatrix} \hat{M} & 0 \\ 0 & -\hat{M}^\dagger \end{pmatrix},$$

where

$$\hat{M} = \frac{1}{2}\omega_{0k}\sigma_k - \frac{i}{4}\epsilon_{iks}\omega_{ik}\sigma_s.$$

Here ϵ_{iks} is the three dimensional totally antisymmetric symbol, $\epsilon_{123} = +1$. In consequence,

$$S(\hat{L}) = \begin{pmatrix} \Lambda & 0 \\ 0 & (\Lambda^\dagger)^{-1} \end{pmatrix}, \tag{5.20}$$

where

$$\Lambda = \pm \exp \hat{M}. \tag{5.21}$$

Furthermore, $\det\Lambda = 1$ because $\mathrm{tr}\hat{M} = 0$. One can show that the set of all matrices Λ given by formula (5.21) together with their inverses and their products coincides with the group of all 2×2 complex matrices with unit determinant, denoted by $SL(2, C)$. It is the smallest connected group containing all such products.

Condition (5.11) in the spinor representation (5.19) written for $S(\hat{L})$ of the form (5.20) is equivalent to the following two relations

$$\Lambda^\dagger\tilde{\sigma}^\mu\Lambda = L^\mu{}_\nu\tilde{\sigma}^\nu, \quad \Lambda^{-1}\sigma^\mu(\Lambda^\dagger)^{-1} = L^\mu{}_\nu\sigma^\nu, \tag{5.22}$$

where $(\sigma^\mu) = (\sigma_0, -\sigma_k)$, $(\tilde{\sigma}^\mu) = (\sigma_0, \sigma_k)$, and $k = 1, 2$, and 3. The two relations (5.22) are equivalent to each other. With the help of the identity

$$\mathrm{Tr}(\tilde{\sigma}^\mu\tilde{\sigma}^\nu) = 2\delta_{\mu\nu},$$

the first of relations (5.22) gives

$$L^\mu{}_\nu = \frac{1}{2}\mathrm{Tr}(\Lambda^\dagger\tilde{\sigma}^\mu\Lambda\tilde{\sigma}^\nu). \tag{5.23}$$

Formulas (5.22) and (5.23) relate the Lorentz matrix $\hat{L} = (L^\mu{}_\nu)$ to the $SL(2, C)$ matrix Λ. Using (5.22) and (5.23) one can prove that the matrix $(L^\mu{}_\nu)$ obtained from formula (5.23) belongs to $L_+^\uparrow$ for any $\Lambda \in SL(2, C)$.

In the quantum mechanical context, apart from the invariance of the Dirac equation under Poincaré transformations, one also has to check that the scalar product (5.6) is invariant. Only then may one say that the quantum mechanics of the Dirac particle is Poincaré invariant. It turns out that the scalar product is indeed invariant, but we skip the proof.

5.2 The Dirac Field

All spinor fields have a rather peculiar property: continuous rotations around a certain fixed axis in the space by an angle which increases from 0 to 2π do not reproduce the initial field when the rotation angle becomes equal to 2π. The initial field is obtained for the angle equal to 4π. For example, let us take

$$\hat{\omega} = \phi \begin{pmatrix} 0 & 0 & 0 & 0 \\ 0 & 0 & -1 & 0 \\ 0 & 1 & 0 & 0 \\ 0 & 0 & 0 & 0 \end{pmatrix}.$$

Then,

$$\hat{L}(\phi) = \exp \hat{\omega} = \begin{pmatrix} 1 & 0 & 0 & 0 \\ 0 & \cos\phi & -\sin\phi & 0 \\ 0 & \sin\phi & \cos\phi & 0 \\ 0 & 0 & 0 & 1 \end{pmatrix},$$

and in the spinor representation of the Dirac matrices

$$\hat{M} = \frac{i}{2}\phi\,\sigma_3, \quad \Lambda(\phi) = \exp \hat{M} = \cos\frac{\phi}{2}\sigma_0 + i\sin\frac{\phi}{2}\sigma_3.$$

It is clear that $\hat{L}(\phi)$ represents a rotation by angle ϕ around the third axis. For $\phi = 0$ we obtain $\Lambda = \sigma_0$. Taking the sign $+$ in formula (5.21) we have $S(\phi = 0) = I_4$. Let us now increase ϕ in a continuous manner to 2π. Then, $\Lambda(\phi) \to -\sigma_0$, and in consequence $S(2\pi) = -I_4$. Increasing ϕ further, we obtain $S = I_4$ for $\phi = 4\pi$. The formulas for $\hat{L}(\phi)$ and $\Lambda(\phi)$ used above have been obtained by writing the exponential function as the series, $\exp x = \sum_{l=0}^{\infty} x^l/l!$, and noticing that

$$\hat{\omega}^{2l} = \phi^{2l}(-1)^l \begin{pmatrix} 0 & 0 & 0 & 0 \\ 0 & 1 & 0 & 0 \\ 0 & 0 & 1 & 0 \\ 0 & 0 & 0 & 0 \end{pmatrix}, \quad (\sigma_3)^{2l} = \sigma_0,$$

where $l = 0, 1, 2, \ldots$. Furthermore, the odd powers $\hat{\omega}^{2l+1}$ can be calculated by writing them as the product

$$\hat{\omega}\hat{\omega}^{2l} = \phi^{2l+1}(-1)^l \begin{pmatrix} 0 & 0 & 0 & 0 \\ 0 & 0 & -1 & 0 \\ 0 & 1 & 0 & 0 \\ 0 & 0 & 0 & 0 \end{pmatrix}.$$

An analogous trick is used in order to calculate $(\sigma_3)^{2l+1}$. Finally, we recognize the series expansions for the sine and cosine functions.

The classical Dirac field is represented by $\psi(x) = (\psi^\alpha)$, where $\alpha = 1, 2, 3, 4$ and ψ^α are complex numbers. By definition, under the Poincaré transformations $x' = \hat{L}x + a$

$$\psi'(x') = S(\hat{L})\psi(x),$$

as in the previous section. The important difference is that now $\psi(x)$ is not interpreted as a wave function with the probabilistic interpretation. In particular, there is no need to introduce a scalar product. The Dirac equation (5.1) is obtained as the Euler–Lagrange equation for the following Lagrangian

$$\mathcal{L} = \frac{i}{2}(\overline{\psi}\gamma^\mu \partial_\mu \psi - \partial_\mu \overline{\psi}\gamma^\mu \psi) - m\overline{\psi}\psi. \tag{5.24}$$

As the independent dynamical variables we may take $\text{Re}\psi^\alpha$, $\text{Im}\psi^\alpha$ or, equivalently, ψ^α, $\overline{\psi}_\alpha$. In the following discussion we use the latter choice. Then the Euler–Lagrange equation corresponding to $\overline{\psi}$ has the form (5.1), while functional derivatives with respect to ψ^α, $\alpha = 1, 2, 3, 4$, give

$$i\partial_\mu \overline{\psi}\gamma^\mu + m\overline{\psi} = 0. \tag{5.25}$$

When we relate $\overline{\psi}$ with ψ using formula (5.7), this last equation becomes equivalent to the Dirac equation (5.1). In order to check this, we notice that formulas (5.3) and (5.4) imply that

$$(\gamma^0)^\dagger = A^\dagger A \gamma^0 (A^\dagger A)^{-1}, \quad (\gamma^i)^\dagger = -A^\dagger A \gamma^i (A^\dagger A)^{-1}. \tag{5.26}$$

Taking the Hermitian conjugate of the Dirac equation (5.1), eliminating $(\gamma^\mu)^\dagger$ with the help of the formulas given above, multiplying the resulting equation by $A^\dagger A \gamma^0$, and finally anti-commuting γ^0 with γ^i, we obtain equation (5.25).

Lagrangian (5.24) has real values, and it is invariant with respect to Poincaré transformations. In order to check this latter property, it is convenient to first derive the transformation law of the field $\overline{\psi}$. Formulas (5.7) and (5.9) give

$$\overline{\psi'}(x') = \psi(x)^\dagger (S(\hat{L}))^\dagger A^\dagger A \gamma^0.$$

Next, using formulas (5.17) and (5.26) on the r.h.s. of this formula we obtain

$$\overline{\psi'}(x') = \overline{\psi}(x) S^{-1}(\hat{L}). \tag{5.27}$$

The invariance of the Lagrangian follows from (5.9), (5.27) and (5.11).

Lagrangian (5.24) is also invariant with respect to global $U(1)$ transformations of the form

$$\psi'(x) = \exp(i\alpha)\psi(x), \quad \overline{\psi'}(x) = \exp(-i\alpha)\overline{\psi}(x), \tag{5.28}$$

where $\alpha \in [0, 2\pi)$. Noether's theorem applied to this internal continuous symmetry gives the conserved current

$$j^\mu(x) = \overline{\psi}(x)\gamma^\mu\psi(x). \tag{5.29}$$

A model with local $U(1)$ symmetry can be obtained from Lagrangian (5.24) by replacing the ordinary derivatives with covariant ones, as described in the previous chapter.

The energy-momentum tensor for the Dirac field can be calculated from the following formula

$$T^\mu_{\ \nu} = -\mathcal{L}\delta^\mu_\nu - \frac{\partial\mathcal{L}}{\partial(\psi^\alpha_{,\mu})}\mathcal{D}_\nu\psi^\alpha - \frac{\partial\mathcal{L}}{\partial(\overline{\psi}_{\alpha,\mu})}\mathcal{D}_\nu\overline{\psi}_\alpha, \tag{5.30}$$

where the Lie derivatives have the form

$$\mathcal{D}_\nu\psi^\alpha = -\partial_\nu\psi^\alpha, \quad \mathcal{D}_\nu\overline{\psi}_\alpha = -\partial_\nu\overline{\psi}_\alpha.$$

These formulas follow from Noether's theorem applied to the translational symmetry of the Dirac Lagrangian (5.24). In particular, the energy density of the Dirac field is equal to

$$T^0_{\ 0} = -\frac{1}{2}i(\overline{\psi}\gamma^k\partial_k\psi - \partial_k\overline{\psi}\gamma^k\psi) + m\overline{\psi}\psi. \tag{5.31}$$

It is not bounded from below. Hence, the classical Dirac field model is not acceptable from a physical viewpoint. It turns out that the remedy consists in quantizing the Dirac field, see the next chapter. The same is also true for the Weyl and Majorana fields discussed below.

5.3 The Weyl Fields

The Dirac field can be decomposed into two so called Weyl fields. This decomposition is Lorentz invariant. It yields an interesting new perspective on the Dirac field. Definition of the Weyl fields involves the γ_5 matrix introduced as follows

$$\gamma_5 = \frac{i}{4!}\epsilon_{\mu\nu\lambda\sigma}\gamma^\mu\gamma^\nu\gamma^\lambda\gamma^\sigma, \tag{5.32}$$

where $\epsilon_{\mu\nu\lambda\sigma}$ is the four dimensional antisymmetric symbol, $\epsilon_{0123} = +1$. Because Dirac matrices with different index values anticommute,

$$\gamma_5 = i\gamma^0\gamma^1\gamma^2\gamma^3. \tag{5.33}$$

This formula is useful when checking that

$$\gamma_5\gamma^\mu + \gamma^\mu\gamma_5 = 0, \quad (\gamma_5)^2 = I_4. \tag{5.34}$$

The γ_5 matrix is Hermitian in the Dirac representation, as well as in all representations of the Dirac matrices which are unitarily equivalent to the Dirac representation, i.e., when the matrix A in formula (5.4) is unitary. Finally, as follows from formulas (5.26),

$$(\gamma_5)^\dagger = A^\dagger A \gamma_5 (A^\dagger A)^{-1}. \tag{5.35}$$

Let us introduce two matrices

$$P_\pm = \frac{1}{2}(I_4 \pm \gamma_5). \tag{5.36}$$

They have the following properties

$$P_+ + P_- = I_4, \;\; (P_\pm)^2 = P_\pm, \;\; P_+P_- = 0 = P_-P_+. \tag{5.37}$$

The Weyl fields ψ_R and ψ_L are defined as follows

$$\psi_R(x) = P_+\psi(x), \;\; \psi_L(x) = P_-\psi(x), \tag{5.38}$$

where ψ is the Dirac field. ψ_R and ψ_L are eigenvectors of γ_5, namely

$$\gamma_5\psi_R(x) = \psi_R(x), \;\; \gamma_5\psi_L(x) = -\psi_L(x). \tag{5.39}$$

It is clear that

$$\psi(x) = \psi_R(x) + \psi_L(x). \tag{5.40}$$

The letters R or L stand for 'right-handed' or 'left-handed', respectively. These traditional names for the Weyl fields refer to the helicity of the particles which appear in the quantum versions of models with these fields.

The decomposition (5.40) of the Dirac field into Weyl fields is interesting because it is preserved under Poincaré transformations (5.9). If we decompose $\psi'(x')$,

$$\psi'(x') = \psi'_R(x') + \psi'_L(x'), \;\; \psi'_{R,L}(x) = P_\pm\psi'(x),$$

then

$$\psi'_R(x') = S(\hat{L})\psi_R(x), \;\; \psi'_L(x') = S(\hat{L})\psi_L(x). \tag{5.41}$$

These formulas are a consequence of the fact that the matrices $S(\hat{L})$ and γ_5 commute,

$$S(\hat{L})\gamma_5 = \gamma_5 S(\hat{L}). \tag{5.42}$$

This very important property of the γ_5 matrix follows from its definition (5.32) and relation (5.11),

$$S^{-1}(\hat{L})\gamma_5 S(\hat{L}) = \frac{i}{4!}\epsilon_{\mu\nu\lambda\sigma} L^{\mu}{}_{\delta} L^{\nu}{}_{\kappa} L^{\lambda}{}_{\alpha} L^{\sigma}{}_{\beta} \gamma^{\delta}\gamma^{\kappa}\gamma^{\alpha}\gamma^{\beta}$$

$$= \frac{i}{4!} \det \hat{L} \ \ \epsilon_{\delta\kappa\alpha\beta} \, \gamma^{\delta}\gamma^{\kappa}\gamma^{\alpha}\gamma^{\beta} = \gamma_5$$

because $\det \hat{L} = 1$ for Lorentz matrices from the $L_{+}^{\uparrow}$ group. Thus, γ_5 is invariant under such Lorentz transformations. Because of transformation laws (5.41), the Weyl fields can be regarded as independent relativistic spinor fields.

Let us write Lagrangian (5.24) for the Dirac field as a function of the Weyl fields. From now on we use a representation for γ^{μ} that is unitarily equivalent to the Dirac representation, hence γ_5 is Hermitian. The Majorana and spinor representations belong to this class. The Dirac field is eliminated with the help of formula (5.40), while for the conjugate Dirac field $\overline{\psi}$ we first use the following formulas

$$\overline{(\psi_R)} = \overline{\psi} P_{-}, \quad \overline{(\psi_L)} = \overline{\psi} P_{+},$$

next $P_{+}P_{-} = 0, \ \ P_{+}\gamma^{\mu} P_{+} = 0$, and other similar formulas. It turns out that

$$\mathcal{L} = \frac{1}{2} i [\overline{(\psi_R)}\gamma^{\mu}\partial_{\mu}\psi_R - \partial_{\mu}\overline{(\psi_R)}\gamma^{\mu}\psi_R] + \frac{1}{2} i [\overline{(\psi_L)}\gamma^{\mu}\partial_{\mu}\psi_L - \partial_{\mu}\overline{(\psi_L)}\gamma^{\mu}\psi_L]$$
$$- m\overline{(\psi_L)}\psi_R - m\overline{(\psi_R)}\psi_L. \quad (5.43)$$

The Dirac equation (5.1) is split as follows

$$i\gamma^{\mu}\partial_{\mu}\psi_R - m\psi_L = 0, \quad i\gamma^{\mu}\partial_{\mu}\psi_L - m\psi_R = 0. \quad (5.44)$$

The conserved current (5.29) is a sum of two separate terms for ψ_R and ψ_L,

$$j^{\mu} = \overline{(\psi_R)}\gamma^{\mu}\psi_R + \overline{(\psi_L)}\gamma^{\mu}\psi_L. \quad (5.45)$$

The two terms in (5.45) have identical form because the $U(1)$ transformations (5.28) act on ψ_R and ψ_L in exactly the same manner. We see from formula (5.43) and equations (5.44) that the parameter m can be regarded as a measure of the coupling of the ψ_R and ψ_L fields in the Dirac Lagrangian (5.24). In the case of $m = 0$ the Lagrangian is split into two separate parts, each one being relativistically invariant.

The form (5.43) of the Lagrangian for the Dirac field reveals that in the case of $m = 0$ the $U(1)$ symmetry (5.28) is enlarged to $U(1)_R \times U(1)_L$ symmetry, defined by the following transformation laws:

$$U(1)_R: \ \psi'_R(x) = e^{i\omega}\psi_R(x), \ \ \overline{\psi'_R}(x) = e^{-i\omega}\overline{\psi_R}(x), \ \ \psi'_L(x) = \psi_L(x), \quad (5.46)$$

$$U(1)_L: \ \psi'_L(x) = e^{i\delta}\psi_L(x), \ \ \overline{\psi'_L}(x) = e^{-i\delta}\overline{\psi_L}(x), \ \ \psi'_R(x) = \psi_R(x), \quad (5.47)$$

where ω and δ are two independent, real, continuous parameters. Let us replace these parameters by α and β such that

$$\omega = \alpha + \beta, \quad \delta = \alpha - \beta.$$

It is clear that for $\beta = 0$ we obtain the familiar $U(1)$ symmetry (5.28) which exists also when $m \neq 0$. On the other hand, for $\alpha = 0$ we have a new $U(1)$ symmetry, called the chiral symmetry. The chiral transformation of the Dirac field has the form

$$\begin{aligned} \psi'(x) &= e^{i\beta}\psi_R(x) + e^{-i\beta}\psi_L(x) \\ &= e^{i\gamma_5\beta}\psi_R(x) + e^{i\gamma_5\beta}\psi_L(x) = \exp(i\gamma_5\beta)\,\psi(x), \end{aligned} \tag{5.48}$$

and for the conjugate Dirac field

$$\overline{\psi}'(x) \equiv \overline{\psi'}(x) = \overline{\psi}(x)\exp(i\gamma_5\beta). \tag{5.49}$$

Noether's theorem applied to the chiral transformations gives a conserved current of the form

$$j_5^\mu = \overline{\psi_R}\gamma^\mu\psi_R - \overline{\psi_L}\gamma^\mu\psi_L = \overline{\psi}\gamma^\mu\gamma_5\psi. \tag{5.50}$$

Simple calculation with the use of the Dirac equation (5.1) and its conjugate (5.25) shows that

$$\partial_\mu j_5^\mu = 2im\overline{\psi}\gamma_5\psi.$$

Thus, the current j_5^μ is conserved when $m = 0$.

Note that there is the possibility of a new spinor field theory involving just one of Weyl fields, let it be ψ_R, with the Lagrangian

$$\mathcal{L}_R = \frac{i}{2}[\overline{(\psi_R)}\gamma^\mu\partial_\mu\psi_R - \partial_\mu\overline{(\psi_R)}\gamma^\mu\psi_R], \tag{5.51}$$

and with the constraint

$$\gamma_5\psi_R = \psi_R, \tag{5.52}$$

see (5.39). This model is relativistically invariant. It contains half degrees of freedom of the Dirac field. Of course, there also exists the model with ψ_R replaced by ψ_L.

Constraint (5.52) can be explicitly solved. For example, in the spinor representation (5.19) of the Dirac matrices,

$$\gamma_5 = \begin{pmatrix} \sigma_0 & 0 \\ 0 & -\sigma_0 \end{pmatrix}, \tag{5.53}$$

and conditions (5.39) give

$$\psi_R = \begin{pmatrix} \xi \\ 0 \end{pmatrix}, \quad \psi_L = \begin{pmatrix} 0 \\ \zeta \end{pmatrix}, \tag{5.54}$$

where ξ and ζ are arbitrary two-component complex spinors, often called Weyl spinors. The Dirac field ψ can be written as

$$\psi(x) = \begin{pmatrix} \xi(x) \\ \zeta(x) \end{pmatrix}.$$

For this reason, the Dirac field is called a bispinor field. Lagrangian $\mathcal{L}_R$ expressed by the spinor ξ has the form

$$\mathcal{L}_R = \frac{i}{2}(\xi^\dagger \tilde{\sigma}^\mu \partial_\mu \xi - \partial_\mu \xi^\dagger \tilde{\sigma}^\mu \xi). \tag{5.55}$$

There is a problem with a mass term for the ξ field. The term $\xi^\dagger \xi$ is not Lorentz invariant. Lorentz invariant expression $\xi^T \hat{\epsilon} \xi$, see below, simply vanishes because the matrix $\hat{\epsilon}$ is antisymmetric.

Models with only one Weyl field ψ_R or ψ_L are not invariant under the spatial reflection. The spatial reflection P acts on the Dirac field in the same manner as on the bispinor wave function in the quantum mechanics of the Dirac particle, namely

$$P\psi(x^0, \vec{x}) = e^{i\eta}\gamma^0 \psi(x^0, -\vec{x}),$$

where the constant factor $\exp(i\eta)$ is called the intrinsic parity of the field. This definition implies that

$$P\,\psi_R = (P\psi)_L, \quad P\,\psi_L = (P\psi)_R.$$

Hence, the operator P intertwines the spaces of the right- and left-handed fields, while for the invariance we need an operator that acts within one such space. The Dirac field model with Lagrangian (5.24) is invariant under the spatial reflection.

Formulas (5.20) and (5.24) give the Poincaré transformations of Weyl spinors:

$$\xi'(x') = \Lambda \xi(x), \quad \zeta'(x') = (\Lambda^\dagger)^{-1} \zeta(x). \tag{5.56}$$

There exist old conventions about the indices of Weyl spinors, namely

$$\xi = (\xi^\alpha), \ \zeta = (\zeta_{\dot{\alpha}}), \ \xi^* = (\xi^{\dot{\alpha}}), \ \zeta^* = (\zeta_\alpha),$$

where $*$ denotes complex conjugation, and $\alpha, \dot{\alpha} = 1, 2$. These are accompanied by conventions for the indices of the $SL(2, C)$ matrices:

$$\Lambda = (\Lambda^\alpha{}_\beta), \quad (\Lambda^\dagger)^{-1} = (((\Lambda^\dagger)^{-1})_{\dot{\alpha}}{}^{\dot{\beta}}), \quad \Lambda^* = (\Lambda^{\dot{\alpha}}{}_{\dot{\beta}}), \quad (\Lambda^T)^{-1} = (((\Lambda^T)^{-1})_\alpha{}^\beta).$$

For example, the transformation law of the ζ^* spinor is written in the form

$$\zeta'^*_\alpha(x') = ((\Lambda^T)^{-1})_\alpha{}^\beta \zeta^*_\beta(x).$$

The transformation laws of the spinors ξ, ζ, ξ^*, and ζ^* are not independent. The reason is that for any matrix $\Lambda \in SL(2, C)$ the following identity is true

$$(\Lambda^T)^{-1} = \hat{\epsilon}\Lambda\hat{\epsilon}^{-1}, \tag{5.57}$$

where

$$\hat{\epsilon} = i\sigma_2 = \begin{pmatrix} 0 & 1 \\ -1 & 0 \end{pmatrix}.$$

A simple way to check this identity consists in explicit computation of the both sides, and taking into account the fact that $\det\Lambda = 1$. Matrix elements of $\hat{\epsilon}$ are denoted as $\epsilon_{\alpha\beta}$, while matrix elements of $\hat{\epsilon}^{-1}$ as $\epsilon^{\alpha\beta}$. Note that $\xi^\alpha \xi_\alpha = 0$, and $\xi_\alpha \xi'^\alpha = -\xi'_\alpha \xi^\alpha$ because $\epsilon_{\alpha\beta}\epsilon^{\beta\sigma} = \delta^\sigma_\alpha$. Due to identity (5.57) the spinor $\hat{\epsilon}^{-1}\zeta^*$ has the same transformation law as the spinor ξ in the first formula (5.56). Therefore, it should have an upper index without a dot. Complex conjugation adds or removes the dot, $\hat{\epsilon}$ lowers the spinor index and $\hat{\epsilon}^{-1}$ rises it. For instance, if $\zeta = (\zeta_{\dot{\alpha}})$ then $\zeta^* = (\zeta_\alpha)$, or if $\xi = (\xi^\alpha)$ then $\xi_\alpha = \epsilon_{\alpha\beta}\xi^\beta$.

Let us end this section by remarking that the γ_5 matrix exists only when the space-time under consideration has an even dimension. This is a consequence of a theorem about the size of Dirac matrices [5] which says that in d-dimensional space-time γ^μ matrices are quadratic with the number of columns and rows equal to $2^{[d/2]}$, where $[d/2]$ denotes the integer part of $d/2$. For example, when $d = 2$ we may take as Dirac matrices

$$\gamma^0 = \sigma_1, \quad \gamma^1 = \sigma_2.$$

The γ_5 matrix should obey relations (5.34) by definition. The explicit formula (5.32) is valid only when $d = 4$. Let us assume that d is odd, and let us suppose that γ_5 obeying (5.34) exists. Then, the set of $d + 1$ matrices

$$\gamma^0, \gamma^1, \ldots, \gamma^{d-1}, i\gamma_5$$

satisfies all of the requirements for Dirac matrices in $(d+1)$-dimensional space-time. Their size is equal to $2^{[d/2]} = 2^{(d-1)/2}$. On the other hand, the theorem quoted above says that Dirac matrices have the size $2^{[(d+1)/2]} = 2^{(d+1)/2}$, which is larger by a factor of 2. This contradiction shows that γ_5 cannot exist. In consequence, the Weyl fields can also only be defined in an even dimensional space-time.

5.4 The Majorana Field

The fact that the matrices $S(\hat{L})$ are real in the Majorana representation (5.14), suggests that there exists a relativistic, real bispinor field $\psi = (\psi^\alpha)$, $\alpha = 1, 2, 3, 4$, with real components ψ^α. The Poincaré transformations of such bispinors,

$$\psi'(x') = S(\hat{L})\psi(x) \tag{5.58}$$

give bispinors ψ' which also have real components. Moreover, the Dirac equation in the Majorana representation,

$$i\gamma_M^\mu \partial_\mu \psi(x) - m\psi(x) = 0, \tag{5.59}$$

contains matrices $i\gamma_M^\mu$ which have real elements. Therefore this equation is compatible with the assumption that ψ is real. The real field ψ which has transformation law (5.58) and obeys (5.59) is called the Majorana field. It contains half of the degrees of freedom of the Dirac field. Note that the $U(1)$ transformations (5.28) cannot be defined for the Majorana field because they would violate the condition that the field has real values. For the same reason it is not possible to introduce the local $U(1)$ gauge symmetry which would determine coupling of the Majorana field to the Abelian gauge field. In particular, the Majorana field cannot be coupled to the electromagnetic field in the minimal way, that is by replacing ordinary derivatives with the covariant ones.

On the other hand, the current $\overline{\psi}\gamma_M^\mu\psi$ still exists and is conserved, but it should not be interpreted as the current of electric charge. Actually, the presence of this conserved current might seem a paradox, because there is a theorem, known as the inverse Noether theorem, which says that in such a case there exists a corresponding continuous symmetry. However, among the assumptions of that theorem is the very existence of a Lagrangian. All field equations considered in previous sections were of the Lagrange type, that is they could be obtained as Euler–Lagrange equations from certain Lagrangians. So, what is the Lagrangian for the Majorana field? A straightforward attempt to obtain the Lagrangian, just by taking the Dirac Lagrangian (5.24) in the Majorana representation and assuming that ψ is real, fails because it gives $\mathcal{L} = 0$ (Exercise 5.2a). Trying a more general and systematic approach, let us assume that the Lagrangian has the following form

$$\mathcal{L}_1(\psi, \partial_\mu\psi) = A_{\alpha\beta}\psi^\alpha\psi^\beta + B^\mu_{\alpha\beta}\psi^\alpha\partial_\mu\psi^\beta,$$

where $A_{\alpha\beta}$ and $B^\mu_{\alpha\beta}$ are real constants. In the matrix notation,

$$\mathcal{L}_1(\psi, \partial_\mu\psi) = \psi^T \hat{A}\psi + \psi^T \hat{B}^\mu\partial_\mu\psi,$$

where $\hat{A} = (A_{\alpha\beta})$ and $\hat{B}^\mu = (B^\mu_{\alpha\beta})$. The antisymmetric part of the matrix $\hat{A}$ gives a vanishing contribution, hence we may assume that this matrix is symmetric.

Moreover, the symmetric parts $\hat{B}^{\mu}_{s}$ of the matrices $\hat{B}^{\mu}$ may be omitted because they lead to a term which is the total divergence $\partial_{\mu}(\psi^T \hat{B}^{\mu}_S \psi)$. Thus, $\hat{A}^T = \hat{A}$ and $(\hat{B}^{\mu})^T = -\hat{B}^{\mu}$ without any loss in generality. The Euler–Lagrange equation obtained from this Lagrangian has the form

$$\hat{B}^{\mu}\partial_{\mu}\psi + \hat{A}\psi = 0,$$

where we have omitted the overall factor 2. We expect that the Lagrangian $\mathcal{L}_1$ is invariant with respect to the Poincaré transformations (5.58). It turns out that the term $\psi^T \hat{A}\psi$ is invariant only if $\hat{A} = 0$ (Exercise 5.2b). This fact implies that the Lagrangian in the above assumed form can exist only in the massless case, $m = 0$ in (5.59).

In the next step, we compare equation $\hat{B}^{\mu}\partial_{\mu}\psi = 0$ with $\gamma^{\mu}_{M}\partial_{\mu}\psi = 0$. These two equations should have identical sets of solutions. The trivial choice $\hat{B}^{\mu} = i\gamma^{\mu}_{M}$ is wrong, because only γ^0_M is antisymmetric. The Dirac equation can be written in the Schroedinger form $i\partial_0\psi = -i\gamma^0_M\gamma^i_M\partial_i\psi$. The Euler-Lagrange equation can be written in this form provided that the matrix $\hat{B}^0$ is nonsingular. Then, we may write it in the form $i\partial_0\psi = -i(\hat{B}^0)^{-1}\hat{B}^i\partial_i\psi$. The two 'Hamiltonians' should coincide, hence $(\hat{B}^0)^{-1}\hat{B}^i = \gamma^0_M\gamma^i_M$, and $\hat{B}^i = \hat{B}^0\gamma^0_M\gamma^i_M$. The conditions $(\hat{B}^i)^T = -\hat{B}^i$ together with $(\hat{B}^0)^T = -\hat{B}^0$ give the following conditions for the $\hat{B}^0$ matrix

$$\gamma^0_M\gamma^i_M\hat{B}^0 = \hat{B}^0\gamma^0_M\gamma^i_M,$$

where $i = 1, 2$ and 3. They are satisfied only by $\hat{B}^0 = c_0 i\gamma_{5M}$, see Exercise 5.2c. Here c_0 is an arbitrary real constant, and the factor i is present because the matrix $\hat{B}^0$ should be real. In consequence, $\hat{B}^i = c_0 i\gamma_{5M}\gamma^0_M\gamma^i_M$. The matrix γ_{5M} has the form

$$\gamma_{5M} = i\gamma^0_M\gamma^1_M\gamma^2_M\gamma^3_M = i\begin{pmatrix} 0 & \sigma_3 \\ -\sigma_3 & 0 \end{pmatrix}.$$

Note that γ_{5M} is antisymmetric, and $\gamma^2_{5M} = I_4$.

Our Lagrangian can be written in the following form

$$\mathcal{L}_1 == -i\overline{\psi}_M\gamma_{5M}\gamma^{\mu}_M\partial_{\mu}\psi_M,$$

where we have put $c_0 = 1$, and $\overline{\psi}_M = \psi^T_M\gamma^0_M$. It is clear that this Lagrangian is invariant with respect to the Poincaré transformations which, by assumption, involve the proper, ortochronous Lorentz transformations (the space and time reflections are excluded). Moreover, it is invariant also with respect to the chiral transformations of the form (5.48) and (5.49) in which we now have γ_{5M}. The matrix $\exp(i\gamma_{5M}\beta)$ is real and orthogonal. The conserved Noether current that follows from this symmetry is equal to $\overline{\psi}\gamma^{\mu}_M\psi$. In the Sect. 5.3, the chiral symmetry gave the current (5.50) which contains the γ_5 matrix. In the present case γ_{5M} is absent in the current because there is another matrix γ_{5M} coming from the Lagrangian $\mathcal{L}_1$, and $\gamma^2_{5M} = I_4$.

The Majorana field can be introduced in another way, often preferred in the literature. We present it working with the Dirac equation (5.1) in the Dirac representation (5.3). Let us define the charge conjugate Dirac field ψ_c:

$$\psi_c(x) = i\gamma_D^2 \psi^*(x), \tag{5.60}$$

where * denotes complex conjugation and $\psi(x)$ is the Dirac field. The name 'charge conjugate' reflects the fact that if $\psi(x)$ obeys the Dirac equation with an external electromagnetic field $A_\mu(x)$,

$$i\gamma_D^\mu(\partial_\mu + iqA_\mu(x))\psi(x) - m\psi(x) = 0,$$

then $\psi_c(x)$ obeys the equation

$$i\gamma_D^\mu(\partial_\mu - iqA_\mu(x))\psi_c(x) - m\psi_c(x) = 0.$$

The change of sign of the coupling to the external electromagnetic field is interpreted as the change of sign of the electric charge carried by the field. Let us impose on the Dirac field the following condition

$$\psi_c(x) = \psi(x),$$

called the Majorana condition. This condition is invariant under Poincaré transformations because in the Dirac representation

$$\gamma_D^2 S_D^*(\hat{L}) = S_D(\hat{L})\gamma_D^2.$$

Note that the Majorana condition breaks the $U(1)$ symmetry (5.28) of the Dirac Lagrangian. The Majorana condition is satisfied by a bispinor of the form

$$\psi_M(x) = \begin{pmatrix} \xi(x) \\ -i\sigma_2\xi^*(x) \end{pmatrix},$$

where $\xi(x)$ can be an arbitrary two-component complex spinor. Such a bispinor ψ_M is invariant under the charge conjugation. It is called the Majorana field in the Dirac representation. The real valued Majorana field (ψ^α) introduced above in the Majorana representation can be identified with $Re\ \xi$ and $Im\ \xi$.

5.5 Anticommuting (Bi)Spinor Fields

The Dirac, Weyl and Majorana fields are defined by their relativistic transformation laws, and by their time evolution equations or, if available, their Lagrangians. The existence of a Lagrangian is a desired feature, in particular because it helps to quantize the field by applying a canonical quantization method or a path integral approach. Furthermore, one can easily write conserved currents using the Noether theorem. The main advantage of the anticommuting (or Grassmann) versions of the Weyl and Majorana fields is that one can easily write Lagrangians for them also in the massive case. Furthermore, the anticommuting versions of the (bi)spinor fields, including the Dirac field, appear in path integrals for these fields, see Sect. 11.3. In the cases where both versions of the spinor field have Lagrangians, one may choose between them. Because the classical spinor fields mainly serve as a starting point for a construction of quantum fields, and do not have direct physical applications as opposed to, e.g., electromagnetic field, the choice is mainly a matter of convenience.[2] The classical Grassmann field theory should be regarded as an auxiliary theoretical construction which acquires physical meaning only when embedded into a quantum field theory.

Let us start from the Majorana field. In the anticommuting version, $\psi^\alpha(x)$ are anticommuting, that is

$$\{\psi^\alpha(x), \psi^\beta(y)\} = 0 \tag{5.61}$$

for all $\alpha, \beta = 1, 2, 3, 4$, and for all $x, y \in M$. Here $\{A, B\} = AB + BA$. Thus, $\psi^\alpha(x)$ are not real numbers as in the previous section. Nevertheless, one can formulate consistent rules for operating with such 'variables'. The mathematical structure which is relevant here is called a Grassmann algebra, and $\psi^\alpha(x)$ are called its generating elements. Because their number is infinite, the algebra is infinite dimensional. The whole algebra is obtained by first, taking all formal products of the generating elements, and next, by including all formal linear combinations of such products. Because x is a continuous variable, such linear combinations generally have the form of sums over discrete bispinor indices $\alpha, \beta, \ldots$, and integrals over $x, y, z, \ldots$. Various products can be related to each other only by applying the rule (5.61). For example, $\psi^\alpha(x)\psi^\beta(y)$ is not reducible to a linear combination of the generating elements, except for $\alpha = \beta,\ x = y$ when that product is equal to 0 according to (5.61). In the case of the Majorana field we have an infinite number of generating elements. Let us note that in the case of a finite number of generating elements one can construct only a finite number of independent products, because all powers of a single generating element vanish. For example, $\exp(\psi^\alpha(x)) = 1 + \psi^\alpha(x)$ exactly!

One can also define a derivative with respect to the generating element $\psi^\alpha(x)$. It is denoted as

$$\frac{\delta}{\delta\psi^\alpha(x)}, \tag{5.62}$$

and is called the Grassmann derivative. In the first step we just define that

$$\frac{\delta a}{\delta\psi^\alpha(x)} = 0, \quad \frac{\delta\psi^\beta(y)}{\delta\psi^\alpha(x)} = \delta^\beta_\alpha \delta(y - x), \tag{5.63}$$

[2]Let us stress again that one should not confuse the field with a wave function of a single quantum particle. For example, originally the Dirac bispinor was introduced as a wave function of a relativistic electron, not a field, and in the quantum mechanical context it has a probabilistic interpretation. No such interpretation is assumed for the Dirac field.

where a is a number. Now, let F be an element of the Grassmann algebra. It can be written as a linear combination of the generating elements and their products. By definition, the derivative acts on F linearly, that is term by term in that linear combination. Also as a part of the definition, the derivative acts on products of numbers and/or generating elements according to the Leibniz rule, with the modification that the symbol (5.62) of the Grassmann derivative anticommutes with the generating elements, and with other Grassmann derivatives. For example, let us take

$$F = a + b(y)\psi^{\beta}(y) + \int d^4x d^4y \; c(x, y)\psi^{\beta}(x)\psi^{\gamma}(y),$$

where a, $b(y)$ and $c(x, y)$ have complex values. Then

$$\frac{\delta F}{\delta\psi^{\alpha}(z)} = b(y)\delta(y - z)\delta^{\beta}_{\alpha} + \int d^4y \; c(z, y)\delta^{\beta}_{\alpha}\psi^{\gamma}(y) - \int d^4x \; c(x, z)\delta^{\gamma}_{\alpha}\psi^{\beta}(x).$$

Another ingredient in the theory of the anticommuting Majorana field is a conjugation, denoted by *. By definition, this operation has the following properties

$$(AB)^* = B^*A^*, \quad (aA)^* = a^*A^*, \quad (A^*)^* = A,$$

where a is a complex number, a^* denotes its complex conjugate, and A, B are elements of the Grassmann algebra. The assumption in the previous section that the Majorana field has real components is replaced in the Grassmann version by the assumption that

$$(\psi^{\alpha}(x))^* = \psi^{\alpha}(x), \tag{5.64}$$

i.e., the components are selfconjugate.[3]

Now we are prepared to formulate the Grassmann version of the Majorana field. We take the action in the form

$$S = \int d^4x \; \mathcal{L}, \tag{5.65}$$

with the following Lagrangian

$$\mathcal{L} = \frac{i}{2}(\overline{\psi}\gamma^{\mu}_{M}\partial_{\mu}\psi - \partial_{\mu}\overline{\psi}\gamma^{\mu}_{M}\psi) - m\overline{\psi}\psi, \tag{5.66}$$

where $\overline{\psi}_{\alpha} = \psi^{\beta}(\gamma^0_M)_{\beta\alpha}$. $\mathcal{L}$ has the same form as the Dirac Lagrangian (5.24), but now ψ is the anticommuting real Majorana field. Lagrangian (5.66) does not vanish precisely because $\psi^{\alpha}(x)$ do not commute. It is 'real' in the sense that $\mathcal{L}^* = \mathcal{L}$. When checking this it is helpful first to notice that the products $\gamma^0_M\gamma^{\mu}_M$ are symmetric matrices. The Grassmann version of the stationary action principle has the form

[3] Nevertheless we will call them 'real', unless there is a risk of confusion.

$$\frac{\delta S}{\delta \psi^\alpha(x)} = 0, \tag{5.67}$$

where S is regarded as an element of the Grassmann algebra. It gives the Majorana equation (5.59) for the anticommuting field $\psi^\alpha(x)$,

$$i\gamma_M^\mu \partial_\mu \psi(x) - m\psi(x) = 0.$$

This equation should be regarded as a restriction on the generating elements $\psi^\alpha(x)$ of the initial Grassmann algebra. Before solving it, in principle one should first define the Dirac operator $i\gamma_M^\mu\partial_\mu - mI$ in the Grassmann algebra—it is not clear what is meant by the derivatives ∂_μ of the Grassmann elements. We adopt a pragmatical attitude: $\psi^\alpha(x)$ are written in the form of a Fourier transform, so that the x^μ variables appear in ordinary exponential functions. Then the general solution of the Majorana equation can be written in the form

$$\psi^\alpha(x) = \int d^3k \sum_{\lambda=1,2}\sum_{\epsilon=\pm}\Big[e^{-\epsilon i\omega(\vec{k})x^0 + i\vec{k}\vec{x}}\psi_\lambda^{(\epsilon)\alpha}(\vec{k})c_\epsilon^\lambda(\vec{k}) + e^{\epsilon i\omega(\vec{k})x^0 - i\vec{k}\vec{x}}(\psi_\lambda^{(\epsilon)\alpha}(\vec{k}))^*(c_\epsilon^\lambda(\vec{k}))^*\Big], \tag{5.68}$$

where $\omega(\vec{k}) = \sqrt{m^2 + \vec{k}^2}$, and $\psi_\lambda^{(\epsilon)\alpha}(\vec{k})$ are the four independent solutions of the homogeneous matrix equation

$$\left(\epsilon\,\omega(\vec{k})\gamma_M^0 - k^i\gamma_M^i\right)\psi_\lambda^{(\epsilon)}(\vec{k}) - m\psi_\lambda^{(\epsilon)}(\vec{k}) = 0. \tag{5.69}$$

The $c_\epsilon^\lambda(\vec{k})$ present in (5.68) are independent generating elements of a certain Grassmann algebra, which is a subalgebra of the original algebra generated by all $\psi^\alpha(x)$. The Grassmann elements $\psi^\alpha(x)$ given by formula (5.68) are not independent, nevertheless they still anticommute as in (5.61).

The Lagrangian (5.66) for the anticommuting Majorana field does not have the $U(1)$ symmetry because multiplication by a phase factor can violate the reality condition (5.64). Nevertheless, the current $j^\mu = \overline{\psi}\gamma_M^\mu\psi = \psi^\alpha(\gamma_M^0\gamma_M^\mu)_{\alpha\beta}\psi^\beta$, which corresponds to this symmetry in the case of the Dirac Lagrangian (5.24), is conserved—the Majorana equation implies that $\partial_\mu j^\mu = 0$. The explanation of this puzzle is simple: the would-be current j^μ is always equal to zero because ψ^α and ψ^β anticommute with each other and the matrices $\gamma_M^0\gamma_M^\mu$ are symmetric.

The Grassmann version of the Weyl spinor field $\xi(x) = (\xi^\alpha(x)),\quad \alpha = 1, 2,$ allows for a Lorentz invariant mass term, which is not possible in the c-number version with Lagrangian (5.55). In this case the full set of independent generating elements consists of $\xi^\alpha(x)$ and $\xi^{*\dot\alpha}(x)$. The conjugation is defined as follows:

$$(a\xi^\alpha(x))^* = a^*\xi^{*\dot\alpha}(x), \quad (a\xi^{*\dot\alpha}(x))^* = a^*\xi^\alpha(x),$$

where a is a complex number, and a^* is its complex conjugation. The Lagrangian has the form

$$\mathcal{L} = \frac{i}{2}\left(\xi^{*\dot{\alpha}}\tilde{\sigma}^{\mu}_{\dot{\alpha}\beta}\partial_\mu\xi^\beta - \partial_\mu\xi^{*\dot{\alpha}}\tilde{\sigma}^{\mu}_{\dot{\alpha}\beta}\xi^\beta\right) + \frac{m}{2}\left(\xi^\alpha\epsilon_{\alpha\beta}\xi^\beta - \xi^{*\dot{\alpha}}\epsilon_{\alpha\beta}\xi^{*\dot{\beta}}\right).$$

This Lagrangian is 'real' in the sense that it is invariant under conjugation, i.e., $\mathcal{L}^* = \mathcal{L}$. Of course $\mathcal{L}$ is not a number, it is just an element of the Grassmann algebra. The mass term does not vanish because

$$\xi^\alpha\xi^\beta = -\xi^\beta\xi^\alpha, \quad \xi^{*\dot{\alpha}}\xi^{*\dot{\beta}} = -\xi^{*\dot{\beta}}\xi^{*\dot{\alpha}}.$$

The Grassmann version of the Dirac bispinor field in Minkowski space-time is presented at the beginning of Sect. 6.2, and a Euclidean version of it in Sect. 14.2.

Exercises

5.1 Check that

$$\omega_{\mu\nu}(\gamma^\mu\gamma^\nu\gamma^\rho - \gamma^\rho\gamma^\mu\gamma^\nu) = -4\omega^\rho{}_\sigma\gamma^\sigma,$$

where $\omega_{\mu\nu} = -\omega_{\nu\mu}$.
Hint: Use the Dirac relations (5.2) and notice that $\omega_{\mu\nu}\eta^{\mu\nu} = 0$.

5.2 (a) Check that the matrices $\alpha^i_M = \gamma^0_M\gamma^i_M$ are symmetric. Next, prove that the Dirac Lagrangian (5.24) with $\gamma^\mu = \gamma^\mu_M$ vanishes if all components of the bispinor ψ are real numbers.
(b) Check that the term $\psi^T\hat{A}\psi$, where $\hat{A} \neq 0$ and $\hat{A}^T = \hat{A}$, is not invariant with respect to transformations of the form (5.58).
Hints: The invariance requires that $S^T(\hat{L})\hat{A}S(\hat{L}) = \hat{A}$ for all $\hat{L} \in L^\uparrow_+$. Show that this condition is equivalent to

$$(\hat{A}\gamma^\mu_M\gamma^\nu_M)^T = -\hat{A}\gamma^\mu_M\gamma^\nu_M,$$

where $\mu \neq \nu$, $\mu, \nu = 0, 1, 2$ and 3. Next, consider these conditions taking, for instance, $\mu = 1, \nu = 3$; $\mu = 0, \nu = 1$ and $\mu = 0, \nu = 2$. Show that they are satisfied only by $\hat{A} = 0$.
(c) Show that the conditions

$$\gamma^0_M\gamma^i_M\hat{B}^0 = \hat{B}^0\gamma^0_M\gamma^i_M,$$

where $i = 1, 2$ and 3, are satisfied only by $\hat{B}^0 = c_0 i\gamma_{5M}$.
Hint: First compute the three matrices $\gamma^0_M\gamma^i_M$. Write $\hat{B}^0$ in the block form

$$\hat{B}^0 = \begin{pmatrix} a & b \\ -b^T & c \end{pmatrix},$$

where a, b, c are real, 2 by 2 matrices, $a^T = -a,\ c^T = -c$.

5.3 Show, by acting on the Dirac equation (5.1) with the operator $i\gamma^\mu\partial_\mu + mI_4$, that every component of the Dirac spinor satisfies the Klein–Gordon equation.

5.4 Prove that the matrices $\{\Gamma^J\} = \{I_4, \gamma^\mu, \gamma_5, \gamma^\mu\gamma_5, \sigma^{\mu\nu}\}$, where $\sigma^{\mu\nu} = \frac{i}{2}[\gamma^\mu, \gamma^\nu]$, form a basis in the vector space (over the complex number field) of 4×4 matrices.
Hint: Check that $\mathrm{tr}\,(\Gamma^J\Gamma^K)$ does not vanish if and only if $J = K$ and use this to prove that for $\lambda_i \in \mathbb{C}$:

$$\sum_{J=1}^{16} \lambda_J \Gamma^J = 0 \quad \Rightarrow \quad \lambda_J = 0,\ J = 1, \ldots, 16.$$

5.5 Let $\psi_\lambda^{(\pm)\,\alpha}(\vec{k}),\ \lambda = 1, 2$, denote linearly independent solutions of the Dirac equation in the momentum space:

$$\left(\epsilon\,\omega(\vec{k})\gamma^0_{\alpha\beta} - k^i\gamma^i_{\alpha\beta} - m\right)\psi_\lambda^{(\epsilon)\,\beta}(\vec{k}) = 0,$$

where $\omega(\vec{k}) = \sqrt{m^2 + \vec{k}^2}$. Denote

$$u_\lambda^\alpha(\vec{k}) = \psi_\lambda^{(+)\,\alpha}(\vec{k}), \quad v_\lambda^\alpha(\vec{k}) = \psi_\lambda^{(-)\,\alpha}(-\vec{k}).$$

Prove the following identities:

$$\left(\omega(\vec{k})\gamma^0 - k^i\gamma^i - m\right)u_\lambda(\vec{k}) = 0,$$

$$\left(\omega(\vec{k})\gamma^0 - k^i\gamma^i + m\right)v_\lambda(\vec{k}) = 0.$$

5.6 For the Dirac bispinors normalized as

$$u_\lambda(\vec{k})^\dagger u_\sigma(\vec{k}) = \frac{\omega(\vec{k})}{m}\delta_{\lambda\sigma}, \quad v_\lambda(\vec{k})^\dagger v_\sigma(\vec{k}) = \frac{\omega(\vec{k})}{m}\delta_{\lambda\sigma}$$

$(m > 0)$, demonstrate the identities

$$\bar{u}_\lambda(\vec{k})u_\sigma(\vec{k}) = \delta_{\lambda\sigma}, \qquad \bar{v}_\lambda(\vec{k})v_\sigma(\vec{k}) = -\delta_{\lambda\sigma},$$

and

$$\sum_{\lambda=1}^{2} u_\lambda^\alpha(\vec{k})\,\bar{u}_\lambda^\beta(\vec{k}) = \left(\frac{\not{k}+mI_4}{2m}\right)^{\alpha\beta}, \qquad \sum_{\lambda=1}^{2} v_\lambda^\alpha(\vec{k})\,\bar{v}_\lambda^\beta(\vec{k}) = \left(\frac{\not{k}-mI_4}{2m}\right)^{\alpha\beta},$$

where $\not{k} = \omega(\vec{k})\gamma^0 - k^i\gamma^i$.
Hint: In order to check the last two formulas notice that the $u_1(\vec{k})$, $u_2(\vec{k})$, $v_1(\vec{k})$, $v_2(\vec{k})$ form a basis in the vector space of Dirac bispinors. Decompose an arbitrary bispinor in this basis and, using this decomposition, check that the actions of both sides of the identities on such an arbitrary bispinor coincide.

5.7 Derive the Gordon identities

$$\bar{u}_\lambda(\vec{p})\gamma^\mu u_\sigma(\vec{q}) = \frac{1}{2m}\bar{u}_\lambda(\vec{p})\left[(p+q)^\mu + i\sigma^{\mu\nu}(p-q)_\nu\right]u_\sigma(\vec{q}),$$

and

$$\bar{u}_\lambda(\vec{p})\gamma^\mu\gamma_5 u_\sigma(\vec{q}) = \frac{1}{2m}\bar{u}_\lambda(\vec{p})\left[(p-q)^\mu\gamma_5 + i\sigma^{\mu\nu}(p+q)_\nu\gamma_5\right]u_\sigma(\vec{q}).$$

What would the analogous identities for the bispinors $v_\lambda(\vec{k})$ look like?

5.8 Check that $\bar{\psi}(x)\psi(x)$ and $\bar{\psi}(x)\gamma_5\psi(x)$ are scalars under the Poincaré transformations (with $L \in L_+^\uparrow$), while $\bar{\psi}(x)\gamma^\mu\psi(x)$ and $\bar{\psi}(x)\gamma_5\gamma^\mu\psi(x)$ behave like four-vectors.
Remark: $\bar{\psi}(x)\psi(x)$ and $\bar{\psi}(x)\gamma_5\psi(x)$, and similarly $\bar{\psi}(x)\gamma^\mu\psi(x)$ and $\bar{\psi}(x)\gamma_5\gamma^\mu\psi(x)$ behave differently under the reflection $\vec{x} \to -\vec{x}$, but the reflection does not belong to $L_+^\uparrow$. Had we study it we would have discovered that $\bar{\psi}(x)\gamma_5\psi(x)$ is in fact a pseudoscalar and $\bar{\psi}(x)\gamma_5\gamma^\mu\psi(x)$ a pseudovector (like, for instance, a vector product of three dimensional vectors).

Chapter 6
The Quantum Theory of Free Fields

Abstract The canonical quantization of the free, real scalar field. Difficulties with the Schroedinger representation. Inequivalent representations of the canonical commutation relations. The Fock representation. Basic quantum observables: the total energy and momentum of the field. A description of quantum states in terms of particles. The field operator as a generalized function. The classical Dirac field as a system with constraints. The Faddeev–Jackiw method and quantization of the free Dirac field. The Dirac vacuum and the appearance of a free, relativistic, spin 1/2 particle and its antiparticle. Extraction of the physical degrees of freedom of the free electromagnetic field. The canonical quantization of the electromagnetic field and the appearance of a free, massless particle (the photon).

Quantum field theory, that is the quantum theory of systems with an infinite number of degrees of freedom, provides an explanation of rather nontrivial phenomena, including the very fact that the world seems to be built of well-defined quantum particles with intrinsic characteristics like spin, electric charge, and so forth. Also, the fact that particles come in a great number of perfectly identical copies, is explained if we assume that in nature there physically exist certain quantum fields. These fields are the basic physical constituents of the material world whereas the particles are secondary.

Quantum field theory still has some unsolved problems. Among them is the question of how to find an appropriate Hilbert space in which one can have a probabilistic interpretation of the theory, in terms of objects directly accessible to the methods of experimental physics (various 'particles' in most cases). This is a very difficult problem, especially when interactions are present, as opposed to the case of quantum mechanics, where there is no doubt what the pertinent Hilbert space is.

In this chapter we describe three main types of free quantum fields. By definition, 'the free quantum field' means that the evolution equation for the field operator in the Heisenberg picture is linear. The free quantum fields are very well understood from both physical and mathematical viewpoints.

H. Arodź and L. Hadasz, *Lectures on Classical and Quantum Theory of Fields*,
Graduate Texts in Physics, DOI 10.1007/978-3-319-55619-2_6

6.1 The Real Scalar Field

The configuration space of the classical real scalar field consists of smooth, real functions $\phi(\vec{x})$ of the vector $\vec{x} \in R^3$. Field trajectories in this space are represented by the functions $\phi = \phi(t, \vec{x})$ of the time t and $\vec{x}$. The Lagrangian of this field has the form[1]

$$\mathcal{L} = \frac{1}{2}\left(\partial_0\phi\,\partial_0\phi - \partial_i\phi\,\partial_i\phi - m^2\phi^2\right). \tag{6.1}$$

It is a function of $\phi(t, \vec{x})$ and $\partial_0\phi(t, \vec{x})$, which are regarded as independent arguments of $\mathcal{L}$ because there is no relation between them at a given time t. We assume that $m^2 > 0$. An example of a quantum field with $m = 0$ is discussed in Sect. 6.3.

The canonical momentum for the field ϕ is defined as follows

$$\pi(t, \vec{x}) = \frac{\partial\mathcal{L}}{\partial(\partial_0\phi(t, \vec{x}))}. \tag{6.2}$$

This formula corresponds to $p = \partial L/\partial\dot{q}$ known from classical mechanics. In our case

$$\pi(t, \vec{x}) = \partial_0\phi(t, \vec{x}). \tag{6.3}$$

Note that the canonical momentum differs from the density T^{0i} of the conserved momentum,

$$T^{0i} = -\partial_i\phi\,\partial_0\phi.$$

In classical mechanics of a free particle the two momenta coincide.

There is no derivation of quantum theory from the classical one. The reason is that the quantum theory is much more general. Actually, it is the classical theory which is derived from the quantum theory as an approximation that is valid only if certain conditions are satisfied. The historical fact that certain classical theories were discovered a long time before the quantum theories, can, to some extent, be explained by the lack of sufficiently precise experimental equipment which could have allowed physicists in the past to observe microscopic phenomena. Another reason is that the majority of phenomena that we can directly perceive with our senses, can be understood in terms of classical physics with a satisfactory accuracy. Thus, the quantum theory is postulated, not derived. Nevertheless, there exist several so called methods of quantization. In fact, they should be regarded merely as certain heuristic rules for how to arrive at (hopefully) consistent quantum theories. Such rules work in certain cases, while in others they have to be modified or even abandoned.

In this chapter we use the most popular method of quantization called the canonical quantization. It is a straightforward generalization of the method applied when passing from classical to quantum mechanics. Thus, we assume that there exist Hermitian field operators $\hat{\phi}(\vec{x})$ and Hermitian canonical momentum operators $\hat{\pi}(\vec{x})$,

---

[1] $c = \hbar = 1, x^0 = t$.

which obey the following commutation relations: for all $\vec{x}$ and $\vec{y} \in R^3$

$$[\hat{\phi}(\vec{x}), \hat{\phi}(\vec{y})] = 0, \quad [\hat{\pi}(\vec{x}), \hat{\pi}(\vec{y})] = 0, \quad [\hat{\phi}(\vec{x}), \hat{\pi}(\vec{y})] = i\delta(\vec{x} - \vec{y})I, \tag{6.4}$$

where I denotes the identity operator. The field and canonical momentum operators here are considered in the Schroedinger picture. In the natural units $[\hat{\phi}] = \text{cm}^{-1}$ and $[\hat{\pi}] = \text{cm}^{-2}$. The Hermiticity of the field and of the canonical momentum operators is the quantum counterpart of the fact that the classical $\phi(\vec{x})$ and $\pi(\vec{x})$ are real.

We also have to postulate the form of the operators which correspond to observables. We use the heuristic principle of correspondence, which says that the dependence of the quantum observables on the field and canonical momentum operators should resemble the dependence of the corresponding classical observables on the classical field ϕ and on the classical canonical momentum π. Because the classical energy is given by the formula

$$E = \frac{1}{2} \int d^3x \, (\pi^2 + \partial_i\phi \, \partial_i\phi + m^2\phi^2),$$

we postulate that the quantum Hamiltonian in the Schroedinger picture has the form

$$\hat{H} = \frac{1}{2} \int d^3x \, \left(\hat{\pi}^2 + \partial_i\hat{\phi} \, \partial_i\hat{\phi} + m^2\hat{\phi}^2\right). \tag{6.5}$$

Similarly, the momentum operator is postulated as

$$\hat{P}^i = -\frac{1}{2} \int d^3x \left(\hat{\pi} \, \partial_i\hat{\phi} + \partial_i\hat{\phi} \, \hat{\pi}\right), \tag{6.6}$$

where we have taken the Hermitian part of the product of noncommuting operators.

In quantum mechanics, the assumptions made above would be sufficient to define the quantum model, and we could pass to the calculations of the spectrum of the Hamiltonian, evolution of wave packets, and so forth. In quantum field theory much more is needed. The point is that the postulates listed above specify only the algebraic structure of the quantum model. In the case of the quantum mechanics, the analogous algebraic structure (i.e. the commutation relations between the position and momentum operators) together with a formula for the Hamiltonian is essentially sufficient to determine the full quantum model. There is a theorem by J. von Neumann which says that there is just one realization of such commutation relations in a Hilbert space up to unitary equivalence.[2] For example, in the one dimensional case it is sufficient to take the well-known $L^2(R^1)$ space with $\hat{p} = -i\partial/\partial q$ and $\hat{q} = q\cdot$, where the notation $q\cdot$ means that $\hat{q}\psi(q) = q\psi(q)$. Here q is a Cartesian coordinate on R^1. This is the Schroedinger representation of the quantum mechanics.

[2]The theorem actually speaks about the realizations of the so called Weyl relations, which are closely related to the canonical commutation relations, but not equivalent to them. Nevertheless our slightly imprecise description of the theorem captures its meaning.

In the case of field theory there is a problem with finding a Hilbert space realization of the algebraic structure. It can be solved in several models, including the ones presented in this chapter, but in many others it is an open question. In fact, the algebraic structure postulated in (6.4) in conjunction with formulas (6.5), (6.6) leads to two problems. To see them, let us try the straightforward generalization of the Schroedinger representation

$$\hat{\pi}(\vec{x}) = -i\frac{\delta}{\delta\phi(\vec{x})}, \quad \hat{\phi}(\vec{x}) = \phi(\vec{x})\cdot, \tag{6.7}$$

where the dot after $\phi(\vec{x})$ means multiplication of numbers. These operators are supposed to act on complex functionals $\Psi[\phi]$, which are counterparts of the wave functions $\psi(q)$ from quantum mechanics. The configuration space of the real scalar field consists of functions $\phi(\vec{x})$, and the functional $\Psi[\phi]$ is just a complex function on this space, in full analogy with the quantum mechanical wave function $\psi(q)$. In order to mark functionals clearly, we use a square bracket around the argument.

The first problem appears when we try to define a scalar product of the functionals. We need a scalar product because otherwise we would not be able to use the standard probabilistic interpretation of the quantum theory. A formula written by analogy with the scalar product in the space $L^2(R)$, namely

$$\langle\Psi_1|\Psi_2\rangle = \int \prod_{\vec{x}\in R^3} d\phi(\vec{x})\ \Psi_1^*[\phi]\Psi_2[\phi], \tag{6.8}$$

does not have any operational meaning because of the undefined infinite product, and therefore it is useless for calculating probabilities of quantum processes. It turns out that a solution to this problem exists, but it is not straightforward.

The second problem has a more technical character, nevertheless it has to be dealt with. It turns out that Hamiltonian (6.5) and momentum operator (6.6) are not properly defined. As we know, the first functional derivative is a generalized function defined as follows:

$$\lim_{\epsilon\to 0}\frac{\Psi[\phi+\epsilon f]-\Psi[\phi]}{\epsilon} = \int d^3x\ \frac{\delta\Psi[\phi]}{\delta\phi(\vec{x})}\ f(\vec{x}),$$

for arbitrary test functions $f(\vec{x})$ from the space $S(R^3)$. Let us now consider the action of the square of the operator $\hat{\pi}(\vec{x})$ on a functional $\Psi[\phi]$. Because the functional derivative $\delta\Psi[\phi]/\delta\phi(\vec{x})$ is a generalized function of the variable $\vec{x}$, it can not be regarded as a functional of ϕ. The reason is that in order to be a functional, it should have a well-defined numerical value, whereas a generalized function of $\vec{x}$ does not necessarily have any definite numerical value at a given $\vec{x}$, see the Appendix. Hence, second and higher order functional derivatives require a special definition—it is not correct to regard them as functional derivatives of the first functional derivative of $\Psi[\phi]$. The definition is recursive. The n-th ($n \geq 1$) functional derivative $\delta^n\Psi[\phi]/\delta\phi(\vec{x}_1)\delta\phi(\vec{x}_2)\ldots\delta\phi(\vec{x}_n)$ is, by definition, a generalized function of $\vec{x}_1, \ldots, \vec{x}_n$, hence it is a functional on the space of smooth functions $f(\vec{x}_1, \vec{x}_2, \ldots, \vec{x}_n)$ which vanish at infinity. It turns out that it is sufficient to consider functions of the form $f(\vec{x}_1, \vec{x}_2, \ldots, \vec{x}_n) = f_1(\vec{x}_1) f_2(\vec{x}_2) \cdots f_n(\vec{x}_n)$, where the functions $f_i(\vec{x}_i)$ are test functions from $S(R^3)$. The n-th functional derivative acts on such a function f, giving a number $\Psi^{(n)}[f, \phi]$, which in physics literature is often written as the integral

$$\Psi^{(n)}[f, \phi] = \int \prod_{i=1}^{n} d^3x_i \, \frac{\delta^n \Psi[\phi]}{\delta\phi(\vec{x}_1)\delta\phi(\vec{x}_2)\ldots\delta\phi(\vec{x}_n)} f_1(\vec{x}_1) f_2(\vec{x}_2) \cdots f_n(\vec{x}_n).$$

Now, $\Psi^{(n)}[f, \phi]$ for any fixed f may be regarded as a functional of $\phi(\vec{x})$, and we may calculate the first functional derivative of this functional with respect to $\phi(\vec{x})$. The $(n+1)$-st functional derivative of $\Psi[\phi]$ is defined by the formula

$$\int d^3x_{n+1} \frac{\delta \Psi^{(n)}[f, \phi]}{\delta\phi(\vec{x}_{n+1})} f_{n+1}(\vec{x}_{n+1}) \tag{6.9}$$
$$= \int d^3x_{n+1} \prod_{i=1}^{n} d^3x_i \, \frac{\delta^{n+1} \Psi[\phi]}{\delta\phi(\vec{x}_{n+1})\delta\phi(\vec{x}_1)\delta\phi(\vec{x}_2)\ldots\delta\phi(\vec{x}_n)} \prod_{i=1}^{n+1} f_i(\vec{x}_i).$$

It is a generalized function of the $(n+1)$ vectors $\vec{x}_1, \vec{x}_2, \ldots, \vec{x}_{n+1}$.

The trouble with the $\hat{\pi}^2$ operator in Hamiltonian (6.5) is that it contains a second functional derivative in which $\vec{x}_1 = \vec{x}_2 = \vec{x}$, that is

$$\left. \frac{\delta^2 \Psi[\phi]}{\delta\phi(\vec{x}_1)\delta\phi(\vec{x}_2)} \right|_{\vec{x}_1 = \vec{x}_2 = \vec{x}}.$$

In general, such an object does not have mathematical meaning. In particular, it does not have to be a generalized function of $\vec{x}$. For example, let us consider the functional $\Psi_1[\phi] = \int d^3x \, \phi^2(\vec{x})$. Then,

$$\frac{\delta \Psi_1[\phi]}{\delta\phi(\vec{x}_1)} = 2\phi(\vec{x}_1),$$

and

$$\frac{\delta^2 \Psi_1[\phi]}{\delta\phi(\vec{x}_1)\delta\phi(\vec{x}_2)} = 2\delta(\vec{x}_1 - \vec{x}_2).$$

It is clear that the substitution $\vec{x}_1 = \vec{x}_2 = \vec{x}$ gives the meaningless result $\delta(0)$. The product $\hat{\pi}\hat{\phi}$ present in the momentum operator also leads to a mathematically undefined term. Let us calculate

$$\hat{\pi}(\vec{x}_1)\hat{\phi}(\vec{x}_2)\Psi_1[\phi]$$
$$= -i\frac{\delta}{\delta\phi(\vec{x}_1)}(\phi(\vec{x}_2)\Psi_1[\phi]) = -i\delta(\vec{x}_1 - \vec{x}_2)\Psi_1[\phi] - i\phi(\vec{x}_2)\frac{\delta}{\delta\phi(\vec{x}_1)}\Psi_1[\phi].$$

It is clear that the first term on the r.h.s. becomes meaningless if we put $\vec{x}_1 = \vec{x}_2 = \vec{x}$. Thus, the straightforward Schroedinger representation is not good in quantum field theory.

The solution to these two problems: finding the Hilbert space and constructing physically relevant operators in it, is quite intricate. The very fact that it exists is far from trivial. Before presenting the details, let us first sketch the underlying idea. First, we assume that the correct Hamiltonian differs from the one given by formula (6.5) by a term of the form $c_0 I$, where c_0 is a number and I is the identity operator. There is a formulation of quantum dynamics which is insensitive to this difference: the Heisenberg picture. If $\hat{O}_S$ is an operator in the Schroedinger picture, its counterpart in the Heisenberg picture is defined by the formula

$$\hat{O}_H(t) = e^{i\hat{H}t}\hat{O}_S e^{-i\hat{H}t}, \tag{6.10}$$

provided that the Hamiltonian $\hat{H}$ does not depend on time. The terms $c_0 I$ cancel each other on the r.h.s. of this formula. Therefore time evolution of operators in the Heisenberg picture is correct in spite of the fact that Hamiltonian (6.5) is wrong. Also the Heisenberg evolution equation derived with the use of this Hamiltonian,

$$\frac{d\hat{O}_H(t)}{dt} = i[\hat{H}, \hat{O}_H(t)] + \left(\frac{d\hat{O}_S}{dt}\right)_H (t), \tag{6.11}$$

has the correct form. In the second term on the r.h.s. of this formula we first calculate the time derivative in the Schroedinger picture and next we transform the obtained operator to the Heisenberg picture as in formula (6.10). One may ask whether it is possible to obtain a concrete form of such an evolution equation when the Hamiltonian is not defined yet, because formula (6.5) is meaningless. The answer is that we will only use the algebraic operator relations in the form of commutators, and for such limited purposes Hamiltonian (6.5) is as good as the correct one. Thus, our first step in the construction of the quantum model is just the choice of the Heisenberg picture.

In the next step we find a general solution of the evolution equations in the Heisenberg picture. In this way we restrict the set of operators to be constructed in the as yet unknown Hilbert space to the subset which is relevant from a physical viewpoint. It turns out that for operators from this subset one can provide explicit realizations in Hilbert spaces. We will also find the correct form of observables like the Hamiltonian and the total momentum of the field, but they will be defined only for the physically relevant fields and in the chosen Hilbert space, not on the abstract level of algebraic relations (6.4), (6.5), (6.6). It turns out that in such a restricted framework, the correctly defined observables differ from the symbolic expressions like (6.5), (6.6) by terms of the form $c_0 I$, as assumed. We shall also see that in the case of field theory there are infinitely many unitarily inequivalent choices of Hilbert spaces.

Because all observables are built from the field and its canonical momentum, it is sufficient to consider the Heisenberg evolution equations only for these two operators. Both operators do not depend on time in the Schroedinger picture. The fact that the field and the canonical momentum operators are considered in the Heisenberg picture is denoted simply by adding the time argument t. The evolution equations have the form

$$\frac{\partial \hat{\phi}(t, \vec{x})}{\partial t} = i[\hat{H}, \hat{\phi}(t, \vec{x})] = \hat{\pi}(t, \vec{x}), \tag{6.12}$$

$$\frac{\partial \hat{\pi}(t, \vec{x})}{\partial t} = i[\hat{H}, \hat{\pi}(t, \vec{x})] = -m^2 \hat{\phi}(t, \vec{x}) + \Delta \hat{\phi}(t, \vec{x}). \tag{6.13}$$

Here we follow the tradition that the time derivatives of the field and its canonical momentum operators in the Heisenberg picture are denoted as partial derivatives. Δ denotes the Laplacian with respect to $\vec{x}$. In order to obtain the r.h.s.'s of these equations we have used the canonical commutation relations (6.4), and the fact that the Hamiltonian commutes with the exponentials $\exp(\pm i\hat{H}t)$. We have also applied the formula $[AB, C] = A[B, C] + [A, C]B$. For example,

$$\begin{aligned}
&i[\hat{H}, \hat{\pi}(t, \vec{x})] \\
&= e^{it\hat{H}} \frac{i}{2} \int d^3y \left(\frac{\partial \hat{\phi}(\vec{y})}{\partial y^i} \frac{\partial}{\partial y^i} [\hat{\phi}(\vec{y}), \hat{\pi}(\vec{x})] + (\frac{\partial}{\partial y^i} [\hat{\phi}(\vec{y}), \hat{\pi}(\vec{x})]) \frac{\partial \hat{\phi}(\vec{y})}{\partial y^i} \right) e^{-it\hat{H}} \\
&+ \frac{im^2}{2} e^{it\hat{H}} \int d^3y \left(\hat{\phi}(\vec{y}) \, [\hat{\phi}(\vec{y}), \hat{\pi}(\vec{x})] + [\hat{\phi}(\vec{y}), \hat{\pi}(\vec{x})] \, \hat{\phi}(\vec{y}) \right) e^{-it\hat{H}} \\
&= e^{it\hat{H}} \frac{i}{2} \int d^3y \left(\frac{\partial \hat{\phi}(\vec{y})}{\partial y^i} \frac{\partial}{\partial y^i} i\delta(\vec{y} - \vec{x}) + (\frac{\partial}{\partial y^i} i\delta(\vec{y} - \vec{x})) \frac{\partial \hat{\phi}(\vec{y})}{\partial y^i} \right) e^{-it\hat{H}} \\
&+ \frac{im^2}{2} e^{it\hat{H}} \int d^3y \, \hat{\phi}(\vec{y}) 2i\delta(\vec{y} - \vec{x}) e^{-it\hat{H}} = \Delta \hat{\phi}(t, \vec{x}) - m^2 \hat{\phi}(t, \vec{x}).
\end{aligned}$$

Eliminating $\hat{\pi}$ in (6.13) with the help of (6.12) we obtain the Klein–Gordon equation

$$\frac{\partial^2 \hat{\phi}(t, \vec{x})}{\partial t^2} - \Delta \hat{\phi}(t, \vec{x}) + m^2 \hat{\phi}(t, \vec{x}) = 0 \tag{6.14}$$

for the field operator $\hat{\phi}(t, \vec{x})$ in the Heisenberg picture.

The general solution of (6.14) can be found in the same manner as in Chap. 1. It has the following form

$$\hat{\phi}(t, \vec{x}) = \int d^3k \left(e^{-ikx} \tilde{a}(\vec{k}) + e^{ikx} \tilde{b}(\vec{k}) \right), \tag{6.15}$$

where

$$kx = \omega(\vec{k})t - \vec{k}\vec{x}, \quad \omega(\vec{k}) = \sqrt{\vec{k}^2 + m^2},$$

and

$$\tilde{b}(\vec{k}) = \tilde{a}^\dagger(\vec{k})$$

because of the Hermiticity of the field operator. For a later convenience we rescale the $\tilde{a}$ operators,

$$\tilde{a}(\vec{k}) = \frac{\hat{a}(\vec{k})}{\sqrt{2(2\pi)^3\omega(\vec{k})}}.$$

Thus,

$$\hat{\phi}(t, \vec{x}) = \int \frac{d^3k}{\sqrt{2(2\pi)^3\omega(\vec{k})}} \left(e^{-ikx}\hat{a}(\vec{k}) + \text{h.c.}\right), \tag{6.16}$$

where h.c. stands for the Hermitian conjugate of the preceding term.

Solution (6.16) contains an arbitrary, operator valued function $\hat{a}(\vec{k})$, where $\vec{k} \in R^3$. Next, we require that $\hat{\phi}(t, \vec{x})$ together with the canonical conjugate momentum given by formula (6.12) obey the canonical commutation relations (6.4). It turns out that those relations are satisfied provided that

$$[\hat{a}(\vec{k}), \hat{a}(\vec{k}')] = 0, \quad [\hat{a}^\dagger(\vec{k}), \hat{a}^\dagger(\vec{k}')] = 0, \quad [\hat{a}(\vec{k}), \hat{a}^\dagger(\vec{k}')] = \delta(\vec{k} - \vec{k}')I. \tag{6.17}$$

These conditions are a kind of ('canonical') constraint on the operators $\hat{a}(\vec{k})$. Their derivation is rather simple. Using the operators $\hat{P}_{\vec{k}}(t)$ introduced in Sect. 1.3 we may extract the operators $\hat{a}(\vec{k})$,

$$\hat{a}(\vec{k}) = \hat{P}_{\vec{k}}(t)\hat{\phi}(t, \vec{x}). \tag{6.18}$$

Thus,

$$\hat{a}(\vec{k}) = i \int d^3x \left[f^*_{\vec{k}}(t, \vec{x})\hat{\pi}(t, \vec{x}) - \partial_0 f^*_{\vec{k}}(t, \vec{x})\hat{\phi}(t, \vec{x})\right], \tag{6.19}$$

where the time t can be chosen arbitrarily. Formula (6.19) and its Hermitian conjugate are inserted on the l.h.s.'s of the commutation relations (6.17). Next we use the canonical commutation relations (6.4) transformed to the Heisenberg picture:

$$[\hat{\phi}(t, \vec{x}), \hat{\phi}(t, \vec{y})] = 0, \quad [\hat{\pi}(t, \vec{x}), \hat{\pi}(t, \vec{y})] = 0,$$

$$[\hat{\phi}(t, \vec{x}), \hat{\pi}(t, \vec{y})] = i\delta(\vec{x} - \vec{y})I. \tag{6.20}$$

Note that the operators in each commutator are taken at the same time t. For this reason, relations (6.20) are called the equal time canonical commutation relations.

With the help of formula (6.16) and relations (6.17) we can compute the commutators of the field and the canonical momentum operators at arbitrary times. For example,

$$[\hat{\phi}(x), \hat{\phi}(y)] = i\Delta(x-y)I, \tag{6.21}$$

where $\Delta(x-y)$ is the Pauli–Jordan function introduced in Sect. 1.3. For brevity, we use here the four-dimensional notation $x = (x^0, \vec{x})$. Of course, for $x^0 = y^0$ formula (6.21) reduces to the first of the equal time commutation relations (6.20). Taking derivatives of both sides of (6.21) with respect to x^0 or y^0 we obtain commutation relations of the types $[\hat{\phi}(x), \hat{\pi}(y)]$, and $[\hat{\pi}(x), \hat{\pi}(y)]$. The formula quoted at the end of Sect. 1.3 shows that the Pauli–Jordan function vanishes when $(x-y)^2 < 0$. Therefore, all of these commutators vanish if x is spatially separated from y. When a field and its canonical momentum in the Heisenberg picture have this property, the field is called the local quantum field. Our scalar field is an example of such a field.

The total energy and the total momentum of the quantum scalar field have the form of integrals over the whole space R^3, see formulas (6.5), (6.6). If these integrals are replaced by integrals over a compact subset V of R^3 (without changing the integrands) we obtain so called local observables. For example, instead of $\hat{H}$ we take

$$\hat{H}_V = \frac{1}{2}\int_V d^3x \left(\hat{\pi}^2 + \partial_i\hat{\phi}\partial_i\hat{\phi} + m^2\hat{\phi}^2\right).$$

In the case of the local quantum field, such local observables commute with each other if the corresponding sets V do not intersect.

The physically relevant quantum field is given by solution (6.16) of the Heisenberg evolution equations, with the restriction that the operators $\hat{a}$ and $\hat{a}^\dagger$ obey the commutation relations (6.17). It is clear that in order to solve the problem of the existence of a Hilbert space realization, it is sufficient to find such a realization of the operators $\hat{a}$ and $\hat{a}^\dagger$. Let us first remove the mathematical complications introduced by the fact that the vector variable $\vec{k}$ is continuous. For example, due to the presence of the Dirac delta on the r.h.s. of the third relation (6.17), $\hat{a}(\vec{k})$ is a generalized function of $\vec{k}$, and therefore it does not have any definite value for a given $\vec{k}$. In most cases this is not important because $\hat{a}$ and $\hat{a}^\dagger$ appear in integrals over the wave vector $\vec{k}$, but here it would hamper our considerations. We introduce an infinite, discrete set of operators $\hat{a}_i$ and $\hat{a}_i^\dagger$, $i = 1, 2, \ldots$, which are related to $\hat{a}(\vec{k})$ and $\hat{a}^\dagger(\vec{k})$ by the following (invertible) formulas:

$$\hat{a}_i = \int d^3k\, h_i(\vec{k})\hat{a}(\vec{k}), \quad \hat{a}_i^\dagger = \int d^3k\, h_i^*(\vec{k})\hat{a}^\dagger(\vec{k}), \tag{6.22}$$

or

$$\hat{a}(\vec{k}) = \sum_{i=1}^{\infty} h_i^*(\vec{k})\hat{a}_i, \quad \hat{a}^\dagger(\vec{k}) = \sum_{i=1}^{\infty} h_i(\vec{k})\hat{a}_i^\dagger. \tag{6.23}$$

The functions $h_i(\vec{k})$, $i = 1, 2, \ldots$, form a complete, orthonormal set, that is

$$\int d^3k\, h_i^*(\vec{k})h_j(\vec{k}) = \delta_{ij}, \quad \sum_{i=1}^{\infty} h_i^*(\vec{k})h_i(\vec{k}') = \delta(\vec{k} - \vec{k}'). \tag{6.24}$$

The precise form of these functions is not needed here. The operators $\hat{a}_i$ and $\hat{a}_j^\dagger$ obey the following commutation relations

$$[\hat{a}_i, \hat{a}_j] = 0, \quad [\hat{a}_i^\dagger, \hat{a}_j^\dagger] = 0, \quad [\hat{a}_i, \hat{a}_j^\dagger] = \delta_{ij} I, \tag{6.25}$$

which are equivalent to (6.17).

Let us now consider the infinite tensor product of $L^2(R^1)$ spaces,

$$\mathcal{H}^\infty = \bigotimes_{i=1}^{\infty} L^2(R^1).$$

In a slightly imprecise description of this space, its elements have the form of linear combinations of a finite number of formal infinite products

$$f_1(\xi_1) f_2(\xi_2) \ldots, \tag{6.26}$$

where $f_i(\xi_i)$ are elements of the $L^2(R^1)$ space. Such products of functions with different arguments are formal because we do not care about their convergence—we are not interested in their numerical value. Except for convergence, such products have all the properties of products with a finite number of factors. Without any loss of generality we may assume that all f_i appearing in the formal products are normalized, that is

$$\int d\xi\, f_i^*(\xi) f_i(\xi) = 1. \tag{6.27}$$

We need the following auxiliary operators in the Hilbert space $L^2(R^1)$ of functions $f(\xi)$ of one real variable $\xi \in R^1$

$$\hat{\alpha}(\xi) = \frac{1}{\sqrt{2}}(\xi + \frac{d}{d\xi}), \quad \hat{\alpha}^\dagger(\xi) = \frac{1}{\sqrt{2}}(\xi - \frac{d}{d\xi}). \tag{6.28}$$

These operators satisfy the following commutation relation

$$[\hat{\alpha}(\xi), \hat{\alpha}^\dagger(\xi)] = I.$$

Operators $\hat{a}_i$ and $\hat{a}_i^\dagger$ have a realization in the space $\mathcal{H}^\infty$, namely we may take

$$\hat{a}_i = \hat{\alpha}(\xi_i), \quad \hat{a}_i^\dagger = \hat{\alpha}^\dagger(\xi_i). \tag{6.29}$$

Thus, the space $\mathcal{H}^\infty$ is large enough to allow for the realizations of all operators $\hat{a}_i,\ \hat{a}_i^\dagger,\ i = 1, 2, \ldots$. Note that there also exist other realizations. For example, instead of $\hat{\alpha},\ \hat{\alpha}^\dagger$ we may use $\hat{\alpha} + cI,\ \hat{\alpha}^\dagger + c^* I$, where c is a complex number.

It remains to introduce a scalar product such that $\hat{a}_i^\dagger$ is Hermitian conjugate to $\hat{a}_i$. A natural definition for the scalar product $\langle h|h'\rangle$ of the two formal products

$$h = \prod_{i=1}^{\infty} f_i(\xi_i), \quad h' = \prod_{j=1}^{\infty} f_j'(\xi_j)$$

has the form

$$\langle h|h'\rangle = \prod_{i=1}^{\infty} \langle f_i|f_i'\rangle_{L^2}, \tag{6.30}$$

where

$$\langle f_i|f_i'\rangle_{L^2} = \int_{R^1} d\xi\ f_i^*(\xi) f_i'(\xi)$$

is the scalar product in $L^2(R^1)$. The infinite product in (6.30) should be convergent—the scalar product has to have a definite numerical value because it gives the probability amplitude in quantum theory. In order to ensure convergence, we assume that all formal products (6.26) are constructed from the same normalized function $f_0(\xi)$ except for a finite number of factors. This is just the simplest solution, but we will not discuss here other possibilities. Thus, in the product $\prod_{i=1}^{\infty} f_i(\xi_i)$ we have $f_i(\xi_i) = f_0(\xi_i)$ for all $i \geq N$, where N is a natural number (which depends on the product). Let us denote by $\mathcal{H}_{f_0}$ the subset of $\mathcal{H}^\infty$ consisting of all such formal products and of their linear combinations. Because $\langle f_0|f_0\rangle_{L^2} = 1$, the product in (6.30) contains only a finite number of factors which may differ from 1, hence it has a definite numerical value. Assuming that the scalar product is anti-linear in its left argument[3] and linear in its right argument, we can compute the scalar product of any two arbitrary elements of $\mathcal{H}_{f_0}$. The standard mathematical procedure of the completion of $\mathcal{H}_{f_0}$ with respect to the norm provided by the scalar product yields a Hilbert space which we denote also by $\mathcal{H}_{f_0}$.

It is clear that the operators $\hat{a}_i$ and $\hat{a}_i^\dagger$, as well as the finite order polynomials constructed from them, act within $\mathcal{H}_{f_0}$. Therefore, this space is sufficient for the Hilbert space realization of our quantum field $\hat{\phi}$. Note that there are infinitely many subspaces of $\mathcal{H}^\infty$ of the described type. They differ from each other by the choice of $f_0 \in L^2(R^1)$. It turns out that this freedom of choice of the subspace allows for various realizations of the quantum field $\hat{\phi}$ which are truly inequivalent—they lead to different physical predictions.

---

[3] That is $\langle c_1 h_1 + c_2 h_2|h\rangle = c_1^* \langle h_1|h\rangle + c_2^* \langle h_2|h\rangle$, where c_1, c_2 are complex numbers.

After solving the Heisenberg evolution equations (6.12), (6.13), and seeing that the operators $\hat{a}_i$ and $\hat{a}_i^\dagger$ have realizations in the Hilbert spaces constructed above, we are prepared to define observables. Let us begin from the Hamiltonian. Using the formula $\hat{\pi} = \partial_0 \hat{\phi}$ and inserting solution (6.16) for $\hat{\phi}$ in formula (6.5) we obtain (Exercise 6.2)

$$\hat{H} = \frac{1}{2} \int d^3k\, \omega(\vec{k}) \left(\hat{a}(\vec{k})\hat{a}^\dagger(\vec{k}) + \hat{a}^\dagger(\vec{k})\hat{a}(\vec{k}) \right). \tag{6.31}$$

Next, we write $\hat{H}$ in the form

$$\hat{H} = \frac{1}{2} \int d^3k d^3k'\, \delta(\vec{k} - \vec{k}') \sqrt{\omega(\vec{k})} \sqrt{\omega(\vec{k}')} \left(\hat{a}(\vec{k}')\hat{a}^\dagger(\vec{k}) + \hat{a}^\dagger(\vec{k})\hat{a}(\vec{k}') \right),$$

and use the second formula (6.24) (the completeness relation) to eliminate the Dirac delta. Introducing the notation

$$g_i(\vec{k}) = \sqrt{\omega(\vec{k})}\, h_i(\vec{k}), \quad \hat{a}[g_i] = \int d^3k\, g_i(\vec{k})\hat{a}(\vec{k}),$$

we finally obtain

$$\hat{H} = \frac{1}{2} \sum_{i=1}^{\infty} \left(\hat{a}[g_i](\hat{a}[g_i])^\dagger + (\hat{a}[g_i])^\dagger \hat{a}[g_i] \right). \tag{6.32}$$

We know from the discussion given at the beginning of this section that $\hat{H}$ given by formula (6.5) is not properly defined. Formula (6.32) is not equivalent to (6.5) because we have substituted the solution of the Klein–Gordon equation for $\hat{\phi}$. Nevertheless, the problem is still present. Namely, it turns out that $\hat{H}$ given by formula (6.32) has an infinite expectation value in any normalized state $|\psi\rangle$. Let us show this. The operators $\hat{a}[g_i]$ and $(\hat{a}[g_i])^\dagger$ obey the following commutation relations

$$\left[\hat{a}[g_i], (\hat{a}[g_i])^\dagger \right] = c_i I, \tag{6.33}$$

where

$$c_i = \int d^3k\, \omega(\vec{k}) |h_i(\vec{k})|^2.$$

Relations (6.33) follow from the definition of these operators and from the third relation (6.17). We assume that the functions $h_i(\vec{k})$ vanish in the limit $|\vec{k}| \to \infty$ sufficiently quickly to ensure convergence of the integral giving c_i. These integrals are bounded from below by a positive number, namely

$$c_i \geq |m| > 0,$$

because the functions h_i are normalized to 1, and $\omega(\vec{k}) \geq |m| > 0$ (remember that we have assumed $m^2 > 0$). Furthermore,

$$\sum_{i=1}^{\infty} \hat{a}[g_i](\hat{a}[g_i])^\dagger = \sum_{i=1}^{\infty} \left((\hat{a}[g_i])^\dagger \hat{a}[g_i] + c_i I\right),$$

as follows from relation (6.33). Because the operators $(\hat{a}[g_i])^\dagger \hat{a}[g_i]$ are positive definite,[4] the expectation value of each term in the sum on the r.h.s. is bounded from below by $|m|\langle\psi|\psi\rangle$, and the whole sum is divergent. Hence, operator (6.31) has an infinite expectation value in any state $|\psi\rangle \neq 0$. Obviously, such an operator cannot be accepted as the Hamiltonian of a physical system. Note that the infinity appears because the sum in formula (6.32) involves an infinite number of terms. It is a consequence of the fact that the field has an infinite number of degrees of freedom.

In view of the argument given above, it is clear that the term $\hat{a}[g_i](\hat{a}[g_i])^\dagger$ should be removed from the Hamiltonian. This should not be done in an arbitrary way because we could loose the correspondence with the classical theory from which we have started, and we would also have to recalculate the Heisenberg evolution equations (6.12) and (6.13), their solution (6.16) and so on. The best approach consists in a 'soft' modification of the Hamiltonian, such that the new Hamiltonian gives the same Heisenberg evolution equations as before. Such modifications exist. Using commutation relation (6.33) we may write

$$\hat{H} = \sum_{i=1}^{\infty} (\hat{a}[g_i])^\dagger \hat{a}[g_i] + \frac{1}{2} \sum_{i=1}^{\infty} c_i I.$$

The last term on the r.h.s. is infinite. However, we can simply drop it because it is proportional to the identity operator and therefore it does not matter when computing commutators. Thus, we postulate that the quantum Hamiltonian for the real scalar field has the form

$$\hat{H} = \sum_{i=1}^{\infty} (\hat{a}[g_i])^\dagger \hat{a}[g_i] = \int d^3k\, \omega(\vec{k}) \hat{a}^\dagger(\vec{k}) \hat{a}(\vec{k}). \tag{6.34}$$

The procedure applied above, that is the application of the commutation relation with the term proportional to the identity operator omitted, in order to remove the operators $\hat{a}(\vec{k})\hat{a}^\dagger(\vec{k})$, is called the normal ordering. In the normally ordered operator, all operators $\hat{a}^\dagger(\vec{k})$ stand to the left of all operators $\hat{a}(\vec{k})$. Such operators are denoted by two colons, i.e. $:\hat{H}:$. At this stage, one can also replace $\sum_{i=1}^{\infty} c_i$ by a real number $2E_0$. The resulting Hamiltonian differs from (6.34) only by the term $E_0 I$, which gives only a trivial shift of the whole spectrum of the Hamiltonian. However, we shall see

[4]Operator $\hat{A}$ is positive definite if its expectation value $\langle\psi|\hat{A}|\psi\rangle$ in an arbitrary state $|\psi\rangle \neq 0$ is positive.

in Chap. 10, that in the case of the Fock realization, described below, postulates of relativistic invariance imply that we have to put $E_0 = 0$.

The Fock realization is distinguished by the fact that the corresponding Hilbert space $\mathcal{H}_{f_0}$, called the Fock space and denoted by $\mathcal{H}_F$, contains a normalized state $|0\rangle$, called the vacuum state, such that for all $\vec{k} \in R^3$

$$\hat{a}(\vec{k})|0\rangle = 0. \tag{6.35}$$

Condition (6.35) is equivalent to

$$\hat{a}_i|0\rangle = 0 \tag{6.36}$$

where $i = 1, 2, \ldots$, with $\hat{a}_i$ defined by formulas (6.22). In the realization (6.29) condition (6.36) is equivalent to the following equations

$$(\xi_i + \frac{d}{d\xi_i}) f_0(\xi_i) = 0,$$

which have the following normalized solution

$$f_0(\xi_i) = (\pi)^{-1/4} \exp(-\frac{1}{2}\xi_i^2).$$

Therefore the vacuum state has the form

$$|0\rangle = \prod_{i=1}^{\infty} f_0(\xi_i). \tag{6.37}$$

Let us remind ourselves that on the r.h.s. of this formula, we have a formal product of functions—it does not have any numerical value. On the other hand, the scalar product $\langle 0|0\rangle$ has a definite numerical value—definition (6.30) gives

$$\langle 0|0\rangle = 1. \tag{6.38}$$

Let us introduce the infinite ladder of states in the Fock space

$$|0\rangle, |\vec{k}\rangle, \ldots |\vec{k}_1\, \vec{k}_2 \ldots \vec{k}_n\rangle, \ldots, \tag{6.39}$$

where

$$|\vec{k}_1\, \vec{k}_2 \ldots \vec{k}_n\rangle = \frac{1}{\sqrt{n!}} \hat{a}^\dagger(\vec{k}_1)\hat{a}^\dagger(\vec{k}_2) \ldots \hat{a}^\dagger(\vec{k}_n)|0\rangle. \tag{6.40}$$

The scalar products of these states can be computed with the help of commutation relations (6.17) and condition (6.35)—we do not have to use the concrete realization in the space $\mathcal{H}_F$. For example,

$$\langle \vec{k}_1 | \vec{k}_2 \rangle = \langle 0 | \hat{a}(\vec{k}_1) \hat{a}^\dagger(\vec{k}_2) | 0 \rangle = \langle 0 | \hat{a}^\dagger(\vec{k}_2) \hat{a}(\vec{k}_1) | 0 \rangle + \delta(\vec{k}_1 - \vec{k}_2) \langle 0 | 0 \rangle = \delta(\vec{k}_1 - \vec{k}_2),$$

because $\hat{a}(\vec{k}_1)|0\rangle = 0$. A similar calculation with multiple uses of relations (6.17) gives

$$\begin{aligned} &\langle \vec{k}_1 \, \vec{k}_2 \ldots \vec{k}_n | \vec{k}'_1 \, \vec{k}'_2 \ldots \vec{k}'_n \rangle \\ &= \frac{1}{n!} \sum_{\text{permutations}} \delta(\vec{k}_1 - \vec{k}'_{i_1}) \delta(\vec{k}_2 - \vec{k}'_{i_2}) \ldots \delta(\vec{k}_n - \vec{k}'_{i_n}), \end{aligned} \tag{6.41}$$

where $(i_1, i_2, \ldots, i_n)$ is a permutation of the set $(1, 2, \ldots, n)$. The sum is over all such permutations. It arises, because the state $|\vec{k}_1 \, \vec{k}_2 \ldots \vec{k}_n\rangle$ does not depend on the order of the wave vectors $\vec{k}_1, \vec{k}_2, \ldots, \vec{k}_n$, as follows from the fact that the operators $\hat{a}^\dagger(\vec{k})$ present in definition (6.40) commute with each other. Furthermore,

$$\langle \vec{k}_1 \, \vec{k}_2 \ldots \vec{k}_n | \vec{k}'_1 \, \vec{k}'_2 \ldots \vec{k}'_m \rangle = 0 \tag{6.42}$$

if $n \neq m$. By definition, the set of states (6.39) is a basis for the Fock space. Thus, any state $|\psi\rangle$ from $\mathcal{H}_F$ can be written in the form

$$\begin{aligned} |\psi\rangle &= \psi_0 |0\rangle + \int d^3k \, \psi_1(\vec{k}) |\vec{k}\rangle + \cdots \\ &+ \int d^3k_1 \ldots d^3k_n \, \psi_n(\vec{k}_1, \vec{k}_2, \ldots, \vec{k}_n) |\vec{k}_1 \, \vec{k}_2 \ldots \vec{k}_n\rangle + \cdots . \end{aligned} \tag{6.43}$$

Here ψ_0 is a complex number, the probability amplitude for finding the vacuum state $|0\rangle$ in $|\psi\rangle$. By assumption, the functions $\psi_n(\vec{k}_1, \ldots, \vec{k}_n)$ are symmetric in $\vec{k}_1, \ldots, \vec{k}_n$. Note that this is not a restriction on the states $|\psi\rangle$ from $\mathcal{H}_F$. The point is that in any case, only symmetric parts of these functions contribute to the r.h.s. of formula (6.43) because $|\vec{k}_1 \ldots \vec{k}_n\rangle$ are symmetric in $\vec{k}_1, \ldots, \vec{k}_n$. The physical interpretation of the functions ψ_n is given below. Simple calculation in which we use (6.41), (6.42) yields the following formula for the norm of the state $|\psi\rangle$

$$\begin{aligned} ||\psi||^2 &= \langle \psi | \psi \rangle = |\psi_0|^2 + \int d^3k \, |\psi_1(\vec{k})|^2 + \cdots \\ &+ \int d^3k_1 d^3k_2 \ldots d^3k_n \, |\psi_n(\vec{k}_1, \vec{k}_2, \ldots, \vec{k}_n)|^2 + \cdots . \end{aligned} \tag{6.44}$$

The Fock space $\mathcal{H}_F$ consists of all vectors $|\psi\rangle$ of the form (6.43), such that the r.h.s. of formula (6.44) is finite. From a physical viewpoint, vectors from $\mathcal{H}_F$ represent states of the quantum field, in a complete analogy with states of a particle in quantum mechanics. Using the Fock space we can construct a perfect quantum field model which has a beautiful interpretation in terms of relativistic, non-interacting quantum particles.

At this point the construction of the quantum theory of the real scalar field is almost finished. It remains only to introduce operators representing other basic observables of the real scalar field, apart from the energy represented by Hamiltonian (6.34). We find them by following the same steps as in the case of the Hamiltonian. Inserting solution (6.16) into formula (6.6) for the total momentum of the field we obtain

$$\hat{P}^i = \frac{1}{2}\int d^3k\, k^i \left(\hat{a}(\vec{k})\hat{a}^\dagger(\vec{k}) + \hat{a}^\dagger(\vec{k})\hat{a}(\vec{k})\right).$$

Normal ordering gives the operator of the total momentum of the field

$$\hat{P}^i = \int d^3k\, k^i\, \hat{a}^\dagger(\vec{k})\hat{a}(\vec{k}). \tag{6.45}$$

There are six more observables for the scalar field which follow from Noether's theorem. They correspond to Lorentz transformations, which are also symmetries of the classical model (6.1), similarly as the space and time translations led to the total energy and momentum integrals of motion. In the classical model, such integrals of motion have the form

$$M^{\mu\nu} = \int d^3x\, (T^{0\mu}x^\nu - T^{0\nu}x^\mu), \tag{6.46}$$

where $T^{\mu\nu} = \partial^\mu\phi\partial^\nu\phi - \eta^{\mu\nu}\mathcal{L}$ are the components of the symmetric energy-momentum tensor. Repeating the usual steps, that is: replacing ϕ and π by the operators $\hat{\phi}$ and $\hat{\pi}$, inserting solution (6.16), and applying the normal ordering, we obtain six Hermitian operators

$$\begin{aligned}\hat{M}^{rs} &= -\frac{i}{2}\int d^3k \left(k^r \frac{\partial \hat{a}^\dagger(\vec{k})}{\partial k^s}\hat{a}(\vec{k}) - k^r \hat{a}^\dagger(\vec{k})\frac{\partial \hat{a}(\vec{k})}{\partial k^s} \right. \\ &\quad \left. - k^s \frac{\partial \hat{a}^\dagger(\vec{k})}{\partial k^r}\hat{a}(\vec{k}) + k^s \hat{a}^\dagger(\vec{k})\frac{\partial \hat{a}(\vec{k})}{\partial k^r} \right),\end{aligned} \tag{6.47}$$

$$\hat{M}^{0r} = -\frac{i}{2}\int d^3k\, \omega(\vec{k}) \left(\frac{\partial \hat{a}^\dagger(\vec{k})}{\partial k^r}\hat{a}(\vec{k}) - \hat{a}^\dagger(\vec{k})\frac{\partial \hat{a}(\vec{k})}{\partial k^r} \right) - \frac{i}{2}\int d^3k\, \frac{k^r}{\omega(\vec{k})}\hat{a}^\dagger(\vec{k})\hat{a}(\vec{k}), \tag{6.48}$$

where $r, s = 1, 2, 3$, and $\hat{M}^{rs} = -\hat{M}^{sr}$. The operators $\hat{M}^{rs}$ represent the three components of the total angular momentum of the field. The operators $\hat{M}^{0r}$ give the quantum counterpart of the initial position of the center-of-energy of the field. To see this, notice that formula (6.46) in the case $\mu = 0$, $\nu = r$ can be written in the form

$$M^{0r} = \int d^3x\, x^r T^{00} - P^r x^0.$$

The position $\vec{X} = (X^r)$ of the center-of-energy of the field is defined as follows

$$\int d^3x \; x^r T^{00} = E X^r,$$

where $E = \int d^3x \; T^{00}$ is the total energy of the field. Therefore

$$E X^r = M^{0r} + P^r x^0.$$

This formula says that the center-of-energy moves in space with the constant velocity $\vec{P}/E$ along a straight line that passes through the point which has Cartesian coordinates equal to M^{0r}/E.

In the quantum theory of the free real scalar field we have just constructed, one can compute the spectrum of the Hamiltonian and of the total momentum. It turns out that the elements of the basis (6.39) are eigenstates of the Hamiltonian, and of the total momentum of the field. In order to prove this fact, it is convenient to use the following formulas

$$[\hat{H}, \hat{a}^{\dagger}(\vec{k})] = \omega(\vec{k})\hat{a}^{\dagger}(\vec{k}), \quad [\hat{P}^i, \hat{a}^{\dagger}(\vec{k})] = k^i \hat{a}^{\dagger}(\vec{k}), \tag{6.49}$$

which are obtained directly from definitions (6.34), (6.45), and commutation relations (6.17). Let us compute $\hat{H}|\vec{k}_1\,\vec{k}_2\,\ldots\vec{k}_n\rangle$. We insert formula (6.40) and commute operator $\hat{H}$ with the operators $\hat{a}^{\dagger}(\vec{k})$ using (6.49) until it reaches the state $|0\rangle$. Each such commutation yields a term proportional to $\omega(\vec{k}_i)$. The last term, in which $\hat{H}$ acts directly on $|0\rangle$, vanishes because

$$\hat{H}|0\rangle = 0, \tag{6.50}$$

as follows from condition (6.36). Therefore

$$\hat{H}|\vec{k}_1\,\vec{k}_2\,\ldots\vec{k}_n\rangle = \left(\sum_{i=1}^{n}\omega(\vec{k}_i)\right)|\vec{k}_1\,\vec{k}_2\,\ldots\vec{k}_n\rangle. \tag{6.51}$$

A similar calculation gives

$$\hat{P}^i|0\rangle = 0, \quad \hat{P}^i|\vec{k}_1\,\vec{k}_2\,\ldots\vec{k}_n\rangle = \left(\sum_{j=1}^{n} k^i_j\right)|\vec{k}_1\,\vec{k}_2\ldots\vec{k}_n\rangle. \tag{6.52}$$

Because the states (6.39) form a basis in the Fock space, they form the complete set of eigenstates of both $\hat{H}$ and $\hat{P}^i$. Of course, these operators commute with each other

$$[\hat{H}, \hat{P}^i] = 0, \quad [\hat{P}^i, \hat{P}^k] = 0. \tag{6.53}$$

Formulas (6.53) become obvious when we notice that

$$[\hat{a}^\dagger(\vec{k})\hat{a}(\vec{k}), \hat{a}^\dagger(\vec{k}')\hat{a}(\vec{k}')] = 0$$

for arbitrary $\vec{k}$ and $\vec{k}'$, as can be checked by direct calculation with the use of commutation relations (6.17).

Formulas (6.51), (6.52) yield plenty of information about the properties of the quantum model. First, the quantum field in the vacuum state $|0\rangle$ has zero energy and momentum. In consequence, the operator

$$U(b^0, \vec{b}) = \exp(-ib^0\hat{H} + ib^k\hat{P}^k), \tag{6.54}$$

which represents translations in time ($x^0 \to x^0 + b^0$) and space ($\vec{x} \to \vec{x} + \vec{b}$), see Chap. 10, leaves the vacuum state unchanged,

$$U(b^0, \vec{b})|0\rangle = |0\rangle. \tag{6.55}$$

The quantum field in the state $|\vec{k}\rangle$ has total momentum equal to $\vec{k}$ and energy equal to $\omega(\vec{k}) = \sqrt{m^2 + \vec{k}^2}$. Moreover, the Schroedinger equation in the Fock space

$$i\frac{\partial|t\rangle}{\partial t} = \hat{H}|t\rangle \tag{6.56}$$

($\hbar = 1$), for the states of the form

$$|t\rangle_1 = \int d^3k\; \psi_1(t, \vec{k})|\vec{k}\rangle$$

is reduced to the equation

$$i\frac{\partial\psi_1(t, \vec{k})}{\partial t} = \sqrt{m^2 + \vec{k}^2}\;\psi_1(t, \vec{k}), \tag{6.57}$$

which in turn coincides with the Schroedinger equation in the momentum representation for a free relativistic particle with rest mass equal to m. Therefore, the states $|\psi\rangle_1$ can be regarded as quantum states of a relativistic particle with rest mass m. For this reason they are called one-particle states, and they form the so called one-particle sector of the Fock space. Note that this particle has positive energy. The problem of states with negative energy, which is present in the relativistic quantum mechanics of a single particle based on the Klein–Gordon equation, does not appear here.

The total momentum of the quantum field in the states $|\vec{k}_1\ \vec{k}_2\ \ldots \vec{k}_n\rangle$ with $n \geq 2$ is equal to $\sum_{i=1}^n \vec{k}_i$, and the total energy to $\sum_{i=1}^n \omega(\vec{k}_i)$. Therefore these states, as well as their 'linear combinations'

$$|t\rangle_n = \int d^3k_1 d^3k_2 \dots d^3k_n \, \psi_n(\vec{k}_1, \vec{k}_2, \dots, \vec{k}_n, t) \, |\vec{k}_1 \, \vec{k}_2 \, \dots \vec{k}_n\rangle,$$

can be regarded as states of n noninteracting identical relativistic particles with rest mass m. The particles are identical because the n-particle wave function in the momentum representation $\psi(\vec{k}_1, \vec{k}_2, \dots, \vec{k}_n, t)$ is symmetric with respect to permutations of $\vec{k}_1, \vec{k}_2, \dots, \vec{k}_n$. They do not interact with each other because the total energy is equal to the sum of kinetic energies $\omega(\vec{k}_i)$ of the particles—there is no interaction energy. We see from formula (6.43) that the Fock space is decomposed into sectors with fixed numbers of identical, relativistic, noninteracting particles. The fact that the states of the field can be described in terms of quantum particles, is called the particle interpretation of the quantum theory of the free real scalar field. Of course, the Fock space also contains states which are linear combinations of states with various numbers of particles. Such states do not have any concrete number of particles, one may only ask about the probability of finding a chosen number of particles.

The operators $\hat{a}^\dagger(\vec{k})$ and $\hat{a}(\vec{k})$ are called (particle) creation and annihilation operators, respectively. Because

$$\hat{a}^\dagger(\vec{k})|\vec{k}_1 \, \vec{k}_2 \, \dots \vec{k}_n\rangle = \sqrt{n+1}|\vec{k} \, \vec{k}_1 \, \vec{k}_2 \, \dots \vec{k}_n\rangle, \tag{6.58}$$

the creation operator transforms a state from the n-particle sector into a state in the sector with $n+1$ particles. The annihilation operator 'moves' states in the opposite direction, namely

$$\begin{aligned}\hat{a}(\vec{k})|\vec{k}_1 \, \vec{k}_2 \, \dots \vec{k}_n\rangle &= \frac{1}{\sqrt{n}} \Big[\delta(\vec{k}_1 - \vec{k})|\vec{k}_2 \, \vec{k}_3 \, \dots \vec{k}_n\rangle \\ &\quad + \delta(\vec{k}_2 - \vec{k})|\vec{k}_1 \, \vec{k}_3 \, \dots \vec{k}_n\rangle + \dots + \delta(\vec{k}_n - \vec{k})|\vec{k}_1 \, \vec{k}_2 \, \dots \vec{k}_{n-1}\rangle \Big].\end{aligned} \tag{6.59}$$

Formula (6.51) does not depend on the form of $\omega(\vec{k})$. Therefore it is valid also for $\omega(\vec{k}) = 1$,

$$\hat{N}|\vec{k}_1 \, \vec{k}_2 \, \dots \vec{k}_n\rangle = n|\vec{k}_1 \, \vec{k}_2 \, \dots \vec{k}_n\rangle,$$

where

$$\hat{N} = \int d^3k \, \hat{a}^\dagger(\vec{k})\hat{a}(\vec{k}). \tag{6.60}$$

For the obvious reason, the operator $\hat{N}$ is called the particle number operator. It commutes with the Hamiltonian and with the total momentum operator. Therefore the translation operator $U(a^0, \vec{a})$, defined by formula (6.54), does not change the number of particles. In particular, this number is constant in time because $U(a^0, \vec{a} = 0)$ is the time evolution operator (whereas $U(a^0 = 0, \vec{a})$ represents the space translation by the vector $\vec{a}$). This feature of the quantum field is related to the absence of interaction between particles. In general, interactions in relativistic quantum field theories can create or destroy particles.

The operator $U(b^0, \vec{b})$ can be used in order to shift the argument of the field operator,

$$U^{-1}(b^0, \vec{b})\ \hat{\phi}(x^0, \vec{x})\ U(b^0, \vec{b}) = \hat{\phi}(x^0 + b^0, \vec{x} + \vec{b}). \tag{6.61}$$

Here $\hat{\phi}$ has the form (6.16), therefore this formula is equivalent to

$$U^{-1}(b^0, \vec{b})\ \hat{a}(\vec{k})\ U(b^0, \vec{b}) = e^{-ib^0\omega(\vec{k})+i\vec{b}\vec{k}}\ \hat{a}(\vec{k}). \tag{6.62}$$

Probably the easiest way to directly check formula (6.62) is to apply both sides of it to each basis state (6.39) in the Fock space, next using formula (6.59) and the fact that the basis states are eigenstates of $\hat{H}$ and $\hat{P}^i$.

Let us note that the quantum field operator $\hat{\phi}(x)$, given by formula (6.16), should not be regarded as an operator valued function of $x \in M$. Rather, it is a generalized function of x. This means that $\hat{\phi}(x)$ is not an operator in the Fock space for any fixed x. A well-defined operator is obtained when we 'smear' the field with a test function $h(x)$ of the class $S(R^4)$

$$\hat{\phi}[h] = \int d^4x\ h(x)\ \hat{\phi}(x). \tag{6.63}$$

To illustrate this point, let us compute the norm of the state $\hat{\phi}[h]|\psi\rangle$. It should be finite if this state belongs to $\mathcal{H}_F$. The square of the norm is equal to

$$\langle\psi|(\hat{\phi}[h])^2|\psi\rangle$$

(for simplicity we have assumed that the test function has real values). The operator $(\hat{\phi}[h])^2$ can be split into four terms containing $\hat{a}^\dagger\hat{a}^\dagger$, $\hat{a}^\dagger\hat{a}$, $\hat{a}\hat{a}$, $\hat{a}\hat{a}^\dagger$, respectively. Using commutation relations (6.17) we can transform the last term into the sum of a term containing $\hat{a}^\dagger\hat{a}$ and of the term

$$\int \frac{d^3k d^3k'}{2(2\pi)^3\sqrt{\omega(\vec{k})\omega(\vec{k}')}} \int d^4x d^4x'\ h(x)h(x')e^{ikx-ik'x'}\delta(\vec{k}-\vec{k}')\langle\psi|I|\psi\rangle$$

$$= \int \frac{d^3k}{2(2\pi)^3\omega(\vec{k})} \int d^4x d^4x'\ h(x)h(x')e^{ik(x-x')}.$$

This expression is finite because the Fourier transform

$$\tilde{h}(k) = \frac{1}{(2\pi)^2}\int d^4x\ e^{-ikx}h(x)$$

of the test function $h(x)$ is also of the class $S(R^4)$ (in the variable $k \in R^4$). On the other hand, if we try to replace the test function by the Dirac delta, $h(x) \to \delta(x - x_0)$ and $\hat{\phi}[h] \to \hat{\phi}(x_0)$, then we obtain the integral $\int \frac{d^3k}{2(2\pi)^3\omega(\vec{k})}$ which is divergent. It turns out that the remaining three types of terms ($\hat{a}^\dagger\hat{a}^\dagger$, $\hat{a}^\dagger\hat{a}$, $\hat{a}\hat{a}$) can also give finite contributions when $h(x)$ is replaced by $\delta(x - x_0)$. This can be seen by expanding the state $|\psi\rangle$ as in (6.43) and using formulas (6.58), (6.59). Then it becomes clear that there exist normalized states $|\psi\rangle \in \mathcal{H}_F$ such that the expression obtained in this way is finite. We conclude that $\hat{\phi}[h]|\psi\rangle$ belongs to $\mathcal{H}_F$ at least for some $|\psi\rangle \in \mathcal{H}_F$, and that $\hat{\phi}(x)|\psi\rangle$ does not belong to $\mathcal{H}_F$ for any $x \in M$ and any $|\psi\rangle \neq 0$ from $\mathcal{H}_F$. Thus, $\hat{\phi}(x)$ cannot be regarded as an operator in the Fock space,[5] as opposed to the smeared field operator $\hat{\phi}[h]$. For this reason powers of $\hat{\phi}(x)$, e.g. $\hat{\phi}^2(x)$, are meaningless, in general. Such powers are present in formula (6.5), so it is not a surprise that the final form (6.34) of the Hamiltonian is not equal to (6.5).

6.2 The Dirac Field

We know from Sect. 5.4 that there exist two versions of the classical Dirac field: with either complex or Grassmann values. Both are not satisfactory from a physical viewpoint, and both can be used as a starting point for constructing the quantum theory of that field. It turns out that the resulting quantum theory of the Dirac field does not have any flaws. It can be regarded as one of the most remarkable achievements of theoretical physics.

We choose the classical anticommuting Dirac field because then the way to the quantum theory is shorter. The Lagrangian has the usual form (5.24) also for this version of the classical Dirac field,

$$\mathcal{L} = \frac{i}{2}\left(\overline{\psi}\gamma^{\mu}\partial_{\mu}\psi - \partial_{\mu}\overline{\psi}\gamma^{\mu}\psi\right) - m\overline{\psi}\psi, \tag{6.64}$$

where $\psi = (\psi^{\alpha})$, $\overline{\psi} = (\overline{\psi}_{\alpha})$, $\alpha = 1, 2, 3, 4$, and we take the matrices γ^{μ} in the Dirac representation (5.3). In matrix notation, ψ is a column, while $\overline{\psi}$ is a row. Let us stress that ψ^{α}, $\overline{\psi}_{\alpha}$ are independent generating elements of a complex Grassmann algebra—there is no relation of the form $\overline{\psi} = \psi^{\dagger}\gamma^0$ (which holds for the complex Dirac field). In this algebra we define conjugation[6]

$$(\psi^{\alpha})^* = \overline{\psi}_{\beta}(\gamma^0)_{\beta\alpha}, \quad (\overline{\psi}_{\alpha})^* = (\gamma^0)_{\alpha\beta}\psi^{\beta}. \tag{6.65}$$

This conjugation is antilinear, that is $(c_1\phi + c_2\chi)^* = c_1^*\phi^* + c_2^*\chi^*$, where c_1 and c_2 are complex numbers, c_i^* is the complex conjugate of c_i, and ϕ, χ are arbitrary elements of the Grassmann algebra. Moreover, $(\phi\chi)^* = \chi^*\phi^*$. Lagrangian (6.64) is 'real' in the sense that $\mathcal{L}^* = \mathcal{L}$, see Exercise 6.5.

---

[5]Nevertheless, we will use the traditional term 'field operator' for $\hat{\phi}(x)$.

[6]Notice the order: the conjugation is introduced in the algebra, hence it is secondary to it. It is not correct to interpret (6.65) as constraints between the generating elements. Formulas (6.65) say that, e.g., the element conjugate to ψ^{α} is by definition equal to $\overline{\psi}_{\beta}(\gamma^0)_{\beta\alpha}$, but not that ψ^{α} is equivalent to $\overline{\psi}_{\beta}(\gamma^0)_{\beta\alpha}$.

The Euler-Lagrange equations have the general form

$$\frac{\delta S}{\delta \psi^\alpha(x)} = 0, \quad \frac{\delta S}{\delta \overline{\psi}_\alpha(x)} = 0,$$

where

$$S = \int d^4x \, \mathcal{L}.$$

In the present case we obtain

$$i\gamma^\mu \partial_\mu \psi - m\psi = 0, \quad i\partial_\mu \overline{\psi} \gamma^\mu + m\overline{\psi} = 0. \tag{6.66}$$

The conjugation (6.65) interchanges these equations.

We shall again apply the canonical quantization method. In the case of Grassmann valued fields we postulate fundamental relations of the kind (6.4) with the commutators replaced by anticommutators. It turns out that such a heuristic rule yields, after a number of steps, a consistent quantum theory. Unfortunately, there are several problems which were absent in the case of the scalar field. First, the usual definition of canonical momenta leads to the presence of constraints. Namely,

$$\pi_\alpha = \frac{\partial \mathcal{L}}{\partial(\partial_0 \psi^\alpha)} = -\frac{i}{2}\overline{\psi}_\beta (\gamma^0)_{\beta\alpha}, \quad \overline{\pi}^\alpha = \frac{\partial \mathcal{L}}{\partial(\partial_0 \overline{\psi}_\alpha)} = -\frac{i}{2}(\gamma^0)_{\alpha\beta}\psi^\beta. \tag{6.67}$$

The minus sign in the first formula appears because $\partial_0 \psi^\alpha$ is the second factor in the product $i\overline{\psi}\gamma^0\partial_0\psi$. Relations (6.67) show that the canonical variables are not independent. Such relations cannot be carried over to the quantum theory because they contradict the canonical anticommutation relations. For example, the canonical anticommutation relations

$$\{\hat{\psi}^\alpha(\vec{x}), \hat{\pi}_\beta(\vec{y})\} = i\delta^\alpha_\beta \delta(\vec{x} - \vec{y}) I, \quad \{\hat{\psi}^\alpha(\vec{x}), \hat{\overline{\psi}}_\beta(\vec{y})\} = 0.$$

are not compatible with the operator counterpart of the first constraint (6.67), which is obtained just by replacing π_α and $\overline{\psi}_\beta$ by the operators $\hat{\pi}_\alpha$ and $\hat{\overline{\psi}}_\beta$, respectively. Inserting this constraint into the first canonical anticommutation relation and using the second one we obtain the contradiction ($0 = I$). Therefore, in the presence of the constraints, the quantization has to be done in a more refined way. One possibility is to use a generalization of the canonical formalism which applies to systems with constraints and which was invented by Dirac, see, e.g., [6]. However, in the present case one may apply an approach proposed by Faddeev and Jackiw [7]. It gives the same result as the former approach but it is a bit simpler.

The approach by Faddeev and Jackiw is based on the fact that the two actions

$$S = \int_{t_1}^{t_2} dt \int d^3x \, \mathcal{L},$$

$$S' = S + \int_{t_1}^{t_2} dt \int d^3x \, \frac{\partial f(\psi, \overline{\psi})}{\partial t},$$

or, equivalently, the two Lagrangians

$$\mathcal{L}, \;\; \mathcal{L}' = \mathcal{L} + \frac{\partial f(\psi, \overline{\psi})}{\partial t},$$

give equivalent quantum theories, see Sect. 11.1 . Here $f(\psi, \overline{\psi})$ can be an arbitrary differentiable function of ψ and $\overline{\psi}$.

Let us write Lagrangian (6.64) in the form

$$\mathcal{L} = i\overline{\psi}\gamma^0\partial_0\psi - \frac{i}{2}\partial_0(\overline{\psi}\gamma^0\psi) + \frac{i}{2}\left(\overline{\psi}\gamma^i\partial_i\psi - \partial_i\overline{\psi}\gamma^i\psi\right) - m\overline{\psi}\psi. \tag{6.68}$$

According to the remark above, we may abandon the second term on the r.h.s. of this formula. The new Lagrangian has the form

$$\mathcal{L}' = i\overline{\psi}\gamma^0\partial_0\psi + \frac{i}{2}\left(\overline{\psi}\gamma^i\partial_i\psi - \partial_i\overline{\psi}\gamma^i\psi\right) - m\overline{\psi}\psi.$$

This Lagrangian gives the following canonical momentum conjugate to ψ^α

$$\pi'_\alpha(t, \vec{x}) = \frac{\partial \mathcal{L}'}{\partial(\partial_0\psi^\alpha(t, \vec{x}))} = -i\overline{\psi}_\beta(t, \vec{x})(\gamma^0)_{\beta\alpha} \tag{6.69}$$

(the minus sign is correct!).

Lagrangian $\mathcal{L}'$ can be written in the form

$$\mathcal{L}' = \partial_0\psi^\alpha\pi'_\alpha - T^0_{\;0}, \tag{6.70}$$

where

$$T^0_{\;0} = \frac{i}{2}\left(\partial_i\overline{\psi}\gamma^i\psi - \overline{\psi}\gamma^i\partial_i\psi\right) + m\overline{\psi}\psi \tag{6.71}$$

coincides with the density of the energy obtained from Noether's theorem applied to the Lagrangian $\mathcal{L}'$.

Now comes the crucial observation: formula (6.70) has the form of the relation between Lagrangian and Hamiltonian, well-known from the canonical formalism in classical mechanics. This tells us that $\overline{\psi}_\beta(t,\vec{x})$ is not a configurational variable! Instead, it is directly related to the canonical momentum conjugate to ψ^α, as shown by formula (6.69). Now we guess that the right way to construct the quantum version of the model is to postulate the following equal-time anticommutation relations

$$\{\hat{\psi}^\alpha(t,\vec{x}),\hat{\psi}^\beta(t,\vec{y})\}=0,\quad \{\hat{\overline{\psi}}_\alpha(t,\vec{x}),\hat{\overline{\psi}}_\beta(t,\vec{y})\}=0, \tag{6.72}$$

$$\{\hat{\overline{\psi}}_\beta(t,\vec{x}),\hat{\psi}^\alpha(t,\vec{y})\}=(\gamma^0)_{\alpha\beta}\delta(\vec{x}-\vec{y})I, \tag{6.73}$$

where $\hat{\psi}^\alpha$ and $\hat{\overline{\psi}}_\beta$ are Heisenberg picture operators corresponding to the classical Grassmann fields ψ^α and $\overline{\psi}_\beta$. The last anticommutation relation follows from the canonical anticommutation relation[7]

$$\{\hat{\pi}'_\beta(t,\vec{x}),\hat{\psi}^\alpha(t,\vec{y})\}=-i\delta^\alpha_\beta\delta(\vec{x}-\vec{y})I.$$

The obvious candidate for the quantum Hamiltonian of the Dirac field is the operator

$$\hat{H}=\int d^3x\,T^0_{\ 0},$$

where $T^0_{\ 0}$ has the form (6.71) with the classical fields $\psi,\overline{\psi}$ replaced by the corresponding operators. Thus,

$$\hat{H}=\frac{i}{2}\int d^3x\left(\partial_i\hat{\overline{\psi}}\gamma^i\hat{\psi}-\hat{\overline{\psi}}\gamma^i\partial_i\hat{\psi}\right)+m\int d^3x\,\hat{\overline{\psi}}\hat{\psi}. \tag{6.74}$$

Similarly as in the case of the scalar field, this Hamiltonian is understood merely as a formal expression which hopefully gives correct commutators. It turns out that indeed, it gives the correct commutators because it differs from the correct Hamiltonian by a multiple of the identity operator.

Let us repeat the steps which we know from our considerations of the scalar field. We start from the Heisenberg evolution equations

$$\frac{\partial\hat{\psi}^\alpha(t,\vec{x})}{\partial t}=i[\hat{H},\hat{\psi}^\alpha(t,\vec{x})],\quad \frac{\partial\hat{\overline{\psi}}_\alpha(t,\vec{x})}{\partial t}=i[\hat{H},\hat{\overline{\psi}}_\alpha(t,\vec{x})]. \tag{6.75}$$

The commutators present in these equations can be reduced to the basic anticommutators (6.72), (6.73) with the help of the identity

$$[AB,C]=A\{B,C\}-\{A,C\}B. \tag{6.76}$$

[7] The canonical commutation relation (6.4) can be written in the form $[\hat{\pi}(t,\vec{x}),\hat{\phi}(t,\vec{y})]=-i\delta(\vec{x}-\vec{y})I$. We replace [,] by {,} precisely in this version.

Multiplication of (6.75) by $i\gamma^0$, and computation of the commutators gives the Dirac equations for the operators $\hat{\psi}^\alpha$ and $\hat{\bar{\psi}}_\beta$,

$$i\gamma^\mu \partial_\mu \hat{\psi} - m\hat{\psi} = 0, \tag{6.77}$$

$$i\partial_\mu \hat{\bar{\psi}}\gamma^\mu + m\hat{\bar{\psi}} = 0. \tag{6.78}$$

The general solution of (6.77) has the form

$$\hat{\psi}(t,\vec{x}) = \int \frac{d^3p}{(2\pi)^{3/2}} \sum_{s=\pm 1/2} \left[v_s^{(+)}(\vec{p})\hat{a}_s^{(+)}(\vec{p})e^{i(\vec{p}\vec{x}-\omega t)} + v_s^{(-)}(\vec{p})\hat{a}_s^{(-)}(\vec{p})e^{i(\vec{p}\vec{x}+\omega t)} \right], \tag{6.79}$$

where

$$\omega(\vec{p}\,) = \sqrt{m^2 + \vec{p}^{\,2}},$$

and $\hat{a}_s^{\pm}(\vec{p}\,)$ are certain operators.

For each fixed wave vector $\vec{p} \in R^3$, the four bispinors $v_s^{\pm}(\vec{p})$ form a basis for the space of bispinors. The components of them are complex numbers, not Grassmann elements. By definition, the basis bispinors obey the following algebraic equations

$$\left(\pm\omega(\vec{p})\gamma^0 - \gamma^i p^i\right) v_s^{(\pm)}(\vec{p}\,) = m v_s^{(\pm)}(\vec{p}\,),$$

which are equivalent to eigenequation for a matrix Hamiltonian $H_D(\vec{p}\,)$ of the Dirac particle with the fixed momentum $\vec{p}$,

$$H_D(\vec{p}\,)v_s^{(\pm)}(\vec{p}\,) = \pm\omega(\vec{p}\,)v_s^{(\pm)}(\vec{p}\,),$$

where

$$H_D(\vec{p}\,) = \begin{pmatrix} m\sigma_0 & \vec{p}\,\vec{\sigma} \\ \vec{p}\,\vec{\sigma} & -m\sigma_0 \end{pmatrix}$$

in the Dirac representation (5.3). In order to specify the basis bispinors uniquely (up to a normalization) we also demand that

$$\Sigma^3(\vec{p})v_s^{(\pm)}(\vec{p}) = s\; v_s^{(\pm)}(\vec{p}),$$

where $s = \pm 1/2$, and the matrix $\Sigma^3(\vec{p})$ is the operator of the third component of spin in the relativistic quantum mechanics of the Dirac particle. Its form depends on the momentum $\vec{p}$ of the particle. Formula for all three components of the spin operator reads

$$\Sigma^i(\vec{p}) = \frac{1}{2\omega(\vec{p}\,)} \begin{pmatrix} m\sigma_i + p^i \frac{\vec{p}\,\vec{\sigma}}{m+\omega(\vec{p}\,)} & i\epsilon_{ikl}p^k\sigma_l \\ -i\epsilon_{ikl}p^k\sigma_l & m\sigma_i + p^i \frac{\vec{p}\,\vec{\sigma}}{m+\omega(\vec{p}\,)} \end{pmatrix},$$

where $\vec{p}\vec{\sigma} = p^j\sigma_j$, and $i, j, k, l = 1, 2, 3$. The matrices $\Sigma^i(\vec{p}\,)$ commute with $H_D(\vec{p}\,)$, and $[\Sigma^j(\vec{p}\,), \Sigma^k(\vec{p}\,)] = i\epsilon_{jkl}\Sigma^l(\vec{p}\,)$. The matrices $H_D(\vec{p}\,)$, $\Sigma^3(\vec{p}\,)$ form the complete set of commuting Hermitian matrices for the particle.

The basis bispinors obey the following orthogonality and normalization conditions

$$(v_r^{\epsilon}(\vec{p}))^{\dagger} v_s^{\epsilon'}(\vec{p}) = \delta_{rs}\delta_{\epsilon\epsilon'}, \tag{6.80}$$

where the indices ϵ and ϵ' have the values $+$ and $-$. Here $v^{\dagger} = (v^*)^T$, * denotes complex conjugation, ^{T} denotes matrix transposition.

Equation (6.78) can be transformed into (6.77) by Hermitian conjugation and multiplication by γ^0. Therefore, the general solution of (6.78) can be expressed by the general solution (6.79), namely

$$\hat{\overline{\psi}}(t, \vec{x}) = \hat{\psi}^{\dagger}(t, \vec{x})\gamma^0. \tag{6.81}$$

Here † denotes the Hermitian conjugation of the field operator in a certain Hilbert space, yet to be defined. Note that due to the Dirac equations (6.77), (6.78), the two initially independent Dirac fields $\hat{\psi}(x)$ and $\hat{\overline{\psi}}(x)$ have become related by formula (6.81). In a field theoretical jargon one says that these fields are independent 'off-shell', and equivalent to each other 'on-shell'.

Using the inverse Fourier transform and relations (6.80) we express the operators $\hat{a}_s^{(\pm)}(\vec{p})$ by the Dirac field

$$\hat{a}_s^{(\pm)}(\vec{p}) = \frac{1}{(2\pi)^{3/2}} e^{\pm i\omega t}((v_s^{(\pm)}(\vec{p}))^{\dagger})^{\alpha} \int d^3x\, e^{-i\vec{p}\vec{x}}\hat{\psi}^{\alpha}(t, \vec{x}). \tag{6.82}$$

Because of (6.72), (6.73), these operators obey the following algebraic relations

$$\{\hat{a}_s^{(\epsilon)}(\vec{p}), \hat{a}_{s'}^{(\epsilon')}(\vec{p}\,')\} = 0, \;\; \{\hat{a}_s^{(\epsilon)}(\vec{p}), (\hat{a}_{s'}^{(\epsilon')}(\vec{p}\,'))^{\dagger}\} = \delta_{ss'}\delta_{\epsilon\epsilon'}\delta(\vec{p} - \vec{p}\,')I, \tag{6.83}$$

where again $\epsilon, \epsilon' = +, -$. The relation involving two operators $\hat{a}^{\dagger}$ is obtained from the first of relations (6.83) by Hermitian conjugation.

Similarly as in the case of the scalar field, we would like to see Hilbert space realizations of the operators $\hat{a}_s^{(\epsilon)}$. The construction of such realizations is analogous to the one presented in the previous section. We take an orthonormal and complete set of functions $h_i(\vec{p}, \epsilon, s)$ of the continuous variable $\vec{p}$ and of discrete variables ϵ, s,

$$\sum_{\epsilon=\pm, s=\pm 1/2} \int d^3p\, h_i(\vec{p}, \epsilon, s) h_j(\vec{p}, \epsilon, s) = \delta_{ij}, \tag{6.84}$$

$$\sum_{i=1}^{\infty} h_i^*(\vec{p}, \epsilon, s) h_i(\vec{p}\,', \epsilon', s') = \delta_{ss'}\delta_{\epsilon\epsilon'}\delta(\vec{p} - \vec{p}\,').$$

and define

$$\hat{a}_i = \sum_{\epsilon,s} \int d^3p \, h_i(\vec{p}, \epsilon, s) \, \hat{a}_s^{(\epsilon)}(\vec{p}), \quad \hat{a}_i^\dagger = \sum_{\epsilon,s} \int d^3p \, h_i^*(\vec{p}, \epsilon, s)(\hat{a}_s^{(\epsilon)}(\vec{p}))^\dagger, \tag{6.85}$$

where $i = 1, 2, \ldots$. The inverse formulas have the form

$$\hat{a}_s^{(\epsilon)}(\vec{p}) = \sum_{i=1}^{\infty} h_i^*(\vec{p}, \epsilon, s)\hat{a}_i, \quad (\hat{a}_s^{(\epsilon)}(\vec{p}))^\dagger = \sum_{i=1}^{\infty} h_i(\vec{p}, \epsilon, s)\hat{a}_i^\dagger. \tag{6.86}$$

It is clear that it suffices to find realizations of the operators $\hat{a}_i$ and $\hat{a}_i^\dagger$. These operators obey the following anticommutation relations obtained from (6.83)

$$\{\hat{a}_i, \hat{a}_j^\dagger\} = \delta_{ij} I, \tag{6.87}$$
$$\{\hat{a}_i, \hat{a}_j\} = 0, \quad \{\hat{a}_i^\dagger, \hat{a}_j^\dagger\} = 0.$$

We again consider an infinite dimensional linear space $\mathcal{H}^\infty$ spanned by formal, infinite products of functions

$$g_1(x_1)g_2(x_2)\ldots,$$

but in the present case, the functions $g_i(x_i)$ are the first order polynomials in x_i,

$$g_i(x_i) = c_i x_i + d_i,$$

where x_i are Grassmann elements which anticommute with each other, and c_i, d_i are complex numbers. Thus, $x_i^2 = 0$, and $x_i x_j = -x_j x_i$. Let us introduce operators $\hat{\beta}_x$, and $\hat{\beta}_x^\dagger$ acting in the two-dimensional complex space of the first order polynomials $c_1 x + c_2$, where x is a Grassmann element:

$$\hat{\beta}_x(c_1 x + c_2) = c_1, \quad \hat{\beta}_x^\dagger(c_1 x + c_2) = c_2 x.$$

Equivalently, we may write that

$$\hat{\beta}_x = \frac{d}{dx}, \quad \hat{\beta}_x^\dagger = x\cdot,$$

where the dot means that the operator acts as multiplication by x. For example, $x(c_1 x + c_2) = c_2 x$ because $x^2 = 0$. The operators $\hat{\beta}_x$ and $\hat{\beta}_x^\dagger$ obey the following anticommutation relation

$$\{\hat{\beta}_x, \hat{\beta}_x^\dagger\} = I.$$

In the basis formed by the two monomials, namely x and 1, these operators are represented by the matrices

$$\hat{\beta}_x \leftrightarrow \begin{pmatrix} 0 & 0 \\ 1 & 0 \end{pmatrix}, \quad \hat{\beta}_x^\dagger \leftrightarrow \begin{pmatrix} 0 & 1 \\ 0 & 0 \end{pmatrix}.$$

The operators $\hat{a}_i$ and $\hat{a}_j^\dagger$ have the following realization in the space $\mathcal{H}^\infty$:

$$\hat{a}_i = \hat{\beta}_{x_i}, \quad \hat{a}_i^\dagger = \hat{\beta}_{x_i}^\dagger. \tag{6.88}$$

Note that $\hat{\beta}_x^\dagger(cx) = 0$. In consequence, there exist normalizable states $|\psi\rangle$ in $\mathcal{H}^\infty$ such that $\hat{a}_i^\dagger|\psi\rangle = 0$. Such states do not exist in the Fock space of the scalar field.

The space $\mathcal{H}^\infty$ is very large. Our operators can be realized in its subspace $\mathcal{H}$ spanned by formal products which differ only by a finite number of factors. We assume that in all these basis formal products, for sufficiently large i we have by assumption $g_i(x_i) = g_\infty(x_i)$, where $g_\infty(x)$ is a fixed first order polynomial, the same for all elements of $\mathcal{H}$. Similarly as in the case of scalar field, we will not discuss here other possibilities. We also assume that

$$\langle g_\infty | g_\infty \rangle = 1.$$

The scalar product in such space is defined as follows

$$\langle g_1 g_2 \cdot \ldots | g_1' g_2' \cdot \ldots \rangle = \prod_{i=1}^{\infty} \langle g_i | g_i' \rangle, \tag{6.89}$$

where for $g_i = c_i x_i + d_i$ and $g_i' = c_i' x_i + d_i'$

$$\langle g_i | g_i' \rangle = c_i^* c_i' + d_i^* d_i'.$$

Strictly speaking, the linear space $\mathcal{H}$ with the scalar product introduced above should be called the pre-Hilbert space. To obtain the Hilbert space, we have to complete it with respect to the norm given by the scalar product using a standard mathematical procedure.

In the second step towards the quantum theory of the Dirac field we construct basic observables. They are represented by operators in $\mathcal{H}$. The obvious candidate for the quantum Hamiltonian has the form (6.74) with $\hat{\psi}$ and $\hat{\overline{\psi}}$ given by formulas (6.79), (6.81). Because these fields obey the Dirac equations (6.77), (6.78) we may write the Hamiltonian in the form

$$\hat{H} = \frac{i}{2} \int d^3x \, \left(\hat{\psi}^\dagger \partial_t \hat{\psi} - \partial_t \hat{\psi}^\dagger \hat{\psi} \right),$$

and

$$\hat{H} = \sum_{s=\pm 1/2} \int d^3p \, \omega(\vec{p}) \left((\hat{a}_s^{(+)}(\vec{p}))^\dagger \hat{a}_s^{(+)}(\vec{p}) - (\hat{a}_s^{(-)}(\vec{p}))^\dagger \hat{a}_s^{(-)}(\vec{p}) \right), \tag{6.90}$$

where in the last step we have used formula (6.79). Notice that Hamiltonian (6.90) is already normally ordered. This is due to the fact that in formula (6.74) the field $\hat{\bar{\psi}}$ always stands to the left of the field $\hat{\psi}$. At this point we could repeat the construction of the Fock space as in the previous Section. The vacuum state $|0\rangle$ would be defined by the conditions

$$\hat{a}_s^{(\pm)}(\vec{p})|0\rangle = 0,$$

and the complete set of basis states would be generated from it by the operators $(\hat{a}_s^{(\pm)}(\vec{p}))^\dagger$ as in formula (6.40). However, considerations analogous to the ones presented at the end of previous Section show that the contribution to the eigenvalues of the Hamiltonian, associated with the operator $(\hat{a}_s^{(-)}(\vec{p}))^\dagger$, is equal to $-\omega(\vec{p})$, hence it is negative. Therefore Hamiltonian (6.90) cannot be accepted because its eigenvalues extend from $-\infty$ to $+\infty$. Such systems have not been found in nature, hence the quantum Dirac field with Hamiltonian (6.90) is unphysical.

There exists a slight modification of Hamiltonian (6.90) which solves this problem. Using the anticommutation relation (6.83), and dropping the term proportional to the identity operator I we obtain the following operator

$$\hat{H}_D = \sum_{s=\pm 1/2} \int d^3p\, \omega(\vec{p}) \left((\hat{a}_s^{(+)}(\vec{p}))^\dagger \hat{a}_s^{(+)}(\vec{p}) + \hat{a}_s^{(-)}(\vec{p}) (\hat{a}_s^{(-)}(\vec{p}))^\dagger \right), \qquad (6.91)$$

It has non-negative expectation values because

$$\langle\psi|(\hat{a}_s^{(+)}(\vec{p}))^\dagger \hat{a}_s^{(+)}(\vec{p})|\psi\rangle = ||\hat{a}_s^{(+)}(\vec{p})|\psi\rangle||^2 \geq 0,$$

and

$$\langle\psi|\hat{a}_s^{(-)}(\vec{p})(\hat{a}_s^{(-)}(\vec{p}))^\dagger|\psi\rangle = ||(\hat{a}_s^{(-)}(\vec{p}))^\dagger|\psi\rangle||^2 \geq 0.$$

Here $||\cdot||$ denotes the norm defined by the scalar product. Notice that the argument presented in previous section, that one should avoid operators of the form $\hat{a}\hat{a}^\dagger$, is based on commutation relation (6.33)—it does not work here. Moreover, $\hat{H}_D$ and $\hat{H}$ give the same Heisenberg evolution equations because they differ only by a multiple of the identity operator. Therefore, $\hat{H}_D$ is a good candidate for the Hamiltonian of the quantum Dirac field, provided that we can find a Hilbert space in which this Hamiltonian has finite eigenvalues. It turns out that such a Hilbert space exists, as shown below.

Let us define the Dirac vacuum state $|0\rangle_D$. By definition, it is a normalized state that obeys the following conditions

$$\hat{a}_s^{(+)}(\vec{p})|0\rangle_D = 0, \quad (\hat{a}_s^{(-)}(\vec{p}))^\dagger|0\rangle_D = 0 \qquad (6.92)$$

for all $s = \pm 1/2$ and for all $\vec{p} \in R^3$. Such a state can be found in the space $\mathcal{H}^\infty$, because this space contains vectors such that $\hat{a}_i^\dagger|\psi\rangle = 0$, as pointed out below formula (6.88). It is clear that

$$\hat{H}_D|0\rangle_D = 0. \tag{6.93}$$

The Dirac vacuum state is sometimes called the Dirac sea. The reason is that one may heuristically write

$$|0\rangle_D = \text{“}\left(\prod_{s=\pm 1/2}\prod_{\vec{p}\in R^3}(\hat{a}_s^{(-)}(\vec{p}))^\dagger\right)\text{”}|0\rangle.$$

Then, the second condition (6.92) is satisfied because relations (6.83) imply that the square of each operator $(\hat{a}_s^{(-)}(\vec{p}))^\dagger$ vanishes. In view of this “formula” the Dirac vacuum may be regarded as the state in which all negative energy states are occupied, hence the sea of negative energy particles. Because the infinite product over $\vec{p} \in R^3$ is not defined, that “formula” cannot serve as the definition of the Dirac vacuum.

The basis states in the Fock space of the quantum Dirac field, analogous to the ones given by formulas (6.39), (6.40) for the quantum scalar field, are defined as follows

$$\begin{aligned}&|(+)\vec{p}_1 s_1, \vec{p}_2 s_2, \ldots, \vec{p}_M s_M; (-)\vec{q}_1 r_1, \vec{q}_2 r_2, \ldots, \vec{q}_N r_N\rangle \\ &= \frac{1}{\sqrt{M!N!}}\hat{a}_{-r_N}^{(-)}(-\vec{q}_N)\ldots\hat{a}_{-r_1}^{(-)}(-\vec{q}_1)(\hat{a}_{s_M}^{(+)}(\vec{p}_M))^\dagger\ldots(\hat{a}_{s_1}^{(+)}(\vec{p}_1))^\dagger|0\rangle_D,\end{aligned} \tag{6.94}$$

where $r_i, s_j = \pm 1/2$ and $M, N = 0, 1, 2, \ldots$. It is understood that $M = 0$ or $N = 0$ means that operators $(\hat{a}_{s_j}^{(+)}(\vec{p}_j))^\dagger$ or $\hat{a}_{r_i}^{(-)}(-\vec{q}_i)$, respectively, are absent. The reason for using $\hat{a}_{-s}^{(-)}(-\vec{q})$ and not $\hat{a}_s^{(-)}(\vec{q})$ is that the states (6.94) are eigenstates of the operator of the total momentum of the quantum Dirac field, with eigenvalues equal to the sum of the wave vectors $\vec{p}_n$ and $\vec{q}_i$, as discussed below. For a similar reason, we take as the spin indices $-r_i$ instead of r_i because then the states are eigenstates of an operator of the total spin with corresponding eigenvalues equal to $\sum_{n=0}^{N} r_n + \sum_{i=0}^{M} s_i$. It turns out that the states (6.94) are the eigenstates of the Hamiltonian,

$$\begin{aligned}&\hat{H}_D|(+)\vec{p}_1 s_1, \ldots, \vec{p}_M s_M; (-)\vec{q}_1 r_1, \ldots, \vec{q}_N r_N\rangle \\ &= \left(\sum_{i=0}^{N}\omega(\vec{q}_i) + \sum_{j=0}^{M}\omega(\vec{p}_j)\right)|(+)\vec{p}_1 s_1, \ldots, \vec{p}_M s_M; (-)\vec{q}_1 r_1, \ldots, \vec{q}_N r_N\rangle,\end{aligned} \tag{6.95}$$

where $\omega(\vec{k}) = \sqrt{m^2 + \vec{k}^2}$. The derivation of this formula is essentially identical as in the case of the quantum scalar field, formula (6.51). Instead of the first formula (6.49) we now have

$$[\hat{H}_D, (\hat{a}_s^{(+)}(\vec{p}))^\dagger] = \omega(\vec{p})(\hat{a}_s^{(+)}(\vec{p}))^\dagger, \ [\hat{H}_D, \hat{a}_s^{(-)}(\vec{q})] = \omega(\vec{q})\hat{a}_s^{(-)}(\vec{q}). \tag{6.96}$$

The commutators on the r.h.s.'s of formulas (6.96) have been calculated with the help of identity (6.76). We have also used the basic anticommutation relations (6.83).

It turns out that the states (6.94) are also eigenstates of the total momentum of the Dirac field. Starting from Noether's theorem applied to the spatial translations, and using formulas (6.79), (6.81) we obtain the following operator

$$\hat{P}^i = \sum_{s=\pm 1/2} \int d^3p\; p^i \left[(\hat{a}_s^{(+)}(\vec{p}))^\dagger \hat{a}_s^{(+)}(\vec{p}) + (\hat{a}_s^{(-)}(\vec{p}))^\dagger \hat{a}_s^{(-)}(\vec{p}) \right],$$

where $i = 1, 2, 3$. Next, we apply the same modification as we did for the Hamiltonian: we anticommute the operators in the second term and drop the generated term proportional to the identity operator. Furthermore, in the second term on the r.h.s. we change the integration variable $\vec{p} \to -\vec{p}$ and the summation index $s \to -s$. The resulting operator has the form

$$\hat{P}_D^i = \sum_{s=\pm 1/2} \int d^3p\; p^i \left((\hat{a}_s^{(+)}(\vec{p}))^\dagger \hat{a}_s^{(+)}(\vec{p}) + \hat{a}_{-s}^{(-)}(-\vec{p}) (\hat{a}_{-s}^{(-)}(-\vec{p}))^\dagger \right). \quad (6.97)$$

It is adopted as the operator of the total momentum of the quantum Dirac field. Calculations similar to the case of the Hamiltonian show that

$$\hat{P}_D^k |(+)\vec{p}_1 s_1, \ldots, \vec{p}_M s_M; (-)\vec{q}_1 r_1, \ldots, \vec{q}_N r_N\rangle =$$
$$= \left(\sum_{n=1}^{N} q_n^k + \sum_{j=1}^{M} p_i^k \right) |(+)\vec{p}_1 s_1, \ldots, \vec{p}_M s_M; (-)\vec{q}_1 r_1, \ldots, \vec{q}_N r_N\rangle. \quad (6.98)$$

Notice that the vectors $\vec{q}_n$ enter the eigenvalues of $\hat{P}_D^k$ with a plus sign, precisely because we have $-\vec{p}$ in the second term in formula (6.97).

We have seen that the sectors '+' and '−' give identical contributions to the eigenvalues of the observables $\hat{H}_D$ and $\hat{P}_D^k$. In fact, this is also true for the remaining six observables $\hat{M}_{\mu\nu}$ related to Poincaré symmetry.[8] However, these two sectors do differ when we take into account the internal $U(1)$ symmetry of the Lagrangians $\mathcal{L}$ or $\mathcal{L}'$. This global symmetry group acts on the classical Grassmannian fields ψ and $\overline{\psi}$ as follows

$$\psi'(x) = e^{i\alpha}\psi(x), \quad \overline{\psi'}(x) = e^{-i\alpha}\overline{\psi}(x).$$

[8] It turns out that eigenstates of $\hat{M}_{\mu\nu}$ are not given by the basis states (6.94), but by certain integrals over the wave vectors $\vec{p}, \vec{q}$ and linear combinations over the indices s_i, r_i. Nevertheless, the sectors '+', '−' give identical contributions to the eigenvectors and to the corresponding eigenvalues.

The corresponding total conserved charge has the form

$$Q = e \int d^3x \, \overline{\psi}\gamma^0\psi,$$

where e is a constant. This expression suggests that in the quantum theory, charge is represented by the following operator

$$\hat{Q} = e \int d^3x \, \hat{\overline{\psi}}\gamma^0\hat{\psi} = e \int d^3x \hat{\psi}^\dagger(t,\vec{x})\hat{\psi}(t,\vec{x})$$
$$= e\sum_s \int d^3p \, \left((\hat{a}_s^{(+)}(\vec{p}))^\dagger \hat{a}_s^{(+)}(\vec{p}) + (\hat{a}_s^{(-)}(\vec{p}))^\dagger \hat{a}_s^{(-)}(\vec{p})\right).$$

One can easily check that this operator commutes with $\hat{H}_D$, hence its eigenvalues and expectation values are constant in time. Nevertheless, this operator is not satisfactory because it has an infinite expectation value in the Dirac vacuum $|0\rangle_D$. Therefore, we perform the by now standard manipulation, consisting in anticommuting the two operators $(\hat{a}_s^{(-)}(\vec{p}))^\dagger$, $\hat{a}_s^{(-)}(\vec{p})$ and dropping the term proportional to the identity operator (such an operation does not have an effect on commutation relations with any other operators). In this way we obtain the correct total $U(1)$ charge operator

$$\hat{Q}_D = e\sum_s \int d^3p \, \left((\hat{a}_s^{(+)}(\vec{p}))^\dagger \hat{a}_s^{(+)}(\vec{p}) - \hat{a}_{-s}^{(-)}(-\vec{p})(\hat{a}_{-s}^{(-)}(-\vec{p}))^\dagger\right) \tag{6.99}$$

In the second term on the r.h.s. we have changed the summation index $s \to -s$ and the integration variable $\vec{p} \to -\vec{p}$ in order to have the same operators as in formula (6.94). The operator $\hat{Q}_D$ has a form analogous to those of the Hamiltonian and the total momentum operators. It is clear that the basis states (6.94) are its eigenstates with the eigenvalues equal to

$$Q = e(M - N). \tag{6.100}$$

Thus, the '−' and '+' states have $U(1)$ charges of the opposite sign.

The Fock space of the quantum Dirac field is spanned by the basis states (6.94). States from this space have all the properties of quantum states of non-interacting particles of rest mass m and spin 1/2. Moreover, there are two species of the particles which differ by the value of their $U(1)$ charge, which can be equal to $+e$ or $-e$. One of the species is called the particle, the other one its antiparticle.[9] Thus, the operators

$$\hat{a}_s(\vec{p}) = \hat{a}_s^{(+)}(\vec{p}), \quad \hat{a}_s^\dagger(\vec{p}) = (\hat{a}_s^{(+)}(\vec{p}))^\dagger \tag{6.101}$$

[9] Let us note that strictly speaking it is not correct to identify them with the real world electron and positron. Such identification would be correct if we could switch off the electromagnetic, weak and gravitational interactions. Nevertheless, the electrons and positrons can approximately be described by the above constructed quantum theory when the interactions are negligibly small.

are the particle annihilation and creation operators, while

$$\hat{d}_s(\vec{p}) = (\hat{a}^{(-)}_{-s}(-\vec{p}))^{\dagger}, \quad \hat{d}^{\dagger}_s(\vec{p}) = \hat{a}^{(-)}_{-s}(-\vec{p}) \tag{6.102}$$

are the antiparticle annihilation and creation operators, respectively. The field operator (6.79) can now be written in the form

$$\hat{\psi}(t,\vec{x}) = \int \frac{d^3p}{(2\pi)^{3/2}} \sum_{s=\pm 1/2} \left[v^{(+)}_s(\vec{p})\hat{a}_s(\vec{p})e^{i(\vec{p}\vec{x}-\omega t)} + v^{(-)}_{-s}(-\vec{p})\hat{d}^{\dagger}_s(\vec{p})e^{i(\omega t-\vec{p}\vec{x})} \right]. \tag{6.103}$$

It has the following commutation relation with the total charge operator $\hat{Q}_D$

$$[\hat{Q}_D, \hat{\psi}(t,\vec{x})] = -e\,\hat{\psi}(t,\vec{x}). \tag{6.104}$$

Using this relation, one can easily prove that the state $\hat{\psi}|\phi\rangle$ has a total $U(1)$ charge which is by e smaller than the $U(1)$ charge of the state $|\phi\rangle$.

Multiparticle wave functions in the momentum representation are defined by expanding a general state vector from the Fock space into the basis vectors (6.94). For example, states describing two particles and two antiparticles have the form

$$|\phi\rangle^{(2,2)} = \sum_{s_1,s_2}\sum_{r_1,r_2} \int d^3p_1 d^3p_2 d^3q_1 d^3q_2\; \phi^{(2,2)}(\vec{p}_1 s_1, \vec{p}_2 s_2; \vec{q}_1 r_1, \vec{q}_2 r_2)$$
$$|(+)\vec{p}_1 s_1, \vec{p}_2 s_2; (-)\vec{q}_1 r_1, \vec{q}_2 r_2\rangle.$$

Because the operators $\hat{a}^{(-)}$ (or $(\hat{a}^{(+)})^{\dagger}$) in the definition (6.94) anticommute, we may assume without loss of generality that the wave function is antisymmetric with respect to the arguments $\vec{q}_1 r_1$ and $\vec{q}_2 r_2$ (or $\vec{p}_1 s_1$ and $\vec{p}_2 s_2$).

On the other hand, the behavior of the wave function under an interchange of whole groups of variables, for example

$$(\vec{q}_1 r_1, \vec{q}_2 r_2) \leftrightarrow (\vec{p}_1 s_1, \vec{p}_2 s_2),$$

is not fixed. Such operations are related to the so called charge conjugation, which is represented by the transformation

$$\hat{a}^{(-)}_{-r}(-\vec{q}) \leftrightarrow (\hat{a}^{(+)}_r(\vec{q}))^{\dagger}, \quad (\hat{a}^{(-)}_{-r}(-\vec{q}))^{\dagger} \leftrightarrow \hat{a}^{(+)}_r(\vec{q}). \tag{6.105}$$

It commutes with the Hamiltonian $\hat{H}_D$, hence it is a symmetry of the quantum theory. Formula (6.105) determines the transformation of the basis states (6.94). The corresponding transformations of general states are defined by writing the states as linear combinations of the basis states. In general, the charge conjugation symmetry does not imply any particular symmetry of a concrete wave function.

Finally, let us have a look at the single particle and antiparticle sectors in the Fock space. The pertinent state vectors have the form

$$|\phi\rangle_1^{(\pm)} = \sum_s \int d^3p\, \phi_1^{(\pm)}(\vec{p}s)|(\pm)\vec{p}s\rangle, \tag{6.106}$$

where $\phi_1^{(\pm)}(\vec{p}s)$ is the single particle (antiparticle) wave function in the momentum representation. The index s describes the spin degrees of freedom of the particle. Time evolution of a single state is governed by the Schroedinger equation

$$i\partial_t|\phi(t)\rangle_1^{(\pm)} = \hat{H}_D|\phi(t)\rangle_1^{(\pm)}. \tag{6.107}$$

Using formulas (6.91), (6.105), and anticommutation relation (6.83), we obtain the Schroedinger equation for the single particle or antiparticle wave functions

$$i\partial_t\phi_1^{(\pm)}(t, \vec{p}s) = \omega(\vec{p})\phi_1^{(\pm)}(t, \vec{p}s). \tag{6.108}$$

Thus both particle and antiparticle have positive energies equal to $\omega(\vec{p})$. The problem of unbounded from below, negative energies of the Dirac particle, present in relativistic quantum mechanics, is absent here. One may say, that in a sense the negative energy states have been transformed into positive energy states of the antiparticle.

6.3 The Electromagnetic Field

The following construction of the quantum theory of the free electromagnetic field is based on the results of Sects. 1.2 and 4.1. It is very similar to the quantum theory of the real scalar field presented in Sect. 6.1. Therefore we shall discuss only the main points.

We consider the free electromagnetic field without any external sources. Its Lagrangian has the form

$$\mathcal{L} = -\frac{1}{4}F_{\mu\nu}F^{\mu\nu}, \tag{6.109}$$

where $F_{\mu\nu} = \partial_\mu A_\nu - \partial_\nu A_\mu$. Moreover, we use the Coulomb gauge condition, that is

$$A_0 = 0, \quad \nabla\vec{A} = 0. \tag{6.110}$$

These conditions eliminate spurious degrees of freedom which do not contribute to the physically relevant quantities, like the electric or magnetic fields. On the other hand, they are not Lorentz invariant. It is important to realize that this fact does not necessarily destroy the Lorentz invariance of the theory of the electromagnetic field. This is because the gauge conditions do not influence the physical degrees of freedom. In fact, it turns out that the Lorentz invariance is not broken, but it is no longer explicit. We shall not discuss this rather complicated issue here.

The condition $\nabla\vec{A} = 0$ can be explicitly solved. Let us write $\vec{A}$ in the form of the Fourier transform

$$\vec{A}(t,\vec{x}) = \frac{1}{(2\pi)^{3/2}} \sum_{\alpha=1}^{3} \int d^3k\, \vec{e}_\alpha(\vec{k}) a_\alpha(t,\vec{k}) e^{i\vec{k}\vec{x}}, \tag{6.111}$$

where $\vec{e}_\alpha(\vec{k})$ are fixed real vectors, called polarization vectors. They are normalized as follows

$$\vec{e}_\alpha(\vec{k})\vec{e}_\beta(\vec{k}) = \delta_{\alpha\beta}. \tag{6.112}$$

We assume that $\vec{k} \neq 0$, and we take

$$\vec{e}_3 = \frac{\vec{k}}{|\vec{k}|}. \tag{6.113}$$

We shall see that photons with $\vec{k} \to 0$ give a vanishing contribution to the total energy and momentum of the electromagnetic field. For this reason the assumption that $\vec{k} \neq 0$, or equivalently that $a_\alpha(t, \vec{k} = 0) = 0$, is consistent with the physics of the electromagnetic field. We also assume that

$$\vec{e}_\alpha(-\vec{k}) = \vec{e}_\alpha(\vec{k}) \;\text{ for }\; \alpha = 1, 2. \tag{6.114}$$

The complex number $a_\alpha(t,\vec{k})$ is called the amplitude of the mode $(\vec{k},\alpha)$ of the electromagnetic field. The fact that $\vec{A}$ is real is equivalent to the conditions

$$a^*_\alpha(t,-\vec{k}) = a_\alpha(t,\vec{k}) \tag{6.115}$$

for $\alpha = 1, 2$, and $a^*_3(t,-\vec{k}) = -a_3(t,\vec{k})$, where $*$ denotes the complex conjugation. From $\nabla\vec{A} = 0$ we obtain the condition

$$\vec{k}\vec{e}_\alpha(\vec{k})a_\alpha(t,\vec{k}) = 0 \tag{6.116}$$

This condition is automatically satisfied[10] for $\vec{k} = 0$, and it implies, for $\vec{k} \neq 0$, that

$$a_3(t,\vec{k}) = 0. \tag{6.117}$$

Thus, the space of vector potentials $\vec{A}$ that is compatible with the Coulomb gauge condition is parameterized by the Fourier amplitudes $a_{1,2}(t,\vec{k})$ which obey conditions (6.115).

---

[10] We assume that the vector potential $\vec{A}$ vanishes at the spatial infinity sufficiently quickly to ensure finiteness of the integral $\int d^3x\, \vec{A}(t,\vec{x})$. Then $\vec{e}_\alpha(\vec{k})a_\alpha(t,\vec{k})$ is finite at $\vec{k} = 0$.

Let us express the Lagrange function L, defined as

$$L = \int d^3x \mathcal{L},$$

by the Fourier amplitudes introduced above. Because $A_0(t, \vec{x}) = 0$, the Lagrangian has the form

$$\mathcal{L} = \frac{1}{2}\partial_0 A^i \partial_0 A^i - \frac{1}{4} F_{ik} F^{ik}. \tag{6.118}$$

Inserting here formula (6.111) and using conditions (6.112), (6.115) we obtain[11]

$$L = \frac{1}{2} \sum_{\alpha=1}^{2} \sum_{i=1}^{2} \int d^3k \left[\dot{a}^i_\alpha(t, \vec{k}) \dot{a}^i_\alpha(t, \vec{k}) - \vec{k}^2 a^i_\alpha(t, \vec{k}) a^i_\alpha(t, \vec{k}) \right]. \tag{6.119}$$

Here we have split $a_\alpha(t, \vec{k})$ into real and imaginary parts

$$a_\alpha(t, \vec{k}) = a^1_\alpha(t, \vec{k}) + i a^2_\alpha(t, \vec{k}).$$

Conditions (6.115) imply that the Fourier amplitudes are not independent,

$$a^1_\alpha(t, \vec{k}) = a^1_\alpha(t, -\vec{k}), \quad a^2_\alpha(t, \vec{k}) = -a^2_\alpha(t, -\vec{k}). \tag{6.120}$$

In order to write the Lagrangian in terms of independent Fourier amplitudes, let us restrict the wave vectors $\vec{k} = (k^1, k^2, k^3)$ to W, where W is the subset of R^3 such that $k^3 \geq 0$. Thus, as the independent dynamical variables, traditionally called the modes of the electromagnetic field, we take $a^{1,2}_\alpha(t, \vec{k})$ where $\vec{k} \in W$ and $\alpha = 1, 2$. Strictly speaking, we still have some double counting of the modes with $k^3 = 0$, but these modes actually do not contribute to L because the plane $k^3 = 0$ has zero volume in R^3.

The Lagrange function written in terms of the independent Fourier amplitudes has the from

$$L = \sum_{\alpha=1}^{2} \sum_{i=1}^{2} \int_W d^3k \left[\dot{a}^i_\alpha(t, \vec{k}) \dot{a}^i_\alpha(t, \vec{k}) - \vec{k}^2 a^i_\alpha(t, \vec{k}) a^i_\alpha(t, \vec{k}) \right]. \tag{6.121}$$

The canonical momenta associated with $a^i_\alpha(t, \vec{k})$, where $\vec{k} \in W$, are given by the functional derivatives

$$\pi^\alpha_i(t, \vec{k}) = \frac{\delta L}{\delta \dot{a}^i_\alpha(t, \vec{k})} = 2\dot{a}^i_\alpha(t, \vec{k}).$$

[11] Let us remember that we use the convention that the arrow denotes vectors with upper indices. Thus, $\vec{A} = (A^i)$ and $F_{ik} = -\partial_i A^k + \partial_k A^i$.

The classical Hamiltonian corresponding to L has the form

$$H = \sum_{\alpha=1}^{2}\sum_{i=1}^{2}\int_W d^3k \left[\frac{1}{4}\pi_i^\alpha(t,\vec{k})\pi_i^\alpha(t,\vec{k}) + \vec{k}^2 a_\alpha^i(t,\vec{k})a_\alpha^i(t,\vec{k})\right]. \tag{6.122}$$

This form of the theory of the classical electromagnetic field is a convenient starting point for constructing the corresponding quantum model. The questions of Hilbert space, the choice of realization of operators, etc., are settled in full analogy with the case of the real scalar field. Therefore we shall omit detailed discussion of these points.

We postulate the equal-time canonical commutation relations

$$\left[\hat{\pi}_i^\alpha(t,\vec{k}), \hat{a}_\beta^j(t,\vec{k}')\right] = -i\delta_{\alpha\beta}\delta_{ij}\delta(\vec{k}-\vec{k}')I, \quad \left[\hat{\pi}_i^\alpha(t,\vec{k}), \hat{\pi}_j^\beta(t,\vec{k}')\right] = 0,$$

$$\left[\hat{a}_\alpha^i(t,\vec{k}), \hat{a}_\beta^j(t,\vec{k}')\right] = 0, \tag{6.123}$$

where $\vec{k} \in W$, and the quantum Hamiltonian (to be changed into the normal ordered one later on)

$$\hat{H} = \sum_{\alpha=1}^{2}\sum_{i=1}^{2}\int_W d^3k \left[\frac{1}{4}\hat{\pi}_i^\alpha(t,\vec{k})\hat{\pi}_i^\alpha(t,\vec{k}) + \vec{k}^2\, \hat{a}_\alpha^i(t,\vec{k})\hat{a}_\alpha^i(t,\vec{k})\right].$$

By assumption, the operators $\hat{a}_\alpha^i$ and $\hat{\pi}_i^\alpha$ are Hermitian. The Heisenberg evolution equations have the form

$$\dot{\hat{a}}_\alpha^i(t,\vec{k}) = \frac{1}{2}\hat{\pi}_i^\alpha(t,\vec{k}), \quad \dot{\hat{\pi}}_i^\alpha(t,\vec{k}) = -2\vec{k}^2\, \hat{a}_\alpha^i(t,\vec{k}). \tag{6.124}$$

Thus, the operators $\hat{a}_\alpha^i$ obey the following equation

$$\ddot{\hat{a}}_\alpha^i(t,\vec{k}) = -\vec{k}^2\hat{a}_\alpha^i(t,\vec{k}). \tag{6.125}$$

Its general Hermitian solution has the form

$$\hat{a}_\alpha^i(t,\vec{k}) = \frac{1}{\sqrt{2|\vec{k}|}}\left[e^{i|\vec{k}|t}(\hat{d}_\alpha^i(\vec{k}))^\dagger + e^{-i|\vec{k}|t}\hat{d}_\alpha^i(\vec{k})\right]. \tag{6.126}$$

The factor $1/\sqrt{2|\vec{k}|}$ has been introduced for later convenience. Let us introduce the following operators

$$\hat{a}_\alpha(\vec{k}) = \hat{d}_\alpha^1(\vec{k}) + i\hat{d}_\alpha^2(\vec{k}), \quad \hat{a}_\alpha^\dagger(\vec{k}) = (\hat{d}_\alpha^1(\vec{k}))^\dagger - i(\hat{d}_\alpha^2(\vec{k}))^\dagger,$$

$$\hat{a}_\alpha(-\vec{k}) = \hat{d}^1_\alpha(\vec{k}) - i\hat{d}^2_\alpha(\vec{k}), \quad \hat{a}^\dagger_\alpha(-\vec{k}) = (\hat{d}^1_\alpha(\vec{k}))^\dagger + i(\hat{d}^2_\alpha(\vec{k}))^\dagger,$$

where $\vec{k} \in W$. The canonical commutation relations (6.123) are equivalent to

$$[(\hat{d}^i_\alpha(\vec{k}))^\dagger, \hat{d}^j_\beta(\vec{k}')] = -\frac{1}{2}\delta_{ij}\delta_{\alpha\beta}\delta(\vec{k} - \vec{k}'), \quad [\hat{d}^i_\alpha(\vec{k}), \hat{d}^j_\beta(\vec{k}')] = 0,$$

where $\vec{k} \in W$. Simple calculation gives

$$[\hat{a}_\alpha(\vec{k}), \hat{a}^\dagger_\beta(\vec{k}')] = \delta_{\alpha\beta}\delta(\vec{k} - \vec{k}'), \quad [\hat{a}_\alpha(\vec{k}), \hat{a}_\beta(\vec{k}')] = 0 \tag{6.127}$$

for all $\vec{k} \in R^3$. These commutators are essentially the same as in the case of the scalar field, except for the index α. The field operator can now be written in the form

$$\hat{\vec{A}}(t, \vec{x}) = \sum_{\alpha=1}^{2} \int_{R^3} \frac{d^3k}{\sqrt{2(2\pi)^3|\vec{k}|}} \vec{e}_\alpha(\vec{k}) \left[e^{-i|\vec{k}|t + i\vec{k}\vec{x}} \hat{a}_\alpha(\vec{k}) + \text{h.c.} \right], \tag{6.128}$$

where h.c. stands for the Hermitian conjugation of the preceding term.

The Hamiltonian expressed by the operators $\hat{a}_\alpha$ and $\hat{a}^\dagger_\alpha$ has the form

$$\hat{H} = \frac{1}{2}\sum_{\alpha=1}^{2} \int_{R^3} d^3k \; |\vec{k}| \left(\hat{a}^\dagger_\alpha \hat{a}_\alpha + \hat{a}_\alpha \hat{a}^\dagger_\alpha \right).$$

At this point we can recognize the same mathematical structures as in the case of the real scalar field. Therefore we repeat the steps from there. The Hamiltonian is changed to the normally ordered one,

$$\hat{H} = \sum_{\alpha=1}^{2} \int_{R^3} d^3k \; |\vec{k}| \; \hat{a}^\dagger_\alpha(\vec{k}) \hat{a}_\alpha(\vec{k}), \tag{6.129}$$

and the Hilbert space is spanned by the basis states

$$|0\rangle, \; |\vec{k}\alpha\rangle = \hat{a}^\dagger_\alpha(\vec{k})|0\rangle, \ldots \; . \tag{6.130}$$

The vacuum state $|0\rangle$ is defined by the condition

$$\hat{a}_\alpha(\vec{k})|0\rangle = 0 \tag{6.131}$$

for all $\vec{k} \in R^3$ and $\alpha = 1, 2$. We see that the states of this quantum field can be regarded as states of particles, called photons, with the two polarizations corresponding to $\alpha = 1, 2$. These polarizations are called transverse because the corresponding polarization vectors $\vec{e}_\alpha(\vec{k})$ are perpendicular to $\vec{k}$. The single particle basis state $|\vec{k}\alpha\rangle$

is an eigenstate of the Hamiltonian (6.129), with energy equal to $|\vec{k}|$ which coincides with the energy of a free relativistic particle with vanishing rest mass. Thus, photons are massless. They do not interact with each other, because all multiparticle eigenstates of the Hamiltonian have eigenvalues equal to the sum of the energies of the participating photons. The single photon wave function in the momentum representation $\phi^1_\alpha(t, \vec{k})$ obeys the Schroedinger equation

$$i\partial_t \phi^1_\alpha(t, \vec{k}) = |\vec{k}|\, \phi^1_\alpha(t, \vec{k}),$$

which follows from the general Schroedinger equation

$$i\partial_t |\phi\rangle = \hat{H}|\phi\rangle,$$

if we restrict $|\phi\rangle$ to the single photon sector, where

$$|\phi\rangle = \sum_{\alpha=1}^{2} \int d^3k\, \phi^1_\alpha(t, \vec{k})|\vec{k}\alpha\rangle.$$

The operators $\hat{a}^\dagger_\alpha(\vec{k})$ commute with each other. Therefore the n-photon basis states $|\vec{k}_1\alpha_1, \vec{k}_2\alpha_2, \ldots, \vec{k}_n\alpha_n\rangle$, as well as the corresponding n-photon wave function $\phi_n(\vec{k}_1\alpha_1, \vec{k}_2\alpha_2, \ldots, \vec{k}_n\alpha_n\rangle$, is symmetric with respect to permutations of the variables $\vec{k}_i\alpha_i,\ \vec{k}_j\alpha_j$. Thus, the free photons are massless bosons.

Exercises

6.1 We have shown in the text that relations (6.17) follow from the canonical commutation relations (6.20). Prove also that the converse is true: (6.20) follows from (6.17).

6.2 Show that Hamiltonian (6.5) can be written in the form (6.31) if $\hat{\phi}$ is given by solution (6.16).
Hints: 1. Obtain $\hat{\phi}(\vec{x})$ and $\hat{\pi}(\vec{x})$ in the Schroedinger picture by putting $t = 0$ in the pertinent formulas in the Heisenberg picture.
2. Use the integrals

$$\int d^3x\, e^{i(\vec{k}\pm\vec{k}')\vec{x}} = (2\pi)^3\delta(\vec{k} \pm \vec{k}'), \qquad \int d^3k\, k^i\hat{a}(\vec{k})\hat{a}(-\vec{k}) = 0.$$

6.3 Prove that the operators

$$\hat{L}^k = \frac{1}{2}\epsilon_{krs}\hat{M}^{rs},$$

where $\hat{M}^{rs}$ are given by formula (6.47), obey the commutation relations

$$[\hat{L}^k,\ \hat{L}^s] = i\epsilon_{ksp}\hat{L}^p,$$

characteristic for quantum angular momentum.

6.4 Find the wave functional $\Psi_0[\phi]$ for the vacuum state $|0\rangle \in \mathcal{H}_F$ of the free real scalar field.
Hints: 1. Use formula (6.19) with $t = 0$ and (6.7) in order to find the Schroedinger representation of the operators $\hat{a}(\vec{k}\,)$.
2. Find a Gaussian type functional that obeys the equation $\hat{a}(\vec{k}\,)\Psi_0[\phi] = 0$, where $\vec{k} \in R^3$.

6.5 Check that Lagrangian (6.64) and the current $j^\mu = \overline{\psi}\gamma^\mu\psi$ are real, that is that $\mathcal{L}^* = \mathcal{L}$ and $(j^\mu)^* = j^\mu$.
Hint: Matrices γ^μ in the Dirac representation have the following property

$$\gamma^\mu = \gamma^0(\gamma^\mu)^\dagger\gamma^0.$$

6.6 Using relation (6.104) prove that the state $\hat{\psi}|\phi\rangle$ has total $U(1)$ charge equal to $Q - e$ if the state $|\phi\rangle$ has charge Q.

6.7 The canonical momentum conjugate to A^i is given by the following formula $\pi_i = \partial\mathcal{L}/\partial(\partial_0 A^i) = \partial_0 A^i$, where $\mathcal{L}$ is given by (6.118). Using (6.128) obtain the equal time commutation relation

$$[\hat{A}^i(t,\vec{x}),\ \hat{\pi}_j(t,\vec{y})] = i\left(\delta_{ij} - \frac{\partial_i\partial_j}{\Delta}\right)\delta(\vec{x}-\vec{y}).$$

Check that the non-canonical form of the r.h.s. is consistent with the Coulomb gauge condition.
Hint: Use the Fourier representation of the Dirac delta. For example, $\Delta^{-1}\delta(\vec{x}-\vec{y}) = -(2\pi)^{-3}\int d^3k\ |\vec{k}|^{-2}\exp(i\vec{k}(\vec{x}-\vec{y}))$.

Chapter 7
Perturbative Expansion in the ϕ_4^4 Model

Abstract Problems with an exact construction of the quantum ϕ_4^4 model. The interaction picture. The Gell-Mann–Low formula for Greeen's functions. The generating functional for Green's functions. The exponential Wick formula. The Feynman free propagator. Regularized Feynman diagrams in four-momentum space. Normal ordered interactions. Cancelation of vacuum bubbles.

We have seen three examples of quantum fields. On the one hand they are extremely important because they show the main features of quantum fields, for example, the appearance of quantum particles. On the other hand, we have obtained only noninteracting particles, and this fact obviously reduces the relevance of the discussed fields for a description of physical phenomena. It is necessary to find quantum field theories (in the literature, rather modestly called 'models') which give interacting particles. Unfortunately, it turns out that this is not an easy task. The level of completeness of the analysis of the quantum fields presented in the previous chapter remains as yet an unreachable ideal in the case of models with interactions. Generally speaking, one is forced either to consider very special models, often of little physical relevance, or to resort to a perturbative expansion. This latter possibility is widely used in most applications of quantum field theory. It is neither simple nor satisfactory from a theoretical viewpoint: it leads to rather cumbersome calculations, and the perturbative series has rather bad convergence properties. Nevertheless, the perturbative approach is a very popular and important tool with many spectacular applications in particle physics and statistical mechanics.

In this chapter we present a derivation of the standard perturbative expansion in powers of interaction. On the basis of a set of assumptions we shall obtain concrete, sensible, approximate formulas for Green's functions. The rules for constructing such perturbative formulas are quite precise. The main ideas of the perturbative expansion are presented here in the example of the ϕ_4^4 model, that is a real scalar field ϕ in the four-dimensional space-time with a self interaction term of the form ϕ^4. We have chosen this relatively simple model in order to get rid of "kinematical" complications which appear when there are several fields or several coupling constants.

H. Arodź and L. Hadasz, *Lectures on Classical and Quantum Theory of Fields*,
Graduate Texts in Physics, DOI 10.1007/978-3-319-55619-2_7

7.1 The Gell-Mann–Low Formula

We consider a relatively simple model, which, on the classical level is defined by the Lagrangian

$$\mathcal{L} = \frac{1}{2}\partial_\mu\phi\,\partial^\mu\phi - \frac{m_0^2}{2}\phi^2 - \frac{\lambda_0}{4!}\phi^4. \tag{7.1}$$

Here m_0^2 and λ_0 are finite, positive constants. In principle, they can be determined experimentally, by measuring certain physical quantities which are calculable in the model and therefore depend on these constants.

Let us first try the same steps as in the case of the free fields. The energy corresponding to (7.1) is given by the formula

$$E = \int d^3x \left(\frac{1}{2}\partial_0\phi\,\partial_0\phi + \frac{1}{2}\partial_i\phi\,\partial_i\phi + \frac{m_0^2}{2}\phi^2 + \frac{\lambda_0}{4!}\phi^4\right). \tag{7.2}$$

The canonical momentum conjugate to ϕ is defined, as always, as

$$\pi(t,\vec{x}) \equiv \frac{\partial\mathcal{L}}{\partial\phi_{,0}(t,\vec{x})}. \tag{7.3}$$

In the present case it is equal to

$$\pi(t,\vec{x}) = \partial_0\phi(t,\vec{x}). \tag{7.4}$$

With the same motivation as for the free real scalar field (Sect. 6.1), we introduce the Hermitian operators $\hat{\phi}(t,\vec{x})$ and $\hat{\pi}(t,\vec{x})$ in the Heisenberg picture, and postulate the equal-time canonical commutation relations

$$\begin{aligned}\left[\hat{\phi}(t,\vec{x}),\hat{\pi}(t,\vec{y})\right] &= i\delta(\vec{x}-\vec{y})I,\\ \left[\hat{\phi}(t,\vec{x}),\hat{\phi}(t,\vec{y})\right] = 0 &= \left[\hat{\pi}(t,\vec{x}),\hat{\pi}(t,\vec{y})\right],\end{aligned} \tag{7.5}$$

as well as the quantum Hamiltonian

$$\hat{H} = \int d^3x \left[\frac{1}{2}\hat{\pi}^2(t,\vec{x}) + \frac{1}{2}\partial_i\hat{\phi}(t,\vec{x})\partial_i\hat{\phi}(t,\vec{x}) + \frac{m_0^2}{2}\hat{\phi}^2(t,\vec{x}) + \frac{\lambda_0}{4!}\hat{\phi}^4(t,\vec{x})\right]. \tag{7.6}$$

The Heisenberg evolution equation[1]

$$\partial_t\hat{O}(t) = i\left[\hat{H},\hat{O}(t)\right] \tag{7.7}$$

[1] We assume here that the considered operators do not depend on time in the Schroedinger picture.

gives

$$\partial_t \hat{\phi}(t, \vec{x}) = i[\hat{H}, \hat{\phi}(t, \vec{x})], \quad \partial_t \hat{\pi}(t, \vec{x}) = i[\hat{H}, \hat{\pi}(t, \vec{x})]. \tag{7.8}$$

Because the Hamiltonian is constant in time, the time t on the r.h.s. of formula (7.6) can be chosen arbitrarily. Therefore, we can compute the commutators on the r.h.s. of (7.8) using the equal time commutators (7.5). We obtain

$$\partial_t \hat{\phi}(t, \vec{x}) = \hat{\pi}(t, \vec{x}),$$

and

$$\partial_t \hat{\pi}(t, \vec{x}) = \Delta \hat{\phi}(t, \vec{x}) - m_0^2 \hat{\phi}(t, \vec{x}) - \frac{\lambda_0}{3!} \hat{\phi}^3(t, \vec{x}),$$

where Δ denotes the three-dimensional Laplacian. It follows from these equations that the operator $\hat{\phi}(t, \vec{x})$ obeys the equation

$$(\partial_t^2 - \Delta + m_0^2) \hat{\phi}(t, \vec{x}) + \frac{\lambda_0}{3!} \hat{\phi}^3(t, \vec{x}) = 0. \tag{7.9}$$

Notice that this equation has the same form as the classical equation (3.25), except that instead of the classical field $\phi(t, \vec{x})$ there is the field operator $\hat{\phi}(t, \vec{x})$.

We have seen in Sect. 6.1 that in the case of the free scalar field it was necessary to replace the 'naive' Hamiltonian (6.5) by the correct one (6.34). Nevertheless, Hamiltonian (6.5) gave the correct evolution equation (6.14). One should expect that also in the present case the 'naive' Hamiltonian (7.6), as well as evolution equation (7.9), do not have a mathematical meaning. The reason is that they involve products of the type

$$\hat{\phi}(x_1) \ldots \hat{\phi}(x_n)|_{x_1 = \ldots = x_n = x},$$

where $n = 2, 3, 4$. Here we have used the four-dimensional notation $x_i = (t_i, \vec{x}_i)$. If $\hat{\phi}(x)$ is a generalized function of x, as suggested by the example of the free quantum scalar field, such products are not defined in general. Yet another difficulty is the nonlinearity of (7.9)—because of it, we would not be able to find its general solution, even if we managed to define the $\hat{\phi}^3(x)$ term.

Because we do not know how to define and solve the Heisenberg evolution equation (7.9), we may try to use the interaction picture in which time evolution is split between states and operators in such a way that the operators evolve as in the free field model. Let us quote the main formulas—their derivations can be found in textbooks on quantum mechanics. The Hamiltonian $\hat{H}$ does not depend on time, hence it has the same form in both the Schroedinger and Heisenberg pictures. Let us split it into the free part $\hat{H}_{0S}$ and the interaction part $\hat{V}_S$, both taken here in the Schroedinger picture marked by the subscript S:

$$\hat{H} = \hat{H}_{0S} + \hat{V}_S,$$

where

$$\hat{H}_{0S} = \frac{1}{2}\int d^3x \left[\hat{\pi}^2(\vec{x}) + \partial_i\hat{\phi}(\vec{x})\partial_i\hat{\phi}(\vec{x}) + m_0^2\hat{\phi}^2(\vec{x})\right], \quad \hat{V}_S = \frac{\lambda_0}{4!}\int d^3x\ \hat{\phi}^4(\vec{x}).$$

Similarly as in the case of free fields, the question of the powers of the field operator will be addressed later. In general, these operators separately depend on time in the Heisenberg picture, while $\hat{H} = \hat{H}_0(t) + \hat{V}(t)$ is constant. For brevity, the Heisenberg picture is denoted just by the presence of the time argument.

Time evolution of states in the interaction picture is given by the unitary operator $U_I(t, t_0)$,

$$|t\rangle_I = U_I(t, t_0)|t_0\rangle_I,$$

where

$$U_I(t, t_0) = e^{i\hat{H}_{0S}t}e^{-i\hat{H}(t-t_0)}e^{-i\hat{H}_{0S}t_0}. \tag{7.10}$$

Operator $\hat{O}_S$ from the Schroedinger picture is represented in the interaction picture by the operator

$$\hat{O}_I(t) = e^{i\hat{H}_{0S}t}\hat{O}_S e^{-i\hat{H}_{0S}t}, \tag{7.11}$$

and in the Heisenberg picture by

$$\hat{O}(t) = e^{i\hat{H}t}\hat{O}_S e^{-i\hat{H}t}. \tag{7.12}$$

Comparing the last two formulas we obtain the relation

$$\hat{O}_I(t) = U_I(t, 0)\hat{O}(t)U_I(0, t). \tag{7.13}$$

The operator $U_I(t, t_0)$ can also be written in the Dyson form

$$U_I(t, t_0) = T\exp(-i\int_{t_0}^{t} dt'\ \hat{V}_I(t')). \tag{7.14}$$

The r.h.s. of this formula is understood as the series

$$T\exp(-i\int_{t_0}^{t} dt'\ \hat{V}_I(t')) = I + \sum_{n=1}^{\infty}\frac{(-i)^n}{n!}T\left(\int_{t_0}^{t} dt'\ \hat{V}_I(t')\right)^n,$$

where T denotes the chronological, or time ordering. It is defined as follows:

$$T\left(\int_{t_0}^{t} dt'\ \hat{V}_I(t')\right)^n = \int_{t_0}^{t} dt_1 \ldots \int_{t_0}^{t} dt_n T\left(\hat{V}_I(t_1)\hat{V}_I(t_2)\ldots\hat{V}_I(t_n)\right),$$

where

$$T\left(\hat{V}_I(t_1)\hat{V}_I(t_2)\ldots\hat{V}_I(t_n)\right)$$
$$=\sum_P \Theta(t_{i_1}-t_{i_2})\Theta(t_{i_2}-t_{i_3})\ldots\Theta(t_{i_{n-1}}-t_{i_n})\hat{V}_I(t_{i_1})\hat{V}_I(t_{i_2})\ldots\hat{V}_I(t_{i_n}).$$

The sum is over the set of all permutations $(t_1, t_2, \ldots, t_n) \to (t_{i_1}, t_{i_2}, \ldots, t_{i_n})$, and Θ denotes the step function.

The operator $\hat{O}_I(t)$ obeys the following evolution equation

$$\frac{d\hat{O}_I(t)}{dt} = i[\hat{H}_{0S}, \hat{O}_I(t)],$$

obtained from definition (7.11) by differentiation with respect to time. In particular,

$$\frac{d\hat{\phi}_I(t,\vec{x})}{dt} = ie^{it\hat{H}_{0S}}[\hat{H}_{0S},\ \hat{\phi}_S(\vec{x})]e^{-it\hat{H}_{0S}} = \hat{\pi}_I(t,\vec{x}),$$

and

$$\frac{d\hat{\pi}_I(t,\vec{x})}{dt} = ie^{it\hat{H}_{0S}}[\hat{H}_{0S},\ \hat{\pi}_S(\vec{x})]e^{-it\hat{H}_{0S}} = \Delta\hat{\phi}_I(t,\vec{x}) - m_0^2\hat{\phi}_I(t,\vec{x}).$$

These two equations imply that $\hat{\phi}_I(t,\vec{x})$ obeys the following equation

$$\left(\frac{\partial^2}{\partial t^2} - \Delta + m_0^2\right)\hat{\phi}_I(t,\vec{x}) = 0. \tag{7.15}$$

It coincides with the operator Klein–Gordon equation, known from Chap. 6. As shown there, its general solution has the form

$$\hat{\phi}_I(t,\vec{x}) = \int \frac{d^3k}{\sqrt{2(2\pi)^3\omega(\vec{k})}}\left(e^{-ikx}\hat{a}_I(\vec{k}) + \text{h.c.}\right), \tag{7.16}$$

where $k^0 = \omega(\vec{k})$.

Canonical commutation relations do not change their form under similarity transformations, hence $\hat{\phi}_I(t,\vec{x})$ and $\hat{\pi}_I(t,\vec{x})$ have equal-time commutation relations of the form (7.5). Similarly as in the case of the free scalar field, one can show that $\hat{a}_I(\vec{k})$ and $\hat{a}_I^\dagger(\vec{k}')$ have the following commutation relations

$$\left[\hat{a}_I(\vec{k}), \hat{a}_I^\dagger(\vec{k}')\right] = \delta(\vec{k}-\vec{k}')I,\quad \left[\hat{a}_I(\vec{k}), \hat{a}_I(\vec{k}')\right] = 0. \tag{7.17}$$

The form of solution (7.16), as well as commutation relations (7.17), are the same as in the case of the free scalar field. Therefore, it is quite natural to consider the Fock space with the basis

$$|0_I\rangle,\ \hat{a}_I^{\dagger}(\vec{k})|0_I\rangle,\ \frac{1}{\sqrt{2}}\hat{a}_I^{\dagger}(\vec{k})\hat{a}_I^{\dagger}(\vec{k}')|0_I\rangle,\ \ldots, \tag{7.18}$$

where the state $|0_I\rangle$ is defined by the condition

$$\hat{a}_I(\vec{k})|0_I\rangle = 0$$

for all $\vec{k} \in R^3$.

In the next step, we insert into the Hamiltonian $\hat{H}_{0I}$ the solution (7.16) for $\hat{\phi}_I$, and $d\hat{\phi}_I/dt$ for $\hat{\pi}_I$. Then, the Hamiltonian is expressed by the 'creation' and 'annihilation' operators $\hat{a}_I^{\dagger}(\vec{k})$ and $\hat{a}_I(\vec{k})$. In order to obtain a well-defined operator $\hat{H}_{0I}$ in the Fock space we apply the normal ordering : :, as discussed in the previous chapter.

The problem with the definition of the interaction operator $\hat{V}_I$ is more severe. It is not to be solved merely by normal ordering—a more drastic modification of the interaction, in the literature called a regularization, is needed in order to convert it into a well-defined operator in the Fock space spanned by the basis vectors (7.18). We shall denote such a regularized interaction by $\hat{V}_{Ig}$ in the interaction picture, and by $\hat{V}_{Sg}$ in the Schroedinger picture (in order to obtain the Schroedinger picture operator it is sufficient to put $t = 0$ in the interaction or Heisenberg picture operators). The problem is generated by the integral $\int d^3x$ over the infinite space. It turns out that the normal ordered monomial $:\hat{\phi}_I^4(t=0,\vec{x}):$ is a generalized function of $\vec{x}$, see, e.g., Chapt. 8, Sect. 4.A in [8]. Therefore, it may be integrated with a test function $g(\vec{x})$, and

$$\hat{V}_{Sg} = \frac{\lambda_0}{4!}\int d^3x\, g(\vec{x}) : \hat{\phi}_I^4(t=0,\vec{x}) :$$

is a well-defined operator, while $\lambda_0 \int d^3x\ :\hat{\phi}_I^4(t=0,\vec{x}):/4!$ is not because the constant function equal to 1 is not a test function. For Hermiticity of $\hat{V}_{Sg}$ the function $g(\vec{x})$ has to be real-valued (Exercise 7.1).

We do not want to ascribe to the regularizing function $g(\vec{x})$ any physical meaning. Therefore, we should remove it by taking the limit

$$g(\vec{x}) \to 1.$$

There is a hope that such a limit, called the removal of the regularization, can be considered in a mathematically rigorous manner, at least on the level of the measurable quantities, like scattering cross-sections or energies of bound states, and that the results obtained in that limit do not contradict the basic physical requirements, such as the unitarity of the time evolution in the quantum theory or Poincaré invariance. Particularly difficult is the problem of recovering Poincaré invariance, because the presence of the fixed test function $g(\vec{x})$ almost surely breaks that invariance, and therefore it has to reappear 'from nowhere' in that limit. A concrete realization of such a programme in the case of interacting fields in four-dimensional space-time does not yet exist. Anyway, in the following considerations we shall use the regularized interaction Hamiltonian in order to avoid mathematically meaningless formulas.

Note that the states (7.18) are not eigenstates of the full regularized Hamiltonian $\hat{H} =: \hat{H}_{0S} : + \hat{V}_{Sg}$. Therefore, there is little hope that they will become the eigenstates after the regularization is removed. This casts a shadow on the physical meaning of these states. In particular, they can hardly be regarded as particle states with definite numbers of particles, and $\vec{k}$ is not equal to the momentum of any particle. Needless to say, the exact eigenvalues and eigenstates of the Hamiltonian $\hat{H}$ are not known.

To summarize, an explicit construction of the quantum ϕ_4^4 model is beyond our reach. This model is not exceptional in this respect. In fact, we do not know the explicit construction of any physically important model with (self)coupled quantum fields defined in four-dimensional space-time.[2] On the other hand, one can construct so called perturbative quantum field theories which are well-defined at every finite order of an expansion with respect to a pertinent interaction Hamiltonian. It turns out that such surrogate quantum field theories can yield predictions which agree with experimental data amazingly well. Principles applied in the construction of the perturbative quantum field theories turn out to be very fruitful. There is no doubt that perturbative expansion is the indispensable tool for applications of quantum field theory. On the other hand, many physically interesting quantities cannot be reliably calculated within the perturbative approach.

We will not present the full perturbative ϕ_4^4 model. We shall concentrate on the so called Green's functions, often also called the correlation functions, $G^{(n)}(x_1, x_2, \ldots, x_n)$, where n is a natural number and $x_i,\ i = 1, 2, \ldots, n$, are points in Minkowski space-time. The Green's functions are defined as the vacuum expectation value of time ordered products of the quantum fields in the Heisenberg picture,

$$G^{(n)}(x_1, x_2, \ldots, x_n) = \langle 0|T\left(\hat{\phi}(x_1)\hat{\phi}(x_2)\cdots\hat{\phi}(x_n)\right)|0\rangle. \tag{7.19}$$

Here T denotes the time ordering, and $|0\rangle$ is the vacuum state in the model, that is the normalized eigenstate of $\hat{H}$ with the lowest eigenvalue E_0—we assume that such an eigenvalue exists. By shifting the Hamiltonian, $\hat{H} \to \hat{H} - E_0 I$, the eigenvalue is shifted to 0. Then, the vector $|0\rangle$ does not depend on time because $i\partial_t|0\rangle = \hat{H}|0\rangle = 0$. From now on we assume that

$$\hat{H}|0\rangle = 0.$$

[2] In the case of fields defined in two- or three-dimensional space-time the situation is a little bit better.

The Green's functions play a very important role in applications of quantum field theory, in particular in calculations of scattering amplitudes of particles. On the mathematical side, Green's functions are generalized functions of n independent four-vectors x_i. Therefore, in general it does not make sense to ask for the value of such a function at fixed values of all x_i, see the Appendix. Also, one cannot construct a well-defined generalized function of a smaller number of variables, say $x_2, x_3, \ldots, x_n$, just by putting, for example, $x_1 = x_2$. The resulting object is not, in general, a generalized function of $x_2, x_3, \ldots, x_n$. This is analogous to putting $x = y$ in the product $\delta(x)\delta(y)$—the resulting object $(\delta(x))^2$ is not a generalized function of x.

The Gell-Mann–Low formula gives $G^{(n)}$ in terms of the interaction picture field $\hat{\phi}_I$ and the state $|0_I\rangle$. In the first step in the derivation of this formula we express $\hat{\phi}$ by $\hat{\phi}_I$ and perform the time ordering. Let $(i_1, i_2, \ldots, i_n)$ be the permutation of $(1, 2, \ldots, n)$ such that $x^0_{i_1} \geq x^0_{i_2} \geq \ldots \geq x^0_{i_n}$. Then,

$$G^{(n)}(x_1, x_2, \ldots, x_n) = \langle 0|\hat{\phi}(x_{i_1}) \ldots \hat{\phi}(x_{i_n})|0\rangle.$$

Next, we apply the following formulas, which are obtained from (7.13):

$$\hat{\phi}(x_k) = U_I^{-1}(x^0_k, 0)\hat{\phi}_I(x_k)U_I(x^0_k, 0),$$

and

$$U_I(x^0_j, 0)U_I^{-1}(x^0_k, 0) = U_I(x^0_j, x^0_k).$$

The result has the form

$$G^{(n)}(x_1, x_2, \ldots, x_n) = \tag{7.20}$$
$$\langle 0|U_I^{-1}(x^0_{i_1}, 0)\hat{\phi}_I(x_{i_1})U_I(x^0_{i_1}, x^0_{i_2})\hat{\phi}_I(x_{i_2}) \ldots \hat{\phi}_I(x_{i_n})U_I(x^0_{i_n}, 0)|0\rangle.$$

In the second step we eliminate the vacuum state $|0\rangle$ in favor of $|0_I\rangle$. The reason is that we know how the operator $\hat{\phi}_I$ acts on $|0_I\rangle$, while the state $|0\rangle$ is in fact completely unknown. First, we prove the formula

$$\lim_{t\to\pm\infty} \langle\psi|e^{i\hat{H}t}|\chi\rangle = \langle\psi|0\rangle\langle 0|\chi\rangle, \tag{7.21}$$

where $|\psi\rangle$ and $|\chi\rangle$ are vectors from the Hilbert space of the model.

We assume that we have the following completeness relation

$$|0\rangle\langle 0| + \int_{E_1}^{\infty} dE \sum_a |E, a\rangle\langle a, E| = I,$$

where E denotes the eigenvalues of the Hamiltonian, $E_1 > 0$ is the lowest energy eigenvalue above the vacuum energy $E_0 = 0$. The index a denotes a set of other quantum numbers (which are eigenvalues of observables commuting with the Hamiltonian). Let us insert this completeness relation on the l.h.s. of formula (7.21). We obtain

$$\lim_{t\to\pm\infty}\langle\psi|e^{i\hat{H}t}|\chi\rangle = \langle\psi|0\rangle\langle 0|\chi\rangle + \lim_{t\to\pm\infty}\int_{E_1}^{\infty} dE\ e^{iEt} f(E),$$

where

$$f(E) = \sum_a \langle\psi|E, a\rangle\langle a, E|\chi\rangle.$$

The completeness relation implies that

$$\int_{E_1}^{\infty} dE\ f(E) = \langle\psi|\chi\rangle - \langle\psi|0\rangle\langle 0|\chi\rangle < \infty,$$

hence the function $f(E)$ is integrable. Here we use the fact that the states $|\psi\rangle$ and $|\chi\rangle$ have finite scalar products with any vector belonging to the Hilbert space.

The integral

$$\int_{E_1}^{\infty} dE\ e^{iEt} f(E)$$

vanishes in the limits $t \to \pm\infty$ under certain assumptions about $f(E)$. The proof is based on theorems about the asymptotic behavior of Fourier transforms, but we shall not go into the mathematical details of it. Roughly, the integral vanishes because the integrand is the product of $f(E)$ with functions of E, namely $\cos(Et)$ and $\sin(Et)$, which oscillate very quickly in the limit $t \to \pm\infty$. In the end, the integral is a sum of positive and negative contributions which in that limit cancel each other out.

Formula (7.21) implies that

$$\lim_{T\to+\infty}\langle\psi|U_I(0, -T)|0_I\rangle = \lim_{T\to+\infty}\langle\psi|e^{-i\hat{H}T}|0_I\rangle = \langle\psi|0\rangle\langle 0|0_I\rangle.$$

Here we have used the fact that $\hat{H}_0|0_I\rangle = 0$. Let us choose[3]

$$\langle\psi| = \langle 0|T\left(\hat{\phi}(x_1)\ldots\hat{\phi}(x_n)\right).$$

We obtain

$$G^{(n)}(x_1, x_2, \ldots, x_n) = \lim_{T'\to+\infty}\frac{\langle 0|T\left(\hat{\phi}(x_1)\ldots\hat{\phi}(x_n)\right)U_I(0, -T')|0_I\rangle}{\langle 0|0_I\rangle}. \tag{7.22}$$

[3]Here we are simplifying things a little bit. In order to be sure that the state $|\psi\rangle$ belongs to the Hilbert space one should integrate $T\left(\hat{\phi}(x_1)\ldots\hat{\phi}(x_n)\right)$ with a test function $h(x_1, x_2, \ldots, x_n)$. We are assuming that such a 'technical' step is done implicitly.

Similarly,

$$\lim_{T"\to+\infty} \langle 0_I|U_I(T",0)|\chi\rangle = \lim_{T"\to+\infty} \langle 0_I|e^{-i\hat{H}T"}|\chi\rangle = \langle 0_I|0\rangle\langle 0|\chi\rangle.$$

Taking

$$|\chi\rangle = \frac{T\left(\hat{\phi}(x_1)\ldots\hat{\phi}(x_n)\right)U_I(0,-T')|0_I\rangle}{\langle 0|0_I\rangle},$$

we obtain the following formula

$$G^{(n)}(x_1,x_2,\ldots,x_n) = \lim_{T',T"\to+\infty} \frac{\langle 0_I|U_I(T",0)T\left(\hat{\phi}(x_1)\ldots\hat{\phi}(x_n)\right)U_I(0,-T')|0_I\rangle}{\langle 0_I|0\rangle\langle 0|0_I\rangle}. \tag{7.23}$$

We have seen in the derivation of formula (7.20) that

$$T\left(\hat{\phi}(x_1)\ldots\hat{\phi}(x_n)\right) = U_I^{-1}(x^0_{i_1},0)\hat{\phi}_I(x_{i_1})U_I(x^0_{i_1},x^0_{i_2})\hat{\phi}_I(x_{i_2})\ldots.\hat{\phi}_I(x_{i_n})U_I(x^0_{i_n},0).$$

Therefore, the numerator on the r.h.s. of formula (7.23) contains the time ordered product of operators which can be written as

$$T\left(\hat{\phi}_I(x_1)\ldots\hat{\phi}_I(x_n)U_I(\infty,-\infty)\right).$$

The denominator in formula (7.23) is equal to $\langle 0_I|U_I(\infty,-\infty)|0_I\rangle$, as follows from formulas (7.10) and (7.21). Thus, we have derived the following remarkable formula, first obtained by Gell-Mann and Low in 1954,

$$G^{(n)}(x_1,x_2,\ldots,x_n) = \frac{\langle 0_I|T\left(\hat{\phi}_I(x_1)\ldots\hat{\phi}_I(x_n)U_I(\infty,-\infty)\right)|0_I\rangle}{\langle 0_I|U_I(\infty,-\infty)|0_I\rangle}, \tag{7.24}$$

where

$$U_I(\infty,-\infty) = T\exp\left(-i\int_{-\infty}^{+\infty} dt\ \hat{V}_{Ig}(t)\right) = T\exp\left(-i\frac{\lambda_0}{4!}\int d^4x\ g(\vec{x}) :\hat{\phi}_I^4(t,\vec{x}):\right).$$

Formula (7.24) is the starting point for the construction of the perturbative expansion for the Green's functions.

The employed regularization involves only the space coordinates $\vec{x}$. It turns out that the integral over the infinite time interval also needs regularization. Therefore, specifically for the purpose of the perturbative approach we will use a more symmetric regularization. The point is, that in the context of the perturbative calculations of the

Green's functions, it suffices to regularize the expression for $U_I(\infty, -\infty)$ because we shall need only the Gell-Mann–Low formula. The new, symmetric regularization utilizes a real-valued test function $g(x_1, x_2, x_3, x_4)$ which is symmetric with respect to permutations of the four-dimensional variables x_j, $j = 1, 2, 3, 4$. Each x_j denotes a point in Minkowski space-time. As always with test functions, it is also assumed that this function is smooth and that it vanishes quickly (e.g., exponentially) when one or more coordinates $x_i^\mu \to \infty$. The symmetrically regularized $U_I(\infty, -\infty)$ has the form

$$U_I(\infty, -\infty) = T \exp\left(-i\hat{V}_{Ig}[\hat{\phi}_I]\right), \tag{7.25}$$

where now

$$\hat{V}_{Ig}[\hat{\phi}_I] = \frac{\lambda_0}{4!} \int \prod_{i=1}^{4} d^4x_i \; g(x_1, x_2, x_3, x_4)\hat{\phi}_I(x_1)\hat{\phi}_I(x_2)\hat{\phi}_I(x_3)\hat{\phi}_I(x_4). \tag{7.26}$$

With this regularization we do not need to introduce the normal ordering. Notice that the operator $\hat{V}_{Ig}$ is Hermitian, because the function g is real and symmetric with respect to permutations of the four-vectors x_i.

7.2 The Generating Functional for Green's Functions

The generating functional $Z[j]$ for the Green's functions is defined as follows:

$$Z[j] = \langle 0|T \exp\left(i \int d^4x \; j(x)\hat{\phi}(x)\right)|0\rangle, \tag{7.27}$$

where $j(x)$ is a real valued, smooth function, which vanishes quickly at infinity (again a test function), sometimes called the external source. Equivalently, we may also write

$$Z[j] = 1 + \sum_{n=1}^{\infty} \frac{i^n}{n!} \int d^4x_1 \ldots d^4x_n \; j(x_1) \ldots j(x_n) \; G^{(n)}(x_1, x_2, \ldots, x_n). \tag{7.28}$$

This last formula is obtained from the definition (7.27) by writing the exponential function as a series and using the definition (7.19) of $G^{(n)}$. Let us use the Gell-Mann–Low formula (7.24) in each term of the sum in (7.28) and reintroduce the exponential function. In this way we obtain yet another formula for $Z[j]$:

$$Z[j] = \frac{\langle 0_I|T\left(\exp\left(i \int d^4x \; j(x)\hat{\phi}_I(x)\right) U_I(\infty, -\infty)\right)|0_I\rangle}{\langle 0_I|U_I(\infty, -\infty)|0_I\rangle}. \tag{7.29}$$

It is clear from formula (7.28) that

$$G^{(n)}(x_1, \ldots, x_n) = (-i)^n \left. \frac{\delta^n Z[j]}{\delta j(x_1) \ldots \delta j(x_n)} \right|_{j=0}. \tag{7.30}$$

In the ϕ_4^4 model the regularized evolution operator U_I is given by formulas (7.25), (7.26). The numerator in formula (7.29), from now on denoted by $Z_I[j]$, can be written in the form

$$Z_I[j] = \exp\left(-i V_{Ig}\left[-i\frac{\delta}{\delta j}\right]\right) Z_0[j], \tag{7.31}$$

where

$$Z_0[j] = \langle 0_I | T \exp(i \int d^4x \; j(x)\hat{\phi}_I(x)) | 0_I \rangle, \tag{7.32}$$

and

$$V_{Ig}\left[-i\frac{\delta}{\delta j}\right] = \frac{\lambda_0}{4!} \int \prod_{i=1}^{4} d^4x_i \; g(x_1, x_2, x_3, x_4) \frac{\delta^4}{\delta j(x_1) \ldots \delta j(x_4)}. \tag{7.33}$$

Here we have used the fact that each derivative $\delta/\delta j(x)$ gives $i\,\hat{\phi}_I(x)$ inside the T-ordered product. The denominator in (7.29) is equal to $Z_I[j=0]$.

The functional $Z_0[j]$ can be explicitly calculated. The most helpful formula in this task is Wick's formula, which has the form

$$T \exp\left(i \int d^4x \; j(x)\hat{\phi}_I(x)\right) = \tag{7.34}$$
$$\exp\left(-\frac{1}{2} \int d^4x d^4x' \; j(x)\Delta_F(x - x')j(x')\right) \; : \exp\left(i \int d^4x \; j(x)\hat{\phi}_I(x)\right) :,$$

where

$$\Delta_F(x - x') = \frac{1}{(2\pi)^4} \int d^4p \; e^{-ip(x-x')} \frac{i}{p^2 - m_0^2 + i0_+}. \tag{7.35}$$

Because the expectation value of the normal ordered exponential function on the r.h.s. of formula (7.34) in the state $|0_I\rangle$ is equal to 1, we immediately obtain

$$Z_0[j] = \exp\left(-\frac{1}{2} \int d^4x d^4x' \; j(x)\Delta_F(x - x')j(x')\right). \tag{7.36}$$

The (generalized) function Δ_F is called the Feynman, or the causal, free propagator. By taking the derivatives $\delta^2/\delta j(x)\delta j(x')$ of both sides of the Wick formula and putting $j = 0$ we find that

$$\Delta_F(x - x') = \langle 0_I | T\left(\hat{\phi}_I(x)\hat{\phi}_I(x')\right) | 0_I \rangle. \tag{7.37}$$

It follows from this formula that Δ_F is the 2-point Green's function of the free scalar field.

In order to prove Wick's formula (7.34), we use the technique of the auxiliary differential equation. Let us introduce the operator

$$\hat{W}(t) = T \exp\left(i \int_{-\infty}^{t} dx'^0 \int d^3x' \; j(x')\hat{\phi}_I(x')\right),$$

where $x' = (x'^0, \vec{x}\,')$. The l.h.s. of the Wick formula is equal to $\hat{W}(+\infty)$. The operator $\hat{W}(t)$ obeys the following differential equation

$$-i\frac{d\hat{W}(t)}{dt} = \int d^3x \; j(t, \vec{x})\hat{\phi}_I(t, \vec{x})\hat{W}(t), \tag{7.38}$$

and the condition

$$\lim_{t \to -\infty} \hat{W}(t) = I.$$

Equation (7.38) can be written in the form

$$-i\frac{d\hat{W}(t)}{dt} = \left(\hat{A}(t) + \hat{A}^\dagger(t)\right)\hat{W}(t),$$

where

$$\hat{A}^\dagger(t) = \int d^3x \; j(t, \vec{x})\hat{\phi}_I^{(-)}(t, \vec{x}), \quad \hat{A}(t) = \int d^3x \; j(t, \vec{x})\hat{\phi}_I^{(+)}(t, \vec{x}).$$

Here

$$\hat{\phi}_I^{(+)}(t, \vec{x}) = \int \frac{d^3k}{\sqrt{2(2\pi)^3\omega(\vec{k})}} e^{-ikx} \, \hat{a}_I(\vec{k})$$

is the positive frequency part of the field $\hat{\phi}_I$. The negative frequency part is given by $\hat{\phi}_I^{(-)}(t, \vec{x}) = (\hat{\phi}_I^{(+)}(t, \vec{x}))^\dagger$. The operators $\hat{A}^\dagger(t)$ with different values of t commute with each other. This fact is crucial for checking that another operator $\hat{X}(t)$, defined by the formula

$$\hat{X}(t) = \hat{\alpha}(t)\hat{W}(t),$$

where

$$\hat{\alpha}(t) = \exp\left(-i\int_{-\infty}^{t} dt'\, \hat{A}^{\dagger}(t')\right),$$

obeys the following equation

$$-i\frac{d\hat{X}(t)}{dt} = \hat{\alpha}(t)\hat{A}(t)\hat{\alpha}^{-1}(t)\,\hat{X}(t). \tag{7.39}$$

The operators $\hat{A}^{\dagger}(t')$ and $\hat{A}(t)$ have a special property: their commutator is proportional to the identity operator,

$$[\hat{A}^{\dagger}(t'), \hat{A}(t'')] = i\int d^3x'd^3x''\; j(t', \vec{x}\,')j(t'', \vec{x}\,'')\;\Delta^{(-)}(t'-t'', \vec{x}\,'-\vec{x}\,'')I, \tag{7.40}$$

where

$$\Delta^{(-)}(x'-x'') = \frac{i}{2(2\pi)^3}\int \frac{d^3p}{\omega(\vec{p})} e^{i(x'-x'')p}.$$

The r.h.s. of (7.39) can be simplified with the help of the following formula, which is valid for operators $\hat{B}$ and $\hat{C}$

$$e^{\hat{C}}\hat{B}e^{-\hat{C}} = \hat{B} + [\hat{C}, \hat{B}] + \frac{1}{2!}[\hat{C}, [\hat{C}, \hat{B}]] + \frac{1}{3!}[\hat{C}, [\hat{C}, [\hat{C}, \hat{B}]]] + \dots. \tag{7.41}$$

In order to prove (7.41), let us consider $\hat{B}(s) = \exp(s\hat{C})\hat{B}\exp(-s\hat{C})$, where s is a real parameter. Of course, $\hat{B}(0) = \hat{B}$, and $\hat{B}(1)$ coincides with the l.h.s. of formula (7.41). It is obvious that

$$\frac{d\hat{B}(s)}{ds} = [\hat{C}, \hat{B}(s)], \quad \frac{d^2\hat{B}(s)}{ds^2} = [\hat{C}, [\hat{C}, \hat{B}(s)]], \quad \text{etc.} \tag{7.42}$$

On the other hand, the Taylor expansion of $\hat{B}(s)$ around $s = 0$ has the form

$$\hat{B}(s) = \hat{B}(0) + s\hat{B}'(0) + \frac{s^2}{2!}\hat{B}''(0) + \dots.$$

Formula (7.41) follows from this expansion when we replace the derivatives $\hat{B}^{(k)}(0)$ by the commutators in accordance with formulas (7.42), and put $s = 1$.

In our case $\hat{B} = \hat{A}(t)$ and $\hat{C} = -i\int_{-\infty}^{t} dt'\; \hat{A}^{\dagger}(t')$. Because of the special property mentioned above, only the first two terms on the r.h.s. of formula (7.41) do not vanish.

Therefore,

$$-i\frac{d\hat{X}(t)}{dt} = \left(\hat{A}(t) - i\int_{-\infty}^{t} dt' \, [\hat{A}^{\dagger}(t'), \hat{A}(t)]\right)\hat{X}(t), \tag{7.43}$$

where the commutator on the r.h.s. is given by formula (7.40). Equation (7.43) has the following solution

$$\hat{X}(t) = \exp\left(i\int_{-\infty}^{t} dt' \, \hat{A}(t')\right)\exp\left(\int_{-\infty}^{t} dt'' \int_{-\infty}^{t''} dt' \, [\hat{A}^{\dagger}(t'), \hat{A}(t'')]\right),$$

which obeys the condition $\lim_{t\to-\infty}\hat{X}(t) = I$. Now we can compute $\hat{W}(t)$ from the formula $\hat{W}(t) = \hat{\alpha}^{-1}(t)\hat{X}(t)$. In particular, in the limit $t \to +\infty$

$$\hat{W}(\infty) = \exp\left(i\int d^4x \; j(x)\hat{\phi}_I^{(-)}(x)\right)\exp\left(i\int d^4x' \; j(x')\hat{\phi}_I^{(+)}(x')\right)$$
$$\exp\left[i\int d^4x'd^4x'' \; \Theta(t''-t')\, j(x')j(x'')\Delta^{(-)}(x'-x'')\right], \tag{7.44}$$

where $x' = (t', \vec{x}\,')$, $x'' = (t'', \vec{x}\,'')$.

The product of the first two exponentials on the r.h.s. of this formula, is just the normal ordered exponent that is present on the r.h.s. of the Wick formula:

$$\exp\left(i\int d^4x \; j(x)\hat{\phi}_I^{(-)}(x)\right)\exp\left(i\int d^4x' \; j(x')\hat{\phi}_I^{(+)}(x')\right)$$
$$= :\exp\left(i\int d^4x \; j(x)\hat{\phi}_I(x)\right):. \tag{7.45}$$

Therefore, it remains to show that

$$-\frac{1}{2}\int d^4x' \int d^4x'' \; j(x')j(x'')\Delta_F(x'-x'') =$$
$$i\int d^4x'd^4x'' \; \Theta(t''-t')\, j(x')j(x'')\Delta^{(-)}(x'-x''). \tag{7.46}$$

Let us start from formula (7.35) for Δ_F in which $d^4p = d^3p\,dp^0$, and $x - x'$ is replaced by $x' - x''$. The integral over p^0 can be calculated with the help of contour integration in the plane of complex p^0. First, we replace $i0_+$ by $i\epsilon$, where $\epsilon > 0$—the original expression is recovered in the limit $\epsilon \to 0_+$ which we shall take at the very end of the calculation. The integrand has simple poles at $p^0_{\pm} = \pm\sqrt{m_0^2 + \vec{p}^{\,2} - i\epsilon}$. The real line (Im $p^0 = 0$) is completed to a closed contour by including the upper half-circle at infinity if $(x' - x'')^0 < 0$, or the lower half-circle if $(x' - x'')^0 > 0$. In each case only one pole contributes to the integral. We obtain

$$\Delta_F(x'-x'')$$
$$= \int \frac{d^3p}{2(2\pi)^3\omega(\vec{p})} \left[\Theta((x'-x'')^0)e^{-ip(x'-x'')} + \Theta((x''-x')^0)e^{ip(x'-x'')}\right], \tag{7.47}$$

where now in the exponentials $p^0 = \omega(\vec{p})$. In the second term we have changed the integration variable $\vec{p} \to -\vec{p}$. The r.h.s. of formula (7.47) can be rewritten with the $\Delta^{(-)}$ function introduced in Sect. 1.3, namely

$$\Delta_F(x'-x'') = -i\left[\Theta(t'-t'')\Delta^{(-)}(x''-x') + \Theta(t''-t')\Delta^{(-)}(x'-x'')\right] \tag{7.48}$$

($t' = x'^0,\ t'' = x''^0$). Formula (7.46) is obtained by multiplying both sides of formula (7.48) by $j(x')\ j(x'')$, integrating over d^4x', d^4x'', and changing the integration variables, $x' \to x'',\ x'' \to x'$, in the first term on the r.h.s. This completes the derivation of Wick formula (7.34).

7.3 Feynman Diagrams in Momentum Space

We shall consider the Fourier transform of the n-point Green's function,

$$\tilde{G}^{(n)}(k_1, k_2, \ldots, k_n) =$$
$$(2\pi)^{-2n} \int d^4x_1 \ldots d^4x_n\ e^{i(k_1x_1+\ldots+k_nx_n)}\ G^{(n)}(x_1, x_2, \ldots, x_n). \tag{7.49}$$

Comparison with formula (7.30) for $G^{(n)}$ suggests that it would be useful to compute the Fourier transform of the functional derivative $\delta/\delta j(x)$. This can be done as follows. The Fourier transform of the external source $j(x)$ is defined by the formula

$$\tilde{j}(q) = \frac{1}{(2\pi)^2} \int d^4y\ e^{-iqy}\ j(y)$$

(note the minus sign in the exponent). Therefore,

$$\frac{\delta\tilde{j}(q)}{\delta j(x)} = \frac{e^{-iqx}}{(2\pi)^2}.$$

The inverse Fourier transform of the external source has the form

$$j(x) = \frac{1}{(2\pi)^2} \int d^4k\ e^{ikx}\ \tilde{j}(k).$$

A functional $F[j]$ with $j(x)$ expressed by $\tilde{j}(k)$ becomes a functional $\tilde{F}[\tilde{j}]$:

$$F[j] = \tilde{F}[\tilde{j}].$$

Therefore,

$$\frac{1}{(2\pi)^2}\int d^4x\, e^{ikx}\frac{\delta F[j]}{\delta j(x)} = \frac{1}{(2\pi)^2}\int d^4x\, e^{ikx}\int d^4q\, \frac{\delta \tilde{j}(q)}{\delta j(x)}\frac{\delta \tilde{F}[\tilde{j}]}{\delta \tilde{j}(q)}$$

$$= \frac{1}{(2\pi)^4}\int d^4x\, e^{ikx}\int d^4q\, e^{-iqx}\frac{\delta \tilde{F}[\tilde{j}]}{\delta \tilde{j}(q)} = \int d^4q\, \delta(q-k)\frac{\delta \tilde{F}[\tilde{j}]}{\delta \tilde{j}(q)} = \frac{\delta \tilde{F}[\tilde{j}]}{\delta \tilde{j}(k)}.$$

The inverse Fourier transform gives

$$\frac{\delta F[j]}{\delta j(x)} = \frac{1}{(2\pi)^2}\int d^4k\, e^{-ikx}\frac{\delta \tilde{F}[\tilde{j}]}{\delta \tilde{j}(k)}.$$

Perturbative computations of the Green's functions could be based on formulas (7.29), (7.30) in which

$$Z[j] = \frac{Z_I[j]}{Z_I[0]}, \tag{7.50}$$

where $Z_I[j]$ is given by formula (7.31) and $Z_I[0] = Z_I[j=0]$. However, it turns out that it is more convenient to use another, equivalent, formula. First, we pass to the Fourier transforms. Then,

$$\tilde{G}^{(n)}(k_1, k_2, \ldots, k_n) = (-i)^n \left.\frac{1}{\tilde{Z}_I[0]}\frac{\delta^n \tilde{Z}_I[\tilde{j}]}{\delta \tilde{j}(k_1)\ldots\delta \tilde{j}(k_n)}\right|_{\tilde{j}=0}. \tag{7.51}$$

The functional $\tilde{Z}_I[\tilde{j}]$ is given by the formula $Z_I[j] = \tilde{Z}_I[\tilde{j}]$. Formula (7.31), written in terms of the Fourier transforms has the form

$$\tilde{Z}_I[\tilde{j}] = \exp\left(-i\tilde{V}_{Ig}\left[-i\frac{\delta}{\delta \tilde{j}}\right]\right)\tilde{Z}_0[\tilde{j}], \tag{7.52}$$

where

$$\tilde{V}_{Ig}\left[-i\frac{\delta}{\delta \tilde{j}}\right] = V_{Ig}\left[-i\frac{\delta}{\delta j}\right]$$

$$= \frac{\lambda_0}{4!}\int d^4q_1\ldots d^4q_4\, \tilde{g}(q_1, q_2, q_3, q_4)\frac{\delta^4}{\delta \tilde{j}(q_1)\ldots\delta \tilde{j}(q_4)}, \tag{7.53}$$

and

$$\tilde{Z}_0[\tilde{j}] = Z_0[j] = \exp\left[-\frac{i}{2}\int d^4k_1 d^4k_2\, \delta(k_1+k_2)\frac{\tilde{j}(k_1)\tilde{j}(k_2)}{k_1^2 - m_0^2 + i0_+}\right]. \tag{7.54}$$

Formula (7.53) contains the Fourier transform of the regularizing function g,

$$\tilde{g}(q_1, q_2, q_3, q_4) = \frac{1}{(2\pi)^8}\int d^4x_1 \dots d^4x_4\, e^{-iq_1x_1 \dots -iq_4x_4} g(x_1, x_2, x_3, x_4).$$

Note that $\tilde{g}(q_1, \dots, q_4)$ is symmetric with respect to permutations of $q_1, \dots, q_4$. The unregularized interaction

$$\hat{V}_I = \frac{\lambda_0}{4!}\int d^4x\, \hat{\phi}_I^4(x)$$

is obtained when

$$g(x_1, x_2, x_3, x_4) = \int d^4x\, \delta(x_1 - x)\delta(x_2 - x)\delta(x_3 - x)\delta(x_4 - x),$$

or equivalently

$$\tilde{g}(q_1, q_2, q_3, q_4) = \frac{1}{(2\pi)^4}\delta(q_1 + q_2 + q_3 + q_4). \tag{7.55}$$

Of course, such a g is not allowed here because the integral of a product of δ's is not a test function. The return to the unregularized interaction will be possible when we modify our perturbative model in a special way. The procedure for this is called renormalization. It is described in the next chapter.

In the next step toward the perturbative expansion, we replace the variational derivatives $-i\delta/\delta\tilde{j}$ by $\tilde{\beta}$, and $\tilde{j}$ by $-i\delta/\delta\tilde{\beta}$, where $\tilde{\beta}(q)$ is a new test function [10]. This is done with the help of the following trick

$$\left.\frac{\delta^n \tilde{Z}_I[\tilde{j}]}{\delta\tilde{j}(k_1)\dots\delta\tilde{j}(k_n)}\right|_{\tilde{j}=0} = \left.\frac{\delta^n \tilde{Z}_I[\tilde{j}]}{\delta\tilde{j}(k_1)\dots\delta\tilde{j}(k_n)}\right|_{\tilde{j}=0} \left. e^{i\int d^4q \tilde{\beta}(q)\tilde{j}(q)}\right|_{\tilde{\beta}=0}$$

$$= \left.\left(\tilde{Z}_0[-i\frac{\delta}{\delta\tilde{\beta}}]\frac{\delta^n}{\delta\tilde{j}(k_1)\dots\delta\tilde{j}(k_n)}\exp\left(-i\tilde{V}_{Ig}[-i\frac{\delta}{\delta\tilde{j}}]\right) e^{i\int d^4q\tilde{\beta}(q)\tilde{j}(q)}\right)\right|_{\tilde{j}=0=\tilde{\beta}}$$

$$= i^n\left.\left(\tilde{Z}_0[-i\frac{\delta}{\delta\tilde{\beta}}]\left(\tilde{\beta}(k_1)\dots\tilde{\beta}(k_n)\exp(-i\tilde{V}_{Ig}[\tilde{\beta}])\right)\right)\right|_{\tilde{\beta}=0},$$

where

$$\tilde{V}_{Ig}[\tilde{\beta}] = \frac{\lambda_0}{4!}\int d^4q_1 \dots d^4q_4\, \tilde{g}(q_1, q_2, q_3, q_4)\tilde{\beta}(q_1)\tilde{\beta}(q_2)\tilde{\beta}(q_3)\tilde{\beta}(q_4), \tag{7.56}$$

and

$$\tilde{Z}_0\left[-i\frac{\delta}{\delta\tilde{\beta}}\right] = \exp\left(\frac{1}{2}\int d^4p_1 d^4p_2\ \delta(p_1+p_2)\frac{\delta}{\delta\tilde{\beta}(p_1)}\Delta_F(p_1)\frac{\delta}{\delta\tilde{\beta}(p_2)}\right)$$
$$= \exp\left(\frac{1}{2}\int d^4p\ \frac{\delta}{\delta\tilde{\beta}(p)}\Delta_F(p)\frac{\delta}{\delta\tilde{\beta}(-p)}\right), \tag{7.57}$$

with

$$\Delta_F(p) = \frac{i}{p^2 - m_0^2 + i0_+}. \tag{7.58}$$

$\Delta_F(p)$ is called the free or Feynman propagator of the real scalar field in four-momentum space. Thus, finally

$$\tilde{G}^{(n)}(k_1, k_2, \dots, k_n) = \frac{\tilde{Z}_I^{(n)}}{\tilde{Z}_I^{(0)}}, \tag{7.59}$$

where

$$\tilde{Z}_I^{(n)} = \left(\tilde{Z}_0\left[-i\frac{\delta}{\delta\tilde{\beta}}\right]\left(\tilde{\beta}(k_1)\dots\tilde{\beta}(k_n)\exp(-i\tilde{V}_{Ig}[\tilde{\beta}])\right)\right)\Big|_{\tilde{\beta}=0}. \tag{7.60}$$

In the case $n = 0$ the factors $\tilde{\beta}(k_i)$ are absent.

Note that formulas (7.59), (7.60) imply that

$$\tilde{G}^{(n)} = 0 \tag{7.61}$$

for any odd n.

The N-th order perturbative approximation for $\tilde{G}^{(n)}$ with even n is obtained by truncating the series

$$\exp(-i\tilde{V}_{Ig}[\tilde{\beta}]) = \sum_{l=0}^{\infty}\frac{(-i)^l}{l!}\tilde{V}_{Ig}^l[\tilde{\beta}] \tag{7.62}$$

to the first $N+1$ terms. It is clear that the perturbative computation of $\tilde{G}^{(n)}$ involves the following three steps. First, evaluation of the indicated functional derivatives. Next, computation of the integrals over the four-momenta. Finally, removal of the regularization. The latter step will be discussed in the next chapter. Now we shall show how one can facilitate the differentiation using a graphical notation, the famous Feynman diagrams.

We begin by defining a graphical representation of the terms that are present in formula (7.60) for $\tilde{Z}_I^{(n)}$. The factors $\tilde{\beta}(k_1), \dots, \tilde{\beta}(k_n)$ are represented by small crosses

$$\times \quad \times \quad \ldots \quad \times$$
$$k_1 \quad k_2 \qquad k_n \tag{7.63}$$

They are called the external vertices, and k_i the external four-momenta. The functional $\tilde{V}_{Ig}[\tilde{\beta}]$ is called the internal vertex, and it is depicted as

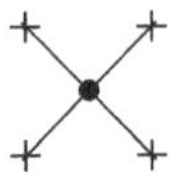

$$\text{(7.64)}$$

The small crosses at the ends of the lines denote the factors $\tilde{\beta}(q)$. The lines emanating from the vertex dot are sometimes called 'legs'.

The exponent in $\tilde{Z}_0[-i\frac{\delta}{\delta\tilde{\beta}}]$, formula (7.57), is depicted as a dumb-bell

$$\bigcirc\!\!-\!\!\!-\!\!\bigcirc \;=\; \frac{1}{2}\int d^4p\, \frac{\delta}{\delta\tilde{\beta}(p)}\Delta_F(p)\frac{\delta}{\delta\tilde{\beta}(-p)}. \tag{7.65}$$

The circles denote the functional derivatives $\delta/\delta\tilde{\beta}$. Thus, formula (7.57) can be presented as

$$\tilde{Z}_0\left[-i\frac{\delta}{\delta\tilde{\beta}}\right] = \sum_{k=0}^{\infty}\frac{1}{k!}(\bigcirc\!\!-\!\!\!-\!\!\bigcirc)^k. \tag{7.66}$$

Non vanishing contributions to $\tilde{Z}_I^{(n)}$ appear only if the number of derivatives exactly matches the number of factors $\tilde{\beta}$, which is equal to $n+4l$ in the l-th order. The l-th order means that we consider contributions which come from the $(l+1)$-th term in the series (7.62) (the term with $l=0$ is the first term). Therefore, $k = 2l + n/2$ dumb-bells are needed. Now let us consider the differentiation in more detail. According to the Leibniz rule each derivative $\delta/\delta\tilde{\beta}$ acts on each factor $\tilde{\beta}$, and

$$\frac{\delta\tilde{\beta}(q)}{\delta\tilde{\beta}(p)} = \delta(q-p).$$

Pictorially, the differentiation removes the circles from the dumb-bells and the crosses from the external or internal vertices. The lines from the dumb-bells either connect two vertices or form a loop at one internal vertex, see, e.g., Figs. 7.1 and 7.2. The remaining expressions $\int d^4p\, \Delta_F(p)$ from the dumb-bells (7.65) we associate with the lines.

The factor $1/2$ can actually be omitted for the following reason. Let us consider the two derivatives from one dumb-bell. Acting on a certain pair of $\tilde{\beta}$'s, say the product $\tilde{\beta}(q)\tilde{\beta}(k)$, they give

$$\frac{1}{2}\int d^4p\; \delta(q-p)\delta(k+p)\Delta_F(p) + \frac{1}{2}\int d^4p\; \delta(q+p)\delta(k-p)\Delta_F(p)$$
$$= \int d^4p\; \delta(q-p)\delta(k+p)\Delta_F(p),$$

because in the second term we may change the integration variable $p \to -p$, and $\Delta_F(p) = \Delta_F(-p)$. Graphically,

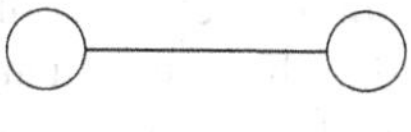

$$\text{O——O} \quad (\times q \quad \times k) \quad = \quad q \bullet\!\!-\!\!\!-\!\!\!-\!\!\bullet\, k,$$

where

$$q \bullet\!\!-\!\!\!-\!\!\!-\!\!\bullet\, k \quad = \quad \int d^4p\; \delta(q+p)\delta(k-p)\Delta_F(p). \tag{7.67}$$

The line is called 'external' if it is attached to at least one external vertex, or 'internal' if both its ends are attached to one or two internal vertices. If $\tilde{\beta}(q)$ (or $\tilde{\beta}(k)$) comes from one of the internal vertices, the corresponding Dirac delta from (7.67) 'eats' the integral over q (or k) present in $\tilde{V}_{Ig}$, see formula (7.56). In consequence, all integrals over $q_1, \ldots, q_4$ from $\tilde{V}_{Ig}$ disappear, and therefore each internal vertex only contributes a factor

$$\frac{-i\lambda_0}{4!}\, \tilde{g}(p_1, p_2, p_3, p_4),$$

where p_i denote the four-momenta from the lines attached to the internal vertex with their signs chosen in accordance with the following rule: the four-momentum p from a line enters the two functions $\tilde{g}$ in the two vertices adjacent to that line with opposite signs: $+p$ in one vertex and $-p$ in the other. Because of invariance of the dumb-bell with respect to the change $p \to -p$, it does not matter in which of the two vertices we take $+p$.

If $\tilde{\beta}$ comes from one of the external vertices, the Dirac delta produced by its differentiation is utilized in order to remove the integral d^4p present in (7.67). Thus, if the line (7.67) is attached to one or two external vertices there is no integral coming from it. In the case of two external vertices, the contribution has the form

$$k_1 \bullet\!\!-\!\!\!-\!\!\bullet\, k_2 \quad = \quad \delta(k_1 + k_2)\Delta_F(k_1). \tag{7.68}$$

Also, note that we have in total $(2l + n/2)!$ contributions obtained by permuting the dumb-bells—the Leibniz rule yields all these terms. Such contributions are equal

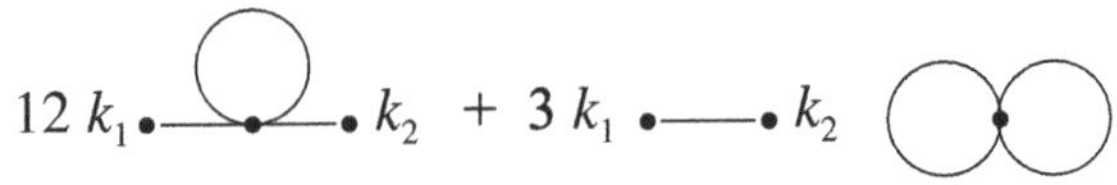

$12\, k_1 \quad k_2 \; + \; 3\, k_1 \quad k_2$

Fig. 7.1 The first order contributions to $\tilde{Z}_I^{(2)}$. The numerical coefficients in front of graphs (12 and 3 in this example) are called the combinatorial factors

to each other, therefore it is sufficient to take one of them and multiply it by the factor $(2l + n/2)!$. This factor exactly cancels the factor $1/(2l + n/2)!$ which appears because we pick only the $k = (2l + n/2)$-th power of the dumb-bell. All other terms in (7.66) give vanishing contributions, either because they have too many or too few derivatives. Therefore, we can forget about the factor $1/(2l + n/2)!$ and about permuting the dumb-bells.

The factor $1/l!$ present in formula (7.62) has to be included as a prefactor in front of each perturbative contribution in the l-th order.

Let us have a look at the perturbative contributions to $\tilde{Z}_I^{(2)}$. All of them have two external vertices (7.63). In the zeroth order the only non vanishing contribution comes from the $k = 1$ term in (7.66), and it is given by formula (7.68). In the first order ($l = 1$) we have one internal vertex (7.64) and three dumb-bells. The resulting contribution has the form presented in Fig. 7.1 (Exercise 7.3).

The closed lines present in Fig. 7.1 appear when a single dumb-bell 'eats' two crosses from one internal vertex. The second term in Fig. 7.1 is the product of terms corresponding to the two subdiagrams: the one given by formula (7.67), and the other given by the two circles. This latter one has the form

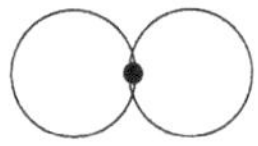

$$= \frac{-i\lambda_0}{4!} \int d^4p\, d^4q\; \tilde{g}(p, -p, q, -q)\Delta_F(p)\Delta_F(q). \qquad (7.69)$$

Note that the expression on the r.h.s. would become meaningless if $\tilde{g}$ was replaced with the Dirac delta (7.35). Apart from the factor $\delta(0)$, the square of the divergent integral $\int d^4p\Delta_F(p)$, would be present. This integral is an example of the so called ultraviolet divergences (UV), to be discussed in the next chapter.

In the second order, we have two internal vertices ($l = 2$), and five dumb-bells—we have to compute the tenth order functional derivative of the product of ten $\tilde{\beta}$'s. The corresponding Feynman diagrams have the form presented in Fig. 7.2.

In order to obtain the second order contribution to $\tilde{Z}_I^{(2)}$, this result has to be multiplied by $1/2!$.

It is clear that the number of diagrams rapidly increases with the order l. A certain reduction of this number occurs when we use the normal ordered interaction $: \tilde{V}_{Ig} :$ instead of $\tilde{V}_{Ig}$. Let us compute the derivatives $(-i)^4\delta^4/\delta j(x_1)\ldots\delta j(x_4)$ of both sides of Wick's formula (7.34) and put $j = 0$ afterwards. We obtain

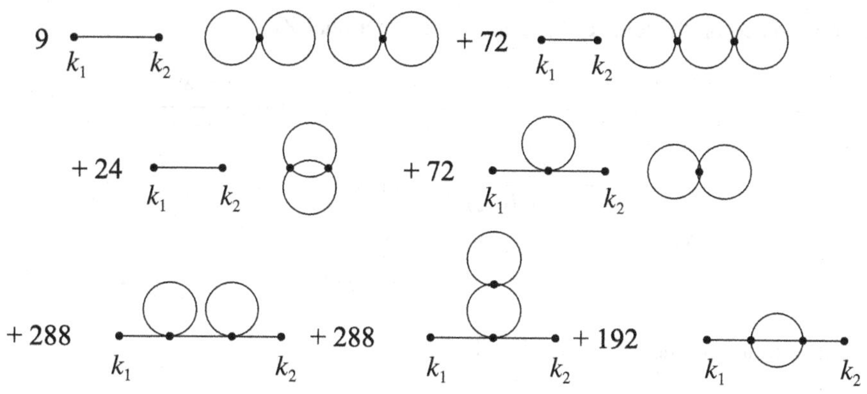

9 k_1 k_2 + 72 k_1 k_2

+ 24 k_1 k_2 + 72 k_1 k_2

+ 288 k_1 k_2 + 288 k_1 k_2 + 192 k_1 k_2

Fig. 7.2 The graphs giving the second order contributions to $\tilde{Z}_I^{(2)}$. The factor $1/2!$ is not included

$$
\begin{aligned}
T\left(\hat{\phi}_I(x_1)\ldots\hat{\phi}_I(x_4)\right) =&: \hat{\phi}_I(x_1)\ldots\hat{\phi}_I(x_4) : + \Delta_F(x_1 - x_2) : \hat{\phi}_I(x_3)\hat{\phi}_I(x_4) : \\
&+ \Delta_F(x_1 - x_3) : \hat{\phi}_I(x_2)\hat{\phi}_I(x_4) : + \ldots + \Delta_F(x_3 - x_4) : \hat{\phi}_I(x_1)\hat{\phi}_I(x_2) : \\
&+ [\Delta_F(x_1 - x_2)\Delta_F(x_3 - x_4) + \Delta_F(x_1 - x_3)\Delta_F(x_2 - x_4) \\
&+ \Delta_F(x_1 - x_4)\Delta_F(x_2 - x_3)]\, I. \qquad (7.70)
\end{aligned}
$$

Analogously, for $i \neq j$

$$
T\left(\hat{\phi}_I(x_i)\hat{\phi}_I(x_j)\right) =: \hat{\phi}_I(x_i)\hat{\phi}_I(x_j) : + \Delta_F(x_i - x_j) I. \qquad (7.71)
$$

It follows from these formulas that

$$
\begin{aligned}
: \hat{\phi}_I(x_1)\ldots\hat{\phi}_I(x_4) :=& T\left(\hat{\phi}_I(x_1)\ldots\hat{\phi}_I(x_4)\right) - \Delta_F(x_1 - x_2) T\left(\hat{\phi}_I(x_3)\hat{\phi}_I(x_4)\right) \\
&- \Delta_F(x_1 - x_3) T\left(\hat{\phi}_I(x_2)\hat{\phi}_I(x_4)\right) - \ldots - \Delta_F(x_3 - x_4) T\left(\hat{\phi}_I(x_1)\hat{\phi}_I(x_2)\right) \\
&+ [\Delta_F(x_1 - x_2)\Delta_F(x_3 - x_4) + \Delta_F(x_1 - x_3)\Delta_F(x_2 - x_4) \\
&+ \Delta_F(x_1 - x_4)\Delta_F(x_2 - x_3)]\, I. \qquad (7.72)
\end{aligned}
$$

Therefore, the modification

$$
T\left(\hat{\phi}_I(x_1)\ldots\hat{\phi}_I(x_4)\right) \to : \hat{\phi}_I(x_1)\ldots\hat{\phi}_I(x_4) :
$$

is represented on the level of the generating functional $Z[j]$ by

$$\frac{\delta^4}{\delta j(x_1)\dots\delta j(x_4)} \to \frac{\delta^4}{\delta j(x_1)\dots\delta j(x_4)} + \Delta_F(x_1-x_2)\frac{\delta^2}{\delta j(x_3)\delta j(x_4)}$$
$$+ \Delta_F(x_1-x_3)\frac{\delta^2}{\delta j(x_2)\delta j(x_4)} + \dots + \Delta_F(x_3-x_4)\frac{\delta^2}{\delta j(x_1)\delta j(x_2)}$$
$$+ \Delta_F(x_1-x_2)\Delta_F(x_3-x_4) + \Delta_F(x_1-x_3)\Delta_F(x_2-x_4)$$
$$+ \Delta_F(x_1-x_4)\Delta_F(x_2-x_3). \quad (7.73)$$

After introducing the $\tilde{\beta}$'s we finally obtain

$$: \tilde{V}_{Ig}[\tilde{\beta}] := \frac{\lambda_0}{4!}\Bigg[\int d^4q_1 \dots d^4q_4\, \tilde{g}(q_1,\dots,q_4)\tilde{\beta}(q_1)\dots\tilde{\beta}(q_4)$$
$$-6\int d^4p\int d^4q_1 d^4q_2\, \Delta_F(p)\tilde{g}(p,-p,q_1,q_2)\tilde{\beta}(q_1)\tilde{\beta}(q_2)$$
$$+3\int d^4p d^4q\, \Delta_F(p)\Delta_F(q)\tilde{g}(p,-p,q,-q)\Bigg]. \quad (7.74)$$

The third term on the r.h.s. of this formula does not depend on $\tilde{\beta}$. Therefore, it cancels out in the quotient $\tilde{Z}_I^{(n)}/\tilde{Z}_0^{(0)}$, and we may omit it. The change to the normal ordered interaction $\tilde{V}_{Ig} \to: \tilde{V}_{Ig}$: is graphically presented in Fig. 7.3.

Thus, in the case of the normal ordered interaction we have two internal vertices, namely

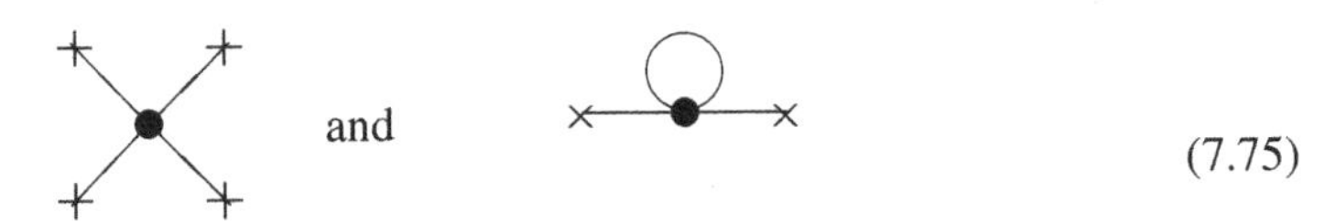

(7.75)

which appear in the combination shown in Fig. 7.3. Due to the presence of the 2-leg internal vertex, we now have new Feynman diagrams, in addition to the former ones with the 4-leg internal vertex. It turns out that the new diagrams exactly cancel all diagrams which have one or more internal lines starting and ending at the same internal vertex. To summarize, in the case of the normal ordered interaction, again only the vertices (7.63), (7.64) are used to construct the diagrams, but there is the

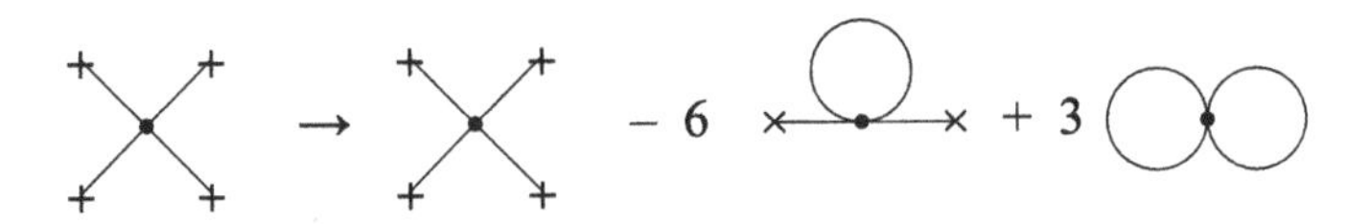

Fig. 7.3 The change to the normal ordered interaction

additional rule that each internal line connects two different 4-leg vertices. It is clear that the net number of Feynman diagrams which have to be taken into account is significantly smaller in the case of the normal ordered interaction. From now on we use the normal ordered interaction unless explicitly stated otherwise.

Another simplification is due to the denominator $\tilde{Z}_I^{(0)}$ in formula (7.59): it turns out that it cancels all the so called 'vacuum bubbles' in the perturbative expansion of the numerator $\tilde{Z}_I^{(n)}$. By vacuum bubbles we mean (sub)diagrams which do not contain any external vertices. Examples can be seen in the first two lines of Fig. 7.2. The first graph in the second line of Fig. 7.2 contains a vacuum bubble which is present also when we take the normal ordering of the interaction. Let us consider a graph Γ with n external and l internal vertices which does not contain any vacuum bubbles among its subdiagrams. Such a graph is a contribution of the l-th order to $\tilde{Z}_I^{(n)}$, that is, a contribution to

$$\left(\frac{(-i)^l}{l!(2l+\frac{n}{2})!}\left(\bigcirc\!\!-\!\!\bigcirc\right)^{2l+\frac{n}{2}}\left[\tilde{\beta}(k_1)\ldots\tilde{\beta}(k_n)\left(:\tilde{V}_{Ig}[\tilde{\beta}]:\right)^l\right]\right)\Big|_{\tilde{\beta}=0}.$$

In the orders $l+m$, where $m>0$, the graph Γ will appear as a subgraph of the larger graphs. Let us consider only such graphs in which Γ is multiplied by vacuum bubbles. These larger graphs are contributions to

$$\left(\frac{(-i)^{l+m}\left(\bigcirc\!\!-\!\!\bigcirc\right)^{2l+2m+\frac{n}{2}}}{(l+m)!(2l+2m+\frac{n}{2})!}\left[\tilde{\beta}(k_1)\ldots\tilde{\beta}(k_n)\left(:\tilde{V}_{Ig}[\tilde{\beta}]:\right)^{l+m}\right]\right)\Big|_{\tilde{\beta}=0}. \tag{7.76}$$

In order to form the subgraph Γ, we have to pick $2l+n/2$ dumb-bells from the full set, which contains $2l+2m+n/2$ of them. This gives $(2l+2m+n/2)!/(2m)!(2l+n/2)!$ possibilities. Similarly, we have to choose l internal vertices for the subgraph Γ out of $l+m$ vertices—there are $(l+m)!/l!m!$ possibilities. Therefore, that part of the expression (7.76) which contains Γ as a subgraph is equal to

$$\Gamma\ \frac{1}{(2m)!}\left(\left(\bigcirc\!\!-\!\!\bigcirc\right)^{2m}\frac{(-i)^m}{m!}(:\tilde{V}_{Ig}:)^m\right)\Big|_{\tilde{\beta}=0}. \tag{7.77}$$

Next, notice that

$$\left(\frac{1}{(2m)!}\left(\bigcirc\!\!-\!\!\bigcirc\right)^{2m}\frac{(-i)^m}{m!}(:\tilde{V}_{Ig}:)^m\right)\Big|_{\tilde{\beta}=0} = \left(\tilde{Z}_0[-i\frac{\delta}{\delta\tilde{\beta}}]\,\frac{(-i)^m}{m!}\,(:\tilde{V}_{Ig}:)^m\right)\Big|_{\tilde{\beta}=0}$$

because the powers of the dumb-bell other than $2m$ give vanishing contributions. Thus, the sum of all contributions of order $l + m$ to $\tilde{Z}_I^{(n)}$ such that they contain the subgraph Γ multiplied by vacuum bubbles is equal to

$$\Gamma \left(\tilde{Z}_0[-i\frac{\delta}{\delta\tilde{\beta}}] \frac{(-i)^m}{m!} (: \tilde{V}_{Ig} :)^m \right)\Big|_{\tilde{\beta}=0} . \tag{7.78}$$

Finally, we sum such contributions from all orders $l + m$, where l is fixed and $m = 1, 2, \ldots$. We also add the initial graph Γ without any accompanying vacuum bubbles by including the $m = 0$ term in the sum. The result is equal to

$$\Gamma \, \tilde{Z}_I^{(0)}. \tag{7.79}$$

The factor $\tilde{Z}_I^{(0)}$ cancels with the denominator in formula (7.59). Thus, we have proven that when computing the perturbative contributions to $\tilde{G}^{(n)}$ from formula (7.59), we may abandon the denominator $\tilde{Z}_I^{(0)}$, as well as the vacuum bubbles in the expansion of the numerator $\tilde{Z}_I^{(n)}$.

Perturbative contributions of the l-th order to the four-point Green's function $\tilde{G}^{(4)}(k_1, k_2, k_3, k_4)$ involve four external vertices (7.63), l internal vertices (7.64), and $2l + 2$ dumb-bells. In the zeroth order we have the Feynman diagrams presented in Fig. 7.4. Analytically, this contribution has the form (Exercise 7.4)

$$\delta(k_1 + k_2)\delta(k_3 + k_4)\Delta_F(k_1)\Delta_F(k_3) + \delta(k_1 + k_3)\delta(k_2 + k_4)\Delta_F(k_1)\Delta_F(k_2) + \delta(k_1 + k_4)\delta(k_2 + k_3)\Delta_F(k_1)\Delta_F(k_2). \tag{7.80}$$

In the first order there is just one diagram, see Fig. 7.5. The corresponding contribution to $\tilde{G}^{(4)}$ is equal to

$$- i\lambda_0\tilde{g}(k_1, k_2, k_3, k_4) \prod_{j=1}^{4} \Delta_F(k_j). \tag{7.81}$$

The diagrammatic representation of the second order contribution to $\tilde{G}^{(4)}$ is shown in Fig. 7.6.

All of the second order contributions presented in Figs. 7.2, 7.6 would contain divergent integrals if $\tilde{g}$ were replaced with the Dirac delta (7.55).

Fig. 7.4 The zeroth order contributions to $\tilde{G}^{(4)}$

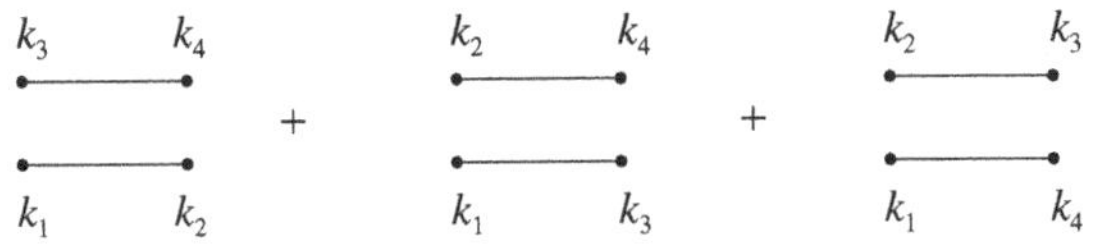

k_3 k_4 / k_1 k_2 + k_2 k_4 / k_1 k_3 + k_2 k_3 / k_1 k_4

Fig. 7.5 The first order contributions to $\tilde{G}^{(4)}$. The second graph is eliminated by the normal ordering prescription

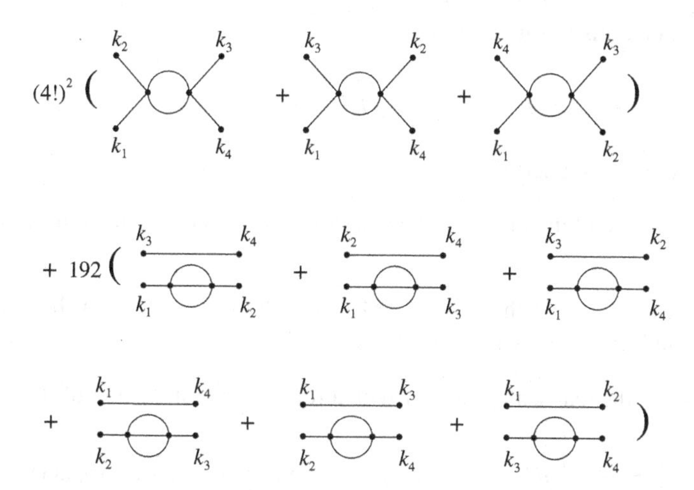

Fig. 7.6 The graphs giving the second order contributions to $\tilde{G}^{(4)}$. The factor 1/2! is not included

Exercises

7.1 Check that the operator $\hat{V}_{Sg} = \frac{\lambda_0}{4!} \int d^3x \; g(\vec{x}) \; : \hat{\phi}_I^4(t = 0, \vec{x}) :$ is Hermitian. Here $g(\vec{x})$ is a real-valued test function.
Hint: Write $\hat{\phi}_I(x)$ in the form $\hat{\phi}_I(x) = \hat{\phi}_I^{(+)}(x) + \hat{\phi}_I^{(-)}(x)$, where $\hat{\phi}_I^{(+)}(x)$ contains the $\hat{a}_I$ part of $\hat{\phi}$.

7.2 Find a general formula for the function $f(E)$ introduced in Sect. 7.1 in the case of the free, real scalar field discussed in Chap. 6. Here

$$|\psi\rangle = \psi_0|0\rangle + \textstyle\sum_{n=1}^{\infty} \int d^3k_1 \dots d^3k_n \; \psi_n(\vec{k}_1, \dots, \vec{k}_n) \; |\vec{k}_1, \dots, \vec{k}_n\rangle,$$
$$|\chi\rangle = \chi_0|0\rangle + \textstyle\sum_{n=1}^{\infty} \int d^3k_1 \dots d^3k_n \; \chi_n(\vec{k}_1, \dots, \vec{k}_n) \; |\vec{k}_1, \dots, \vec{k}_n\rangle,$$

As an example, compute $f(E)$ in the case

$$\psi_n = \chi_n = \delta_{n1}\psi_1(\vec{k}_1),$$

where $\psi_1(\vec{k}) = \exp(-a\vec{k}^2)$, $a > 0$ is a constant.

Answer:

$$f(E) = \sum_{n=1}^{\infty} \int d^3k_1 \dots d^3k_n \; \delta(E - \sum_{i=1}^{n} \omega(\vec{k}_i)) \; \psi_n^*(\vec{k}_1, \dots, \vec{k}_n)\chi_n(\vec{k}_1, \dots, \vec{k}_n),$$

where $\omega(\vec{k}) = \sqrt{m_0^2 + \vec{k}^{\,2}}$.
In the example, this formula gives

$$f(E) = 4\pi \; \Theta(E - m_0) \; E \sqrt{E^2 - m_0^2} \; \exp\left(-2a(E^2 - m_0^2)\right),$$

where Θ is the step function.

7.3 Check the combinatorial coefficients shown in front of the diagrams in Figs. 7.1 and 7.2.

7.4 Check that the zeroth and first order contributions to $\tilde{G}^{(4)}$ in the considered model are indeed given by formulas (7.80), (7.81).

7.5 Consider the real scalar field $\hat{\phi}_I$ with a regularized interaction of the form

$$\hat{V}_{Ig}^{(3)} = \frac{\lambda_0}{3!} \int d^4x_1 d^4x_2 d^4x_3 \; g(x_1, x_2, x_3) \; : \hat{\phi}_I(x_1)\hat{\phi}_I(x_2)\hat{\phi}_I(x_3) :,$$

where $g(x_1, x_2, x_3)$ is a real-valued, symmetric test function, and $\lambda_0 \neq 0$ is a coupling constant.
(a) Construct Feynman diagrams in this model.
(b) Find all diagrams contributing to $\tilde{G}^{(3)}$ in the third order and their combinatorial coefficients.

7.6 In quantum spinor electrodynamics (QED for short), defined by the Lagrangian

$$\mathcal{L} = -\frac{1}{4} F_{\mu\nu} F^{\mu\nu} + \bar{\psi}(i\gamma^\mu \partial_\mu - m_0)\psi - e_0 \bar{\psi}\gamma^\mu A_\mu \psi$$

the generating functional for the Green's functions has the form

$$Z^{\mathrm{QED}}[\eta, \bar{\eta}, J] = \langle 0|T \; \exp\left(i \int d^4x \left(J^\mu(x)\hat{A}_\mu(x) + \bar{\eta}^\alpha(x)\hat{\psi}_\alpha(x) + \hat{\bar{\psi}}_\alpha(x)\eta^\alpha(x)\right)\right) |0\rangle.$$

Here $\hat{A}_\mu(x)$ and $\hat{\psi}_\alpha(x)$, $\hat{\bar{\psi}}_\alpha(x)$ are the electromagnetic and Dirac field operators in the Heisenberg picture. The Green's functions are the vacuum expectation values of time ordered products of these field operators. $J^\mu(x)$ is a classical, commuting source function, while $\bar{\eta}^\alpha(x)$, $\eta^\alpha(x)$ are independent, anticommuting (Grassmann) elements.

(a) Express the Green's functions through functional derivatives of $Z[\eta, \bar{\eta}, J]$; remember, that similarly as Grassmann elements the Grassmann functional derivatives anticommute, for instance

$$\left\{\frac{\delta}{\delta\eta^{\alpha}(x)}, \frac{\delta}{\delta\eta^{\beta}(y)}\right\} = \left\{\frac{\delta}{\delta\eta^{\alpha}(x)}, \frac{\delta}{\delta\bar{\eta}^{\beta}(y)}\right\} = 0.$$

(b) Repeating the steps which led to (7.29), derive the Gell-Mann–Low formula

$$Z^{\text{QED}}[\eta, \bar{\eta}, J] = \frac{\langle 0_I | T \, \exp\left(i \int d^4x \left(J^{\mu}\hat{A}_{I\mu} + \bar{\eta}^{\alpha}\hat{\psi}_{I\alpha} + \hat{\bar{\psi}}_{I\alpha}\eta^{\alpha}\right)\right) U_I(\infty, -\infty)|0_I\rangle}{\langle 0_I | U_I(\infty, -\infty)|0_I\rangle},$$

where

$$U(\infty, -\infty) = T \exp\left(-ie_0 \int d^4x \, \hat{\bar{\psi}}_I(x)\gamma^{\mu}\hat{A}_{I\mu}(x)\hat{\psi}_I(x)\right) \equiv T \exp\left(-i\hat{V}_I^{\text{QED}}[\psi, \bar{\psi}, A]\right).$$

7.7 Derive the Wick formula for the spinor fields

$$T \, \exp\left(i \int d^4x \left(\bar{\eta}(x)\hat{\psi}(x) + \hat{\bar{\psi}}(x)\eta(x)\right)\right) = \exp\left(-\int d^4x \int d^4x' \, \bar{\eta}(x)S_F(x - x')\eta(x')\right) : \exp\left(i \int d^4x \left(\bar{\eta}(x)\hat{\psi}(x) + \hat{\bar{\psi}}(x)\eta(x)\right)\right) :$$

where

$$S_F(x - x') = i \int \frac{d^4k}{(2\pi)^4} \, \mathrm{e}^{-ik(x'-x'')} \frac{\not{k} + m_0}{k^2 - m_0^2 + i0_+} = \langle 0_I | T \left(\hat{\psi}_I(x)\hat{\bar{\psi}}_I(x')\right) |0_I\rangle.$$

7.8 Prove that

$$T \, \exp\left(i \int d^4x \, J^{\mu}(x)\hat{A}_{I\mu}(x)\right) = \exp\left(-\frac{1}{2} \int d^4x \int d^4x' \, J^{\mu}(x)D_F(x - x')_{\mu\nu}J^{\nu}(x')\right) : \exp(\left(i \int d^4x \, J^{\mu}(x)\hat{A}_{I\mu}(x)\right) :$$

where

$$D_F(x-x')_{\mu\nu} = -i\eta_{\mu\nu}\int \frac{d^4k}{(2\pi)^4}\frac{\mathrm{e}^{-ik(x'-x'')}}{k^2+i0_+} = \langle 0_I|T\left(\hat{A}_{I\mu}(x)\hat{A}_{I\nu}(x')\right)|0_I\rangle.$$

7.9 The numerator appearing in the Gell-Mann–Low formula in QED,

$$Z_I^{\mathrm{QED}}[\eta,\bar{\eta},J] \equiv \langle 0_I|T\ \exp\left(i\int d^4x\left(J^\mu\hat{A}_{I\mu}+\bar{\eta}\hat{\psi}_I+\hat{\bar{\psi}}_I\eta\right)\right)U_I(\infty,-\infty)|0_I\rangle,$$

can be rewritten as

$$Z_I^{\mathrm{QED}}[\eta,\bar{\eta},J] = \exp\left(-i\hat{V}^{\mathrm{QED}}\left[\frac{1}{i}\frac{\delta}{\delta\bar{\eta}},-\frac{1}{i}\frac{\delta}{\delta\eta},\frac{1}{i}\frac{\delta}{\delta J}\right]\right)Z_0^{\mathrm{QED}}[\eta,\bar{\eta},J],$$

where

$$Z_0^{\mathrm{QED}}[\eta,\bar{\eta},J]$$
$$= \langle 0_I|T\ \exp\left(i\int d^4x\left(J^\mu(x)\hat{A}_{I\mu}(x)+\bar{\eta}(x)\hat{\psi}_I(x)+\hat{\bar{\psi}}_I(x)\eta(x)\right)\right)|0_I\rangle,$$

and

$$V^{\mathrm{QED}}\left[\frac{1}{i}\frac{\delta}{\delta\bar{\eta}},-\frac{1}{i}\frac{\delta}{\delta\eta},\frac{1}{i}\frac{\delta}{\delta J}\right] = ie_0\int d^4x\,\frac{\delta}{\delta\bar{\eta}(x)}\gamma^\mu\frac{\delta}{\delta\eta(x)}\frac{\delta}{\delta J^\mu(x)}.$$

Using the results of problems 7.7 and 7.8 derive the formula for the momentum space Green's functions in QED, analogous to formula (7.60).

7.10 Find (without calculating the involved integrals over the internal momenta) the perturbative expression for the QED Green's function

$$\tilde{G}_2^{\mathrm{QED}}(p) = \int d^4x\,\mathrm{e}^{ip(x-y)}\langle 0|T\left(\psi(x)\bar{\psi}(y)\right)|0\rangle$$

up to the terms of the order e_0^4.

7.11 Discuss what simplification occurs (i.e., which Feynman diagrams are absent) when we replace the interaction $\hat{V}_I^{\mathrm{QED}}[\psi,\bar{\psi},A]$ with its normal ordered form.

Chapter 8
Renormalization

Abstract General description of ultraviolet divergences in the ϕ_4^4 model. Loop and one-particle irreducible (1PI) diagrams. The superficial degree of divergence. Renormalization of the one-loop contribution to the four-point Green's function (the sunset diagram). The BPHZ subtraction scheme. Lorentz invariant renormalization of the two-point Green's function. The renormalization constants Z_1, Z_3, δm^2 and the multiplicative renormalization.

The perturbative contributions to the Green's functions, discussed in the preceding chapter, contain the regularizing function g or its Fourier transform $\tilde{g}$. Its presence is necessary in order to obtain mathematically meaningful formulas. This is generally true not only for the $:\phi_4^4:$ model, but also for other models of quantum field theory. Apart from mathematical correctness, one would also like to have a physical motivation for the presence and the form of such a function. In some cases this can be provided, and in these cases the regularizing function has a concrete form, and it is called a formfactor. It has a definite physical interpretation. Usually it encodes the fact that the considered quantum particles are not point-like when, for example, they are bound states of more fundamental objects, like nucleons which are bound states of quarks and gluons.

Much more difficult is the case when such a physical justification is not available. This happens when the corresponding quantum particles seem to be truly elementary objects, like, for example, the fundamental particles of the Standard Model—so far, there is no compelling experimental evidence for the existence of some internal structure of quarks, leptons, or gauge vector bosons. In this case the regularizing function should be removed from the theory. The problem is that this cannot be done in a straightforward manner because then we would get mathematically meaningless expressions. The procedure which allows for the removal of the regularizing function g is called renormalization. Renormalization of the perturbative expansion is certainly among the most intricate constructions in theoretical physics. Its main parts were known by 1955, but important contributions were also made around 1970 in connection with the Standard Model.

H. Arodź and L. Hadasz, *Lectures on Classical and Quantum Theory of Fields*, Graduate Texts in Physics, DOI 10.1007/978-3-319-55619-2_8

In this chapter, we outline renormalization in the example of the $:\phi_4^4:$ model. In Sect. 8.1 we carry out a reconnaissance into the problem of ultraviolet (UV) divergences, which would appear if $\tilde{g}$ was replaced by the Dirac delta (7.55). In the subsequent sections these divergences are analyzed in more detail, and finally the problem is solved by adding to the initial interaction so called counterterms.

8.1 Ultraviolet Divergences

We have seen in the preceding chapter that the perturbative contribution to a Green's function, represented by a given graph Γ, contains integrals over the four-momenta associated with the internal lines of the graph. The integrand essentially has the form of a product of the propagators Δ_F and of the $\tilde{g}$ functions.[1] Let us suppose for a while that we substitute for $\tilde{g}$ in the integrand its limiting form (7.55). It is clear that due to the presence of Dirac deltas, a certain number of the integrals can be trivially calculated. Let us eliminate in this manner as many integrations as possible. It can happen that no integrals are left. The corresponding graphs are called tree graphs. Examples are given in Fig. 8.1.

Graphs where some integrals remain present after using all the Dirac deltas are called loop graphs. By definition, the number of independent loops L in the graph Γ is equal to the number of the remaining four-dimensional integrals over the four-momenta, and the four-momenta, over which we still have to integrate are called the loop momenta. Thus, only the graphs with $L \neq 0$ can have the UV divergences—that is the integrals over the loop four-momenta which become divergent when we extend the integration range[2] to the infinite one. The presence of the UV divergences is of course a consequence of the fact that without the regularizing function $\tilde{g}$ the model is mathematically incorrect.

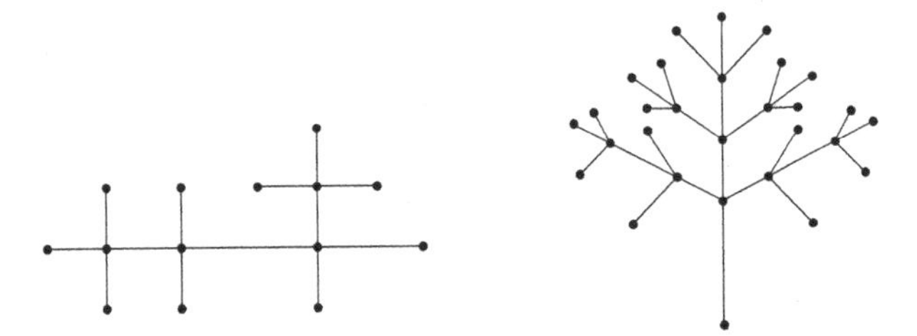

Fig. 8.1 Examples of tree graphs in the $:\phi^4:$ model

[1]In the rather general discussion below, we neglect numerical factors which are present in the perturbative contributions, because they are not important in the qualitative analysis of the UV divergences.

[2]Let us recall that in calculus, the integrals of the type $\int_{-\infty}^{+\infty}$ are defined as the limit of $\int_{M_1}^{M_2}$ when $M_1 \to -\infty,\ \ M_2 \to +\infty$.

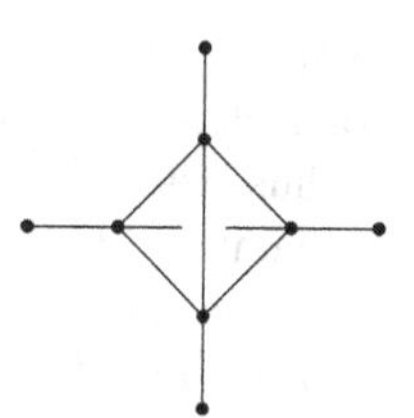

Fig. 8.2 Example of a non planar graph. The horizontal line is continuous in spite of the drawing—it just runs behind the vertical one. This graph has 3 independent loops

It is clear that the calculation of the number of independent loops L can be done separately for each connected component of the graph Γ. Here we use the term 'connected' in the sense known from topology for subsets of R^3. Each connected component is a diagram in its own right, disconnected from the remaining part of the graph Γ. The perturbative contribution corresponding to Γ is equal to the product of the contributions from all of its connected subgraphs. Therefore, from now on we consider only connected graphs.

An explanation is in order as to why we have referred to the topology of figures in R^3, while so far all graphs have been drawn in the plane R^2. There exist graphs which are better presented as figures in the space R^3. If drawn in the plane they would contain superfluous crossings which are not internal vertices (7.64). The graph is called non planar if it is not possible to draw it in the plane without superfluous crossings of lines, under the assumptions that all its lines are continuous and all external lines extend to the infinity.[3] A simple example is given in Fig. 8.2.

Let us consider a connected graph Γ with l internal vertices. We assume that the graph is nontrivial, that is that $l > 0$. It turns out that the Dirac deltas can always be combined to produce at least one delta which does not contain any four-momentum associated with an internal line: that is $\delta(\sum_{i=1}^{n} k_i)$, where k_i are the external four-momenta for the graph. The perturbative contribution of each connected graph Γ is proportional to such δ. In order to show this, let us pick an internal vertex A of Γ—it will serve as the starting point for the following procedure. In the first step we choose one internal line, let us denote it as I_1, attached to that vertex. The four-momentum associated with it is denoted as p. The line I_1 ends at another internal vertex B. Both vertices have their δ's. The four-momentum p appears in both of them, with opposite signs. Thus, we have a product of the form

$$\delta(\sum_{i=1}^{3} q_{Ai} + p)\delta(\sum_{j=1}^{3} q_{Bj} - p),$$

where q_{Ai} and q_{Bj} are the four-momenta associated with the other three lines emanating from A and B, respectively. One of the δ's is used to perform the integral $\int d^4p$ related to the internal line I_1, and to eliminate p from the other δ yielding $\delta(\sum_{j=1}^{3} q_{Aj} + \sum_{j=1}^{3} q_{Bj})$. We may imagine that the two vertices are dragged to each other along the line I_1 and merged, thus producing a six-leg 'vertex' AB proportional

[3] If this assumption is abandoned the graphs can be drawn in the plane, see Exercise 8.1.

to $\delta(\sum_{j=1}^{3} q_{Aj} + \sum_{j=1}^{3} q_{Bj})$. In the second step, we pick another internal vertex C connected to AB by at least one internal line, and we repeat the reasoning from step 1, thus obtaining an effective 'vertex' ABC with 8 legs. We continue this procedure until all l internal vertices of Γ are merged into one 'vertex' which has $2l + 2$ legs. The lines emanating from such an effective 'vertex' can form loops of the type shown in Fig. 7.1—to this kind of 'vertex' the normal ordering prescription does not apply of course—and there are n lines that end at the external vertices. Therefore, the resulting final δ will be just $\delta(\sum_{i=1}^{n} k_i)$ because the two ends of any line forming the loop introduce the zero four-momentum, $q - q = 0$.

The number of loops in the final effective 'vertex' is equal to $(2l + 2 - n)/2$ because only the n external lines are not looped. This number is equal to the number of independent loops L in the graph Γ, hence

$$L = l + 1 - \frac{n}{2}.$$

On the other hand, counting the ends of the n external and I internal lines of the graph Γ we obtain the following relation

$$n + 2I = 4l. \tag{8.1}$$

Note that it implies that n is even. Elimination of n with the help of the latter formula gives

$$L = I - l + 1. \tag{8.2}$$

This formula has a simple heuristic justification: each internal line brings in one four-dimensional integral d^4p, and each internal vertex one δ. One can combine these δ's to produce one that contains only the external momenta, and the remaining $l - 1$ δ's can be used to eliminate integrals. After doing this, the number of remaining four dimensional integrations is equal to $I - l + 1$.

When investigating the UV divergences one may focus on the so called one-particle irreducible (1PI) graphs. By definition, such a graph is connected and, moreover, it is not possible to split it into disconnected parts by cutting one internal line. Furthermore, the Feynman propagators Δ_F are removed from all the external lines. This latter property is marked by removing the dots from the ends of the external lines. Examples of such graphs are given in Fig. 8.3, while Fig. 8.4 shows graphs which are not of the 1PI type.

We may restrict considerations of the UV divergences to the 1PI graphs for the following reasons. First, the external lines of graphs do not introduce any integrations. Moreover, the same is true for each internal line which is the only link between two

Fig. 8.3 Examples of 1PI graphs

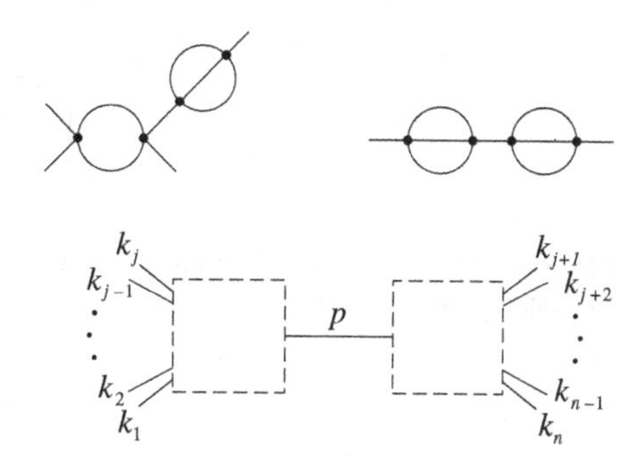

Fig. 8.4 Examples of one-particle reducible graphs

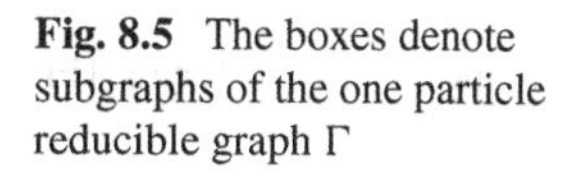

Fig. 8.5 The boxes denote subgraphs of the one particle reducible graph Γ

parts of a non-1PI graph Γ (cutting it would break the graph into disconnected parts). The four-momentum p associated with this line appears in two δ's:

$$\delta(\sum_{i=1}^{j} k_i - p)\, \delta(\sum_{i=j+1}^{n} k_i + p) = \delta(\sum_{i=1}^{j} k_i - p)\, \delta(\sum_{i=1}^{n} k_i),$$

see Fig. 8.5. The first δ on the r.h.s. of this formula eliminates the integral $\int d^p$ associated with the internal line.

Let us have a look at the L four-momentum integrals (in a certain 1PI graph Γ) which are left after using all Dirac δ's (we still imagine that $\tilde{g}$ is replaced by the Dirac δ according to (7.55)). If all components of the loop four-momenta $p_i,\ i = 1, \ldots L$, are restricted to an interval $[-M, M]$ there are no UV divergences.[4] Let us introduce a $4L$-component vector $\underline{w}$: its first four components are equal to p_1, the next four to p_2, and so on. The integration measure $\prod_{i=1}^{L} d^4 p_i$ can be written as $d^{4L}\underline{w}$. The restrictions $-M \le p_i^\mu \le M$ mean that we integrate over the hypercube of size $2M$ with its center located at the origin in the $4L$-dimensional space R^{4L} of vectors $\underline{w}$. As far as the limit $M \to \infty$ is concerned, we may replace the hypercube by the $4L$-dimensional ball of radius M in that space. In spherical coordinates on the R^{4L} space

$$d^{4L}\underline{w} = w^{4L-1}\, dw\, d\Omega,$$

where $d\Omega$ is the solid angle element in that space and w denotes the modulus of $\underline{w}$, $0 \le w \le M$. The integral over the solid angle does not generate any UV divergences by definition—the range of integration over each spherical angle is finite. On the other hand, for large w the integrand behaves like

[4]The integrals may still be divergent for specific values of the external momenta, because the denominators of some propagators can be equal to zero. In order to avoid such divergences, we may replace $i0_+$ in the denominators by $i\epsilon$, where $\epsilon > 0$. The limit $\epsilon \to 0_+$ is taken after we perform the integrations over loop momenta. $\tilde{G}^{(n)}(k_1, k_2, \ldots, k_n)$ is not a smooth function of the external momenta—rather, it is a generalized function of them. Singularities of these functions usually have certain physical meaning. We shall not discuss them because their presence does not jeopardize the existence of the perturbative contributions.

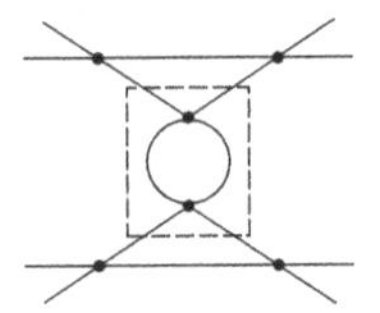

Fig. 8.6 The example of 1PI graph Γ such that $\omega(\Gamma) = -4 < 0$, while $\omega(\gamma) = 0$ for its 1PI subgraph γ shown inside the box

$$w^{4L-1}w^{-2I},$$

where the second factor comes from the propagators of the I internal lines of Γ. Let us introduce the superficial degree of divergence $\omega(\Gamma)$. It is defined for a 1PI graph Γ with L independent loops and I internal lines by the formula

$$\omega(\Gamma) = 4L - 2I. \tag{8.3}$$

It is clear that the integral over w is divergent in the limit $M \to \infty$ when $\omega(\Gamma) \geq 0$. In particular, in the case $\omega(\Gamma) = 0$ we have a logarithmic divergence.[5] Using formulas (8.1), (8.2) we obtain

$$\omega(\Gamma) = 4 - n. \tag{8.4}$$

Thus, in our model the superficial degree of divergence is nonnegative only for 1PI graphs with 2 or 4 external lines.

It turns out that $\omega(\Gamma) < 0$ does not mean that the integral is necessarily convergent. The point is that in the reasoning presented above, we have assumed that the loop four-momenta p_i become infinite in the synchronized manner implied by the limit $w \to \infty$. Actually, we expect that the loop integrals are finite, independently of the way the infinite four-momenta limit is taken. In particular, we may repeat the reasoning presented above for each 1PI subgraph γ of Γ. If $\omega(\gamma) \geq 0$ for one or more such subgraphs we again encounter a UV divergence. An example of such a subgraph is presented in Fig. 8.6. One can prove that the 1PI graph Γ does not have any UV divergences if the superficial degrees of divergence of it and of all its 1PI subgraphs are negative.

To summarize, our preliminary analysis has shown that in the $:\phi_4^4:$ model it suffices to remove the UV divergences from 2- and 4-point 1PI graphs with loops. Such graphs directly appear in the perturbative contributions to $G^{(2)}$ and $G^{(4)}$, and also as subgraphs of graphs with 6 or more external lines. The limit

$$\tilde{g}(q_1, q_2, q_3, q_4) \to \frac{1}{(2\pi)^4}\delta(q_1 + q_2 + q_3 + q_4) \tag{8.5}$$

[5] In some rather special cases the integral can be finite even if $\omega \geq 0$, because the integral over the solid angle Ω can vanish. We shall not consider such exceptions.

will exist if we replace all such potentially UV divergent parts by certain UV convergent terms. The model $:\phi_4^4:$ belongs to the class of so called perturbatively renormalizable field theories. A model is perturbatively renormalizable if the number of external lines n in superficially divergent graphs (1PI graphs with $\omega(\Gamma) \geq 0$) is bounded from above by a finite number n_0—in the case of the $:\phi_4^4:$ model $n_0 = 4$. In certain models $\omega(\Gamma) \geq 0$ only for a finite number of graphs. Such models are called superrenormalizable. Of course, the $:\phi_4^4:$ model is not superrenormalizable.

In nonrenormalizable models the number of external lines in superficially divergent graphs is not bounded from above. We shall see by the end of Sect. 8.4, that the renormalized perturbative expansion in such models contains an arbitrarily large number of constants, whose values are not predicted by the theory—they have to be determined experimentally. It is believed that such models have little predictive power, and therefore they are not popular.

One should note here that Einstein's theory of gravity is nonrenormalizable when quantized in a straightforward, canonical manner. This is one of several obstacles in obtaining a quantum theory of the gravitational field. For that matter, it is not at all obvious that Einstein's theory of gravity should be quantized—it can happen that it is merely an effective theory, that is, an approximate description of effects which in fact are described much better by another, perhaps more general theory which has a satisfactory quantum version. Many theorists are investigating so called superstring models with precisely that goal in mind. At the moment such a deeper theory has not been established, mainly because as yet there are no experimental data to test various proposals.

The (non)renormalizability of a model has a certain connection with the dimensionality of pertinent coupling constants. This can be clearly seen in the example of $:\phi_d^4:$ models, where d is the dimension of space-time. The action functional has the form

$$S = \int d^dx \left(\frac{1}{2}\partial_\mu\phi\partial^\mu\phi - \frac{1}{2}m_0^2\phi^2 - \frac{\lambda_0}{4!}\phi^4 \right). \tag{8.6}$$

In the units $c = 1 = \hbar$ the action S is dimensionless by assumption, $[S] = \text{cm}^0$. Let us take $d = 3$. Then, $[\phi] = \text{cm}^{-1/2}$ and $[\lambda_0] = \text{cm}^{-1}$. In the l-th order of the perturbative expansion, the graphs are proportional to λ_0^l and this constant has the dimension cm^{-l}. In order to have the dimension of the contribution of the whole graph equal to the dimension of $\tilde{G}^{(n)} = \text{cm}^{5n/2}$, negative powers of three-momentum are needed because $[p] = \text{cm}^{-1}$. This suggests better and better convergence of the integrals over the three-momenta as l increases, and therefore superrenormalizability of the model. Indeed, the superficial degree of divergence is equal to

$$\omega = 3L - 2I = 3 - l - \frac{n}{2}$$

(we have used formulas (8.1), (8.2) and the fact that this space-time has three dimensions). It is clear that there is only one case in which we have $\omega \geq 0$: $n = 2,\ l = 2$:

it is the second graph in Fig. 8.3. Note that this graph can appear as a divergent subgraph in other 1PI graphs with $\omega < 0$. Therefore the total number of UV divergent graphs is infinite. The case $n = 2,\ l = 1$ is excluded by the normal ordering. The $:\phi_3^4:$ model is superrenormalizable.

Let us now take $d = 4$. Then $[\lambda_0] = \text{cm}^0$, and the analogous reasoning suggests that the appearance of UV divergences is not related to the order l of the perturbative expansion, apart from the trivial condition $l > 1$. Indeed, we already know that $\omega = 4 - n$.

Adding still one space-time dimension, $d = 5$, gives $[\lambda_0] = \text{cm}^1$. In this case we need positive powers of five-momenta in order to have the right dimension of the perturbative contributions (now $[\phi] = \text{cm}^{-3/2}$ and $[\tilde{G}^{(n)}] = \text{cm}^{7n/2}$). This suggests that UV divergent graphs will appear for any n if l is large enough. Indeed, the formula for the superficial degree of divergence

$$\omega = 5L - 2I = 5 + l - \frac{3}{2}n,$$

shows that $\omega \geq 0$ for an arbitrarily large number of external lines n, if we take sufficiently large order of the perturbative expansion. Hence, the model $:\phi_5^4:$ is nonrenormalizable. In general, increasing the dimensionality of space-time worsens the situation as far as the UV divergences are concerned. Satisfactory models from the perturbative point of view are still possible, but they require very special sets of fields, as well as Lagrangians with symmetries which lead to mutual cancelations of the UV divergent contributions. Examples of such cancelations are given in Chap. 13, where we discuss so called supersymmetric models.

8.2 An Example

The goal of renormalization is to define the limit (8.5) term by term in the perturbative expansion. We already know that this cannot be done in a straightforward manner just by replacing $\tilde{g}$ by the r.h.s. of formula (8.5), because then we would get UV divergent integrals over some loop four-momenta. Below we consider in detail the graph presented in Fig. 8.7. Using this graph, we introduce the main ingredient of renormalization, which is called the subtractions.

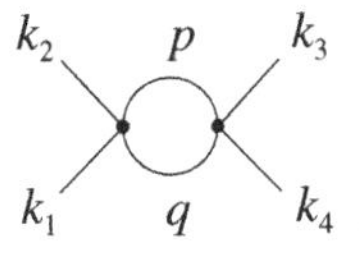

Fig. 8.7 The 1PI graph A_1 renormalized in this section

We will use the following one-parameter family of regularizing functions

$$\tilde{g}(q_1, q_2, q_3, q_4) = \frac{1}{(2\pi)^4}\delta(q_1 + q_2 + q_3 + q_4)\prod_{i=1}^{4}\left(\frac{m_0^2 - M^2}{q_i^2 - M^2 + i\epsilon}\right)^{\frac{N}{2}}, \quad (8.7)$$

where N is a natural number and $\epsilon > 0$. We shall take the limit $\epsilon \to 0_+$ later, when there will be no risk of vanishing denominators. This choice of $\tilde{g}$ is called the Pauli–Villars (P–V) regularization. The limit (8.5) corresponds to $M \to \infty$. It should be noted that the function $\tilde{g}$ given by (8.7) does not belong to the space $S(R^4)$ of test functions. Rather, it is a generalized function. Nevertheless, it vanishes sufficiently quickly when $q_i \to \infty$, so that the loop integrals are finite, see below. This function $\tilde{g}$ has the advantage that it is invariant with respect to Lorentz transformations of the four-momenta. Moreover, it has a simple algebraic form which harmonizes with the form of the free propagator $\Delta_F(q)$.

Because each internal vertex has four legs enumerated by the four-momenta q_i, we may ascribe the P–V factors

$$\left(\frac{m_0^2 - M^2}{q_i^2 - M^2 + i\epsilon}\right)^{N/2}$$

to the legs. Therefore, one may formulate the P-V regularization in an equivalent way, by saying that the internal vertex has the form as if the limit $M \to \infty$ was already taken, that is

$$\times \quad = \frac{\lambda_0}{4!(2\pi)^4}\,\delta(q_1 + q_2 + q_3 + q_4), \quad (8.8)$$

but the free propagators Δ_F, formula (7.58), associated with each internal line are replaced by the P–V regularized propagator Δ_{P-V},

$$\Delta_F(p) \to \Delta_{P-V}(p) = \frac{i}{p^2 - m_0^2 + i\epsilon}\left(\frac{m_0^2 - M^2}{p^2 - M^2 + i\epsilon}\right)^N \quad (8.9)$$

(each internal line has two P–V factors coming from the two legs of the adjacent internal vertices). Actually, this latter formulation of the P–V regularization is the original one. Note that Δ_{P-V} with $N = 1$ may also be written in the following forms

$$\Delta_{P-V}(p) = \frac{i}{p^2 - m_0^2 + i\epsilon} - \frac{i}{p^2 - M^2 + i\epsilon} = -i\int_{m_0^2}^{M^2} \frac{d\lambda}{(p^2 - \lambda + i\epsilon)^2}. \quad (8.10)$$

Formula (8.7), and the substitution (8.9), imply that the factors $[(m_0^2 - M^2)/(q_i^2 - M^2 + i\epsilon)]^{N/2}$ appear also on the external lines of graphs. Because such lines do not play any role as far as the UV divergences are concerned, we may take the limit

$M \to \infty$ for each external line right now. Therefore, the external lines do not introduce any factors in the P–V regularized 1PI graphs (Δ_F's have already been removed).

The Pauli–Villars regularization is sufficient for rendering all 1PI graphs UV finite: the superficial degree of divergence of P–V regularized 1PI graphs is negative. Computation of ω for such a graph, denoted by Γ_{reg}, differs from the one presented in the preceding section only on one point: now the contribution of each internal line behaves like $(p^2)^{-N-1}$. Therefore,

$$\omega(\Gamma_{reg}) = 4L - 2(N+1)I = 4 - 4l + 2I(1-N).$$

It is negative for all $l > 1$, even if we take the lowest possible $N = 1$. When $l = 1$ we would have to take a larger N, e.g., $N = 2$, but luckily the perturbative expansion does not contain any 1PI graphs with $l = 1$ because of the normal ordering. It turns out that expressions which require the P–V regularization with $N = 2$ appear when applying the BPHZ subtraction scheme to the 2-point Green's function, see Sect. 8.4. In the following considerations, we use the regularization with $N = 1$ unless explicitly stated otherwise.

The regularized formula represented by the graph A_1 has the form[6]

$$A_1(k^2; M) = -\frac{\lambda_0^2}{(4!)^2(2\pi)^8}\int d^4p$$
$$\int_{m_0^2}^{M^2} d\lambda_1 \int_{m_0^2}^{M^2} d\lambda_2 \frac{1}{(p^2 - \lambda_1 + i\epsilon)^2[(k+p)^2 - \lambda_2 + i\epsilon]^2},$$

where $k = k_1 + k_2$. Next, we use the identity

$$\frac{1}{a^2 b^2} = \int_0^1 dz\, \frac{6z(1-z)}{[a(1-z) + bz]^4},$$

which is obtained from the simpler identity

$$\frac{1}{ab} = \int_0^1 \frac{dz}{[a(1-z) + bz]^2}$$

by differentiation with respect to a and b. The latter identity can easily be checked by elementary calculation of the integral over z. Thus,

[6]We denote the graph and the formula corresponding to it by the same letter. The presence of P–V regularization is marked by adding the argument M.

$$A_1(k^2; M) = -\frac{\lambda_0^2}{(4!)^2(2\pi)^8} \int d^4p \int_{m_0^2}^{M^2} d\lambda_1 \int_{m_0^2}^{M^2} d\lambda_2$$
$$\int_0^1 dz \frac{6z(1-z)}{[(p^2-\lambda_1+i\epsilon)(1-z)+((k+p)^2-\lambda_2+i\epsilon)z]^4}. \tag{8.11}$$

Let us shift the integration variable p:

$$p = p' - kz, \quad d^4p = d^4p'.$$

The reason for this shift is that the denominator in formula (8.11) depends on p' only through p'^2,

$$[\ldots] = p'^2 + k^2z(1-z) - \lambda_1(1-z) - \lambda_2 z + i\epsilon. \tag{8.12}$$

The expression (8.12) vanishes when

$$p_0' = \pm\sqrt{(\vec{p}\,')^2 + \lambda_1(1-z) + \lambda_2 z - k^2z(1-z) - i\epsilon}.$$

At these points (in the complex p_0' plane) the integrand in the formula (8.11) has poles. Because $\lambda_{1,2} \geq m_0^2$, $\epsilon > 0$, and $z \in [0, 1]$, when

$$k^2 \in (-\infty, 4m_0^2) \tag{8.13}$$

the poles lie close to the real axis, see Fig. 8.8. In this case the integral over p_0' along the contour presented in Fig. 8.8 vanishes. The integrals along the two arcs of the circle vanish when the radius of the circle increases to infinity. Therefore, the integral along the real axis is equal to the integral along the imaginary axis. The integration over imaginary p_0' is equivalent to the integration over real variable p_4, introduced by the formula

$$p_0' = ip_4,$$

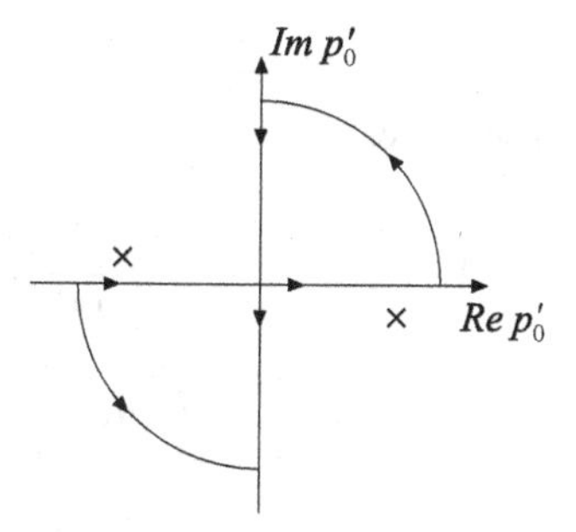

Fig. 8.8 The integration contour in the complex p_0' plane. The position of the poles is marked by the small crosses

$p_4 \in (-\infty, +\infty)$, and

$$dp_0' d^3 p' = i dp_4 d^3 p', \quad p'^2 = -p_4^2 - \vec{p}'^2.$$

Therefore,

$$A_1(k^2; M) = -i \frac{\lambda_0^2}{(4!)^2 (2\pi)^8} \int dp_4 \int d^3 p \int_{m_0^2}^{M^2} d\lambda_1 \int_{m_0^2}^{M^2} d\lambda_2$$
$$\int_0^1 dz \frac{6z(1-z)}{[-p_4^2 - \vec{p}^2 + k^2 z(1-z) - \lambda_1(1-z) - \lambda_2 z + i\epsilon]^4}, \tag{8.14}$$

where we have omitted$'$ in the integration variable $\vec{p}$. The transition from formula (8.11) to (8.14) is called the Wick rotation.

The integration variables $\vec{p}$ and p_4 together form a Euclidean four-momentum $p_E = (\vec{p}, p_4)$, $d^4 p_E = dp_4 d^3 p$. The integrand in (8.14) depends only on $p_E^2 = p_4^2 + \vec{p}^2$. Therefore, we introduce the four-dimensional spherical angles $\Theta_1, \Theta_2, \Theta_3$:

$$p_E = |p_E| \begin{pmatrix} \sin\Theta_1 \sin\Theta_2 \sin\Theta_3 \\ \sin\Theta_1 \sin\Theta_2 \cos\Theta_3 \\ \sin\Theta_1 \cos\Theta_2 \\ \cos\Theta_1 \end{pmatrix},$$

where $0 \le \Theta_1, \Theta_2 \le \pi$ and $0 \le \Theta_3 < 2\pi$. Then,

$$d^4 p_E = |p_E|^3 d|p_E| d\Omega,$$

where the four-dimensional solid angle element has the form

$$d\Omega = \sin^2\Theta_1 \sin\Theta_2 d\Theta_1 d\Theta_2 d\Theta_3.$$

The full solid angle is equal to $2\pi^2$, that is

$$\int d\Omega = 2\pi^2.$$

Therefore,

$$A_1(k^2; M) = -2i\pi^2 \frac{\lambda_0^2}{(4!)^2 (2\pi)^8} \int_0^\infty d|p_E| \int_{m_0^2}^{M^2} d\lambda_1 \int_{m_0^2}^{M^2} d\lambda_2$$
$$\int_0^1 dz \frac{6z(1-z)\,|p_E|^3}{[|p_E|^2 - k^2 z(1-z) + \lambda_1(1-z) + \lambda_2 z - i\epsilon]^4}. \tag{8.15}$$

Because of assumption (8.13) the real part of the expression in the bracket in the denominator does not vanish. Therefore, we may now take the limit $\epsilon \to 0_+$. The integrals over $|p_E|$, λ_1 and λ_2 are elementary. We finally obtain the following formula

$$A_1(k^2; M) = i\pi^2 \frac{\lambda_0^2}{(4!)^2(2\pi)^8} \int_0^1 dz$$

$$\left(\ln \frac{M^2 - k^2 z(1-z)}{M^2(1-z) + m_0^2 z - k^2 z(1-z)} - \ln \frac{M^2 z + m_0^2(1-z) - k^2 z(1-z)}{m_0^2 - k^2 z(1-z)} \right).$$

In the limit $M \to \infty$

$$A_1(k^2; M) = -2i\pi^2 \frac{\lambda_0^2}{(4!)^2(2\pi)^8} \ln \frac{M}{m_0} + \text{(terms finite in the limit } M \to \infty).$$

The logarithmic divergence in the limit $M \to \infty$ is the expected one, because the superficial degree of divergence of graph A_1 is equal to 0.

The divergent term does not depend on k^2. Therefore, the difference

$$A_1(k^2; M) - A_1((\overset{(0)}{k})^2; M),$$

where $\overset{(0)}{k}$ is a fixed four-vector, also remains finite when we remove the regularization. The renormalized contribution of graph A_1 is defined as follows

$$A_1^{ren}(k^2) \overset{df}{=} \lim_{M \to \infty} \left(A_1(k^2; M) - A_1((\overset{(0)}{k})^2; M) \right). \tag{8.16}$$

Note that this definition trivially implies that

$$A_1^{ren}(\overset{(0)}{k}{}^2) \equiv 0. \tag{8.17}$$

This identity is called the renormalization condition.

The four-vector $\overset{(0)}{k}$ is called the subtraction point. In the $:\phi_4^4:$ model the subtraction point is usually given in terms of four four-vectors $\overset{(0)}{k}_i$, $i = 1, 2, 3, 4$, such that

$$\sum_{i=1}^{4} \overset{(0)}{k}_i = 0, \quad (\overset{(0)}{k}_i)^2 = -\mu^2, \quad \overset{(0)}{k}_i \overset{(0)}{k}_j = \frac{1}{3}\mu^2 \ \text{ for } \ i \neq j, \tag{8.18}$$

where μ is a positive parameter with the dimension of mass. This choice for $\overset{(0)}{k}_i$ is called the symmetric subtraction point. For an example of it, see Exercise 8.3. In the case of graph A_1 we have $k = k_1 + k_2$, therefore we take $\overset{(0)}{k} = \overset{(0)}{k}_1 + \overset{(0)}{k}_2$. In consequence,

$$(\overset{(0)}{k})^2 = -\frac{4}{3}\mu^2.$$

It follows from definition (8.16) that

$$A_1^{ren}(k^2) = i\pi^2 \frac{\lambda_0^2}{(4!)^2(2\pi)^8} \int_0^1 dz \ln \frac{m_0^2 - k^2 z(1-z)}{m_0^2 + \frac{4}{3}\mu^2 z(1-z)}. \tag{8.19}$$

Let us recall that this formula is obtained under assumption (8.13). We do not present a calculation of A_1 in the case this assumption is not satisfied.

Formula (8.19) for the renormalized contribution of graph A_1 takes a particularly simple form when $m_0^2 = 0$:

$$A_1^{ren}(k^2)\big|_{m_0^2=0} = i\pi^2 \frac{\lambda_0^2}{(4!)^2(2\pi)^8} \ln\left(-\frac{3k^2}{4\mu^2}\right).$$

In this case, the restriction (8.13) has the form $k^2 < 0$.

The subtraction of $A_1(-\frac{4}{3}\mu^2; M)$ can equivalently be regarded as an ad hoc modification of the interaction by adding to it a new term, called a counterterm. It is chosen in such a way, that the difference

$$A_1(k^2; M) - A_1(-\frac{4}{3}\mu^2; M)$$

appears automatically when calculating the full second order contribution to $\tilde{G}^{(4)}$. The second order contribution in the original model, i.e. without the counterterm, is presented in Fig. 7.6. In order to implement the subtraction we introduce a second internal vertex with four legs, c.f. the vertex (7.64), with a suitably adjusted coefficient. In Fig. 7.6 there are three graphs of the form A_1, which differ from each other only by the external momenta. We need a counterterm for each of them. Because the subtraction is done at the symmetric point, the subtracted terms are identical. Therefore, it is sufficient to add a coefficient of 3 in front of the counterterm for graph A_1. It is convenient to introduce a constant C_1 such that

$$A_1(-\frac{4}{3}\mu^2; M) = -i \frac{\lambda_0^2}{(4!)^2(2\pi)^8} C_1. \tag{8.20}$$

In the limit $M \to \infty$

$$C_1 \cong 2\pi^2 \ln \frac{M}{m_0}.$$

In order to implement the subtractions, it suffices to replace $\tilde{V}_{Ig}[\tilde{\beta}]$ given by formula (7.56) by $\tilde{V}_{Ig}[\tilde{\beta}] + \delta_1 \tilde{V}_{Ig}[\tilde{\beta}]$, where

$$\delta_1 \tilde{V}_{Ig}[\tilde{\beta}] = \frac{\lambda_0^2 C_1}{16(2\pi)^4} \int d^4q_1 d^4q_2 d^4q_3 d^4q_4 \; \tilde{g}(q_1, q_2, q_3, q_4) \tilde{\beta}(q_1) \tilde{\beta}(q_2) \tilde{\beta}(q_3) \tilde{\beta}(q_4). \tag{8.21}$$

Here $\delta_1 \tilde{V}_{Ig}[\tilde{\beta}]$ is the total counterterm for the three graphs from the first line of Fig. 7.6. Note that such a modification of the interaction is equivalent to changing the coupling constant

$$\lambda_0 \to \lambda_0 + \frac{3C_1}{32\pi^4} \lambda_0^2. \tag{8.22}$$

In the second order of the perturbative expansion, we also have the second graph from Fig. 8.3. This graph has two independent loops which share one internal line. In this case the subtractions are more complicated. We shall apply the general BPHZ prescription which is described in the next section.

8.3 BPHZ Subtractions

Let Γ be a 1PI graph in the regularized $:\phi_4^4:$ model with n external four-momenta $k_1, k_2, \ldots, k_n$. The analytical expression corresponding to it has the form

$$A_\Gamma = \delta(\sum_{i=1}^{n} k_i) \int d^4p_1 \ldots d^4p_L \; I_\Gamma(p_1, \ldots, p_L; k_1, \ldots, k_{n-1}; M),$$

where L is the number of independent loops in the graph. A_Γ is finite due to the presence of the regularization, but the existence of the limit $M \to \infty$ requires the subtractions. The integrand I_Γ is the product of the Pauli–Villars regularized propagators Δ_{P-V}, and of numerical factors. The four-momentum k_n has been eliminated from it because $k_n = -\sum_{i=1}^{n-1} k_i$.

The graph Γ can have subgraphs: parts which are graphs of the $:\phi_4^4:$ model in their own right. The subgraphs are denoted by γ, $\gamma \subset \Gamma$. Subgraph γ is called a proper one if $\gamma \neq \Gamma$, and the proper subgraph $\gamma \subset \Gamma$ is called a renormalization part of Γ if it is 1PI and $\omega(\gamma) \geq 0$. Two renormalization parts γ_1 and γ_2 of Γ are disconnected, $\gamma_1 \cap \gamma_2 = \emptyset$, if they do not have any common vertices.

The BPHZ subtractions are defined in terms of Taylor expansions with respect to the external momenta. Let $f(k_1, \ldots, k_{n-1})$ be a function of the external momenta $k_1, \ldots, k_{n-1}$, which is smooth in a vicinity of certain fixed four-momenta $\overset{(0)}{k}_1, \ldots, \overset{(0)}{k}_{n-1}$. It is convenient to use the following notation

$$T_\omega f(k_1, \ldots, k_{n-1}) \stackrel{df}{=} f(\overset{(0)}{k}_1, \ldots, \overset{(0)}{k}_{n-1}) + \left(k_i - \overset{(0)}{k}_i\right)^{\mu_i} \left.\frac{\partial f}{\partial k_i^{\mu_i}}\right|_{k_j = \overset{(0)}{k}_j} + \ldots$$

$$+ \frac{1}{\omega!} \left(k_{i_1} - \overset{(0)}{k}_{i_1}\right)^{\mu_{i_1}} \ldots \left(k_{i_\omega} - \overset{(0)}{k}_{i_\omega}\right)^{\mu_{i_\omega}} \left.\frac{\partial^\omega f}{\partial k_{i_1}^{\mu_{i_1}} \partial k_{i_2}^{\mu_{i_2}} \ldots \partial k_{i_\omega}^{\mu_{i_\omega}}}\right|_{k_j = \overset{(0)}{k}_j},$$

where ω is a nonnegative integer. Thus, $T_\omega f$ denotes the first $\omega + 1$ terms of the Taylor series for f around $\overset{(0)}{k}_i$, $i = 1, \ldots, n-1$. In the case $\omega = 0$ it is just $f(\overset{(0)}{k}_1, \ldots, \overset{(0)}{k}_{n-1})$. In the considerations below $n = 2$ or $n = 4$, because in the model $:\phi_4^4:$ all 1PI graphs with $n > 4$ external lines have $\omega(\Gamma) < 0$. The Taylor expansions are made around the four-momenta from the symmetric point (8.18).

According to the prescription worked out by N.N. Bogoljubov and O.S. Parasiuk, with later contributions by K. Hepp and W. Zimmermann (hence the acronym BPHZ), the subtractions should be done in the following manner. The integrand I_Γ should be replaced by R_Γ which is defined as follows

$$R_\Gamma = \begin{cases} I_\Gamma^{i.s.} & \text{if } \omega(\Gamma) < 0, \\ I_\Gamma^{i.s.} - T_{\omega(\Gamma)} I_\Gamma^{i.s.} & \text{if } \omega(\Gamma) \geq 0, \end{cases} \tag{8.23}$$

where

$$I_\Gamma^{i.s.} = I_\Gamma + \sum_{\{\gamma_1 \ldots \gamma_S : \gamma_i \cap \gamma_k = \emptyset\}} I_{\Gamma/\{\gamma_1 \ldots \gamma_S\}} \prod_{i=1}^{S} \left(-T_{\omega(\gamma_i)} I_{\gamma_i}^{i.s.}\right). \tag{8.24}$$

The sum in the last formula is over all families of disconnected renormalization parts of the graph Γ. One such family is denoted by $\{\gamma_1 \ldots \gamma_S : \gamma_i \cap \gamma_k = \emptyset\}$. The symbol $I_{\Gamma/\{\gamma_1 \ldots \gamma_S\}}$ denotes that part of the integrand I_Γ which does not belong to any of the subgraphs $\gamma_1 \ldots \gamma_S$ from the given family. The superscript $^{i.s.}$ stands for 'internal subtractions'. The internal subtractions are defined recursively, by application of formula (8.24) to γ_i. The graph Γ does not require internal subtractions only if it does not contain any renormalization part. If $\omega(\Gamma) \geq 0$, such a 1PI graph Γ is called primitively divergent.

The renormalized contribution of the graph Γ is defined as follows

$$A_\Gamma^{ren} = \delta(\sum_{i=1}^{n} k_i) \lim_{M \to \infty} \int d^4 p_1 \ldots d^4 p_L \, R_\Gamma(p_1, \ldots, p_L; k_1, \ldots, k_{n-1}; M). \tag{8.25}$$

The main theorem about BPHZ subtractions says that the limit $M \to \infty$ exists, and that A_Γ^{ren} is a generalized function of the external four-momenta. Moreover, the limit $M \to \infty$ commutes with the integrals, that is it can already be taken in the whole integrand R_Γ before the integration. Therefore, in principle, the regularization is not necessary inasmuch as we only consider Feynman diagrams with subtractions. However, investigation of A_Γ^{ren} without a regularization is much harder, because only the convergence of the integral of R_Γ is guaranteed, and not of the integrals of the separate contributions to R_Γ, given by the terms in the sums present in formulas (8.23), (8.24). For this reason, it is convenient to introduce the regularization and to take the limit $M \to \infty$ after the integration over the loop four-momenta. Such an auxiliary regularization is often referred to as the intermediate one.

The fact that the subtractions improve the convergence of the integrals over loop four-momenta has a simple intuitive explanation. All the terms in the Taylor expansion of I_Γ have the same dimensionality. Therefore, the positive dimension introduced by the powers of the external momenta standing in front of the derivatives has to be compensated for by the negative dimension of the derivatives. This means that the terms with derivatives of sufficiently high order necessarily have a negative superficial degree of divergence. Therefore, one may expect that the integrals over the loop four-momenta will remain finite in the limit $M \to \infty$, except for the first $\omega(\Gamma) + 1$ terms of the Taylor expansion. These terms are specifically removed by the subtractions.[7]

Note that formulas (8.23), (8.24), (8.25) can be applied directly to an arbitrary 1PI graph Γ, in any order of the perturbative expansion. We do not need to consider graphs from the lower orders except for the renormalization parts of Γ.

In a particular case where $\omega(\Gamma) < 0$ and all renormalization parts γ_i of Γ are disconnected and primitively divergent, the BPHZ prescription is reduced to independent subtractions for each γ_i. For example, if Γ has only two renormalization parts γ_1 and γ_2, which are disconnected, then we have three families of disconnected renormalization parts $\{\gamma_1\}$, $\{\gamma_2\}$, $\{\gamma_1, \gamma_2\}$, and in consequence

$$\begin{aligned} R_\Gamma &= I_\Gamma - I_{\Gamma/\gamma_1} (T_{\omega(\gamma_1)} I_{\gamma_1}) - I_{\Gamma/\gamma_2} (T_{\omega(\gamma_2)} I_{\gamma_2}) + I_{\Gamma/\{\gamma_1,\gamma_2\}} (T_{\omega(\gamma_1)} I_{\gamma_1})(T_{\omega(\gamma_2)} I_{\gamma_2}) \\ &= I_{\Gamma/\{\gamma_1,\gamma_2\}} (I_{\gamma_1} - T_{\omega(\gamma_1)} I_{\gamma_1})(I_{\gamma_2} - T_{\omega(\gamma_2)} I_{\gamma_2}). \end{aligned}$$

See also Exercise 8.4.

It can happen that all the renormalization parts are nested, that is they form the ordered sequence of subgraphs $\gamma_1 \supset \gamma_2 \supset \ldots \supset \gamma_S$. Then, we have just S one-element families of disconnected renormalization parts, but now the internal subtractions are needed. The BPHZ prescription gives the nested sequence of subtractions. For example, let $\omega(\Gamma) < 0$ and $S = 2$. Then,

$$R_\Gamma = I_\Gamma - I_{\Gamma/\gamma_1} T_{\omega(\gamma_1)} I_{\gamma_1}^{i.s.} - I_{\Gamma/\gamma_2} T_{\omega(\gamma_2)} I_{\gamma_2},$$

where

$$I_{\gamma_1}^{i.s.} = I_{\gamma_1} - I_{\gamma_1/\gamma_2} T_{\omega(\gamma_2)} I_{\gamma_2}.$$

Using the identities

$$I_\Gamma = I_{\Gamma/\gamma_1} I_{\gamma_1}, \quad I_{\Gamma/\gamma_2} = I_{\Gamma/\gamma_1} I_{\gamma_1/\gamma_2}$$

we find that

$$R_\Gamma = I_{\Gamma/\gamma_1} \left(I_{\gamma_1}^{i.s.} - T_{\omega(\gamma_1)} I_{\gamma_1}^{i.s.} \right).$$

In the next section we consider the graph shown in Fig. 8.9. In that case, the renormalization parts are neither disconnected or nested—they overlap—and it is not obvious what is the correct way of making the subtractions. One of the advantages of the BPHZ prescription is that it can be easily applied in such less obvious cases.

---

[7]One may remove more terms than necessary. Such an operation is called an oversubtraction.

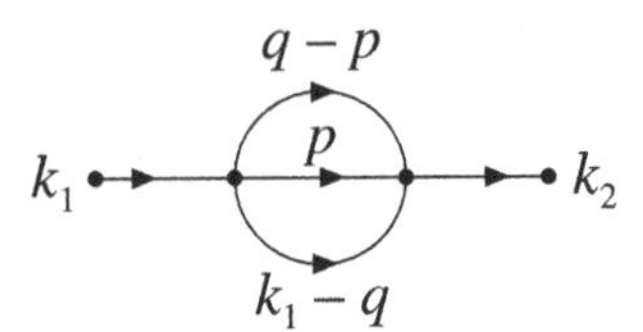

Fig. 8.9 The first nontrivial 1PI graph contributing to $\tilde{G}^{(2)}$. Its 1PI part, obtained by removing Δ_F from the external lines, is denoted as A_2. The meaning of the small arrows is explained in the text

8.4 Renormalization of the 2-Point Green's Function

In the second order of the perturbative expansion for $\tilde{G}^{(2)}$ we have the 1PI graph A_2 presented in Fig. 8.9. When ascribing the four-momenta to the lines of this graph we have taken into account the fact that in the limit (8.5), or even before taking that limit if we use the Pauli–Villars regularization (8.7), the four-momenta are not independent due to the Dirac deltas at the internal vertices. Moreover, we have put arrows on the lines in order to indicate at which end of the line the four-momentum ascribed to the line is taken with a plus sign: the rule is that it is the end the arrow points to. At the other end of the line it is taken with a minus sign. One may imagine that the four-momentum flows along the line, from a source at one end to a sink at the other end—the four-momentum flowing into the sink is counted with the plus sign. In the model $:\phi_4^4:$ we can put an arrow on a given line as we wish because Δ_F and d^4p are not sensitive to the change $p \to -p$.

The nonrenormalized, Pauli–Villars regularized contribution to $\tilde{G}^{(2)}(k_1, k_2; M)$, corresponding to this graph, has the form

$$96\Delta_F(k_1)\Delta_F(k_2)\delta(k_1+k_2)A_2,$$

where

$$A_2 = \int d^4p d^4q \; I_{A_2}(p, q; k_1; M),$$

and

$$I_{A_2}(p, q; k_1; M) = \left(\frac{\lambda_0}{4!(2\pi)^4}\right)^2 \Delta_{P-V}(q-p)\Delta_{P-V}(p)\Delta_{P-V}(k_1-q).$$

The external lines are not regularized.

The graph A_2 has two independent loops. We expect that it is quadratically divergent in the limit $M \to \infty$ because $\omega(A_2) = 2$. The BPHZ prescription can remove the UV divergences from the graph, but it turns out that it violates Lorentz invariance. Therefore, we shall modify the prescription in such a way that the renormalized perturbative contributions to $\tilde{G}^{(2)}$ will be manifestly Lorentz invariant.

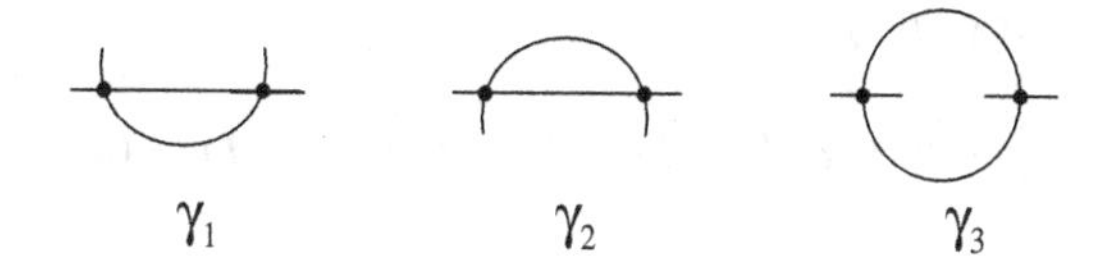

Fig. 8.10 The three renormalization parts of the graph A_2

The graph A_2 contains three renormalization parts $\gamma_i,\ i = 1, 2, 3$, shown in Fig. 8.10. In the limit $M \to \infty$ they are logarithmically divergent, $\omega(\gamma_i) = 0$. We can form three families of disconnected renormalization parts, each family has just one element: $\{\gamma_1\}, \{\gamma_2\}, \{\gamma_3\}$. According to formula (8.24)

$$I_{A_2}^{i.s.} = I_{A_2} - \Delta_{P-V}(q-p)T_0 I_{\gamma_1} - \Delta_{P-V}(k_1-q)T_0 I_{\gamma_2} - \Delta_{P-V}(p)T_0 I_{\gamma_3}. \quad (8.26)$$

As the external four-momenta for subgraph γ_1 we may take $k_1 \equiv p_1$ and $-(q - p) \equiv p_2$ (the four-momenta flowing into the internal vertex on the l.h.s. of the graph). Therefore,

$$I_{\gamma_1} = \left(\frac{\lambda_0}{4!(2\pi)^4}\right)^2 \Delta_{P-V}(p_1 - q)\, \Delta_{P-V}(p_2 + q),$$

and

$$T_0 I_{\gamma_1} = \left(\frac{\lambda_0}{4!(2\pi)^4}\right)^2 \Delta_{P-V}(\overset{(0)}{k_1} - q)\, \Delta_{P-V}(\overset{(0)}{k_2} + q),$$

where $\overset{(0)}{k_i}$ are the four-momenta of the symmetric subtraction point (8.18). For subgraph γ_2, we may take as the external four-momenta $k_1 \equiv p_1,\ q - k_1 \equiv p_2$, hence

$$T_0 I_{\gamma_2} = \left(\frac{\lambda_0}{4!(2\pi)^4}\right)^2 \Delta_{P-V}(\overset{(0)}{k_1} + \overset{(0)}{k_2} - p)\, \Delta_{P-V}(p).$$

In the case of γ_3 the external four-momenta are $k_1 \equiv p_1, -p \equiv p_2$, and

$$T_0 I_{\gamma_3} = \left(\frac{\lambda_0}{4!(2\pi)^4}\right)^2 \Delta_{P-V}(\overset{(0)}{k_1} - q)\, \Delta_{P-V}(\overset{(0)}{k_2} + q).$$

The renormalized contribution of graph A_2 is given by the formulas

$$A_2^{ren} = \lim_{M\to\infty} \int d^4p d^4q\ R_{A_2}(p, q; k_1; M), \quad (8.27)$$

where

$$R_{A_2}(p, q; k_1; M) =$$

$$I^{i.s.}_{A_2}(p, q; k_1; M) - I^{i.s.}_{A_2}(p, q; \overset{(0)}{k}_1; M) - (k_1 - \overset{(0)}{k}_1)^\mu \left. \frac{\partial I^{i.s.}_{A_2}(p, q; k_1; M)}{\partial k_1^\mu} \right|_{k_1 = \overset{(0)}{k}_1}$$

$$- \frac{1}{2}(k_1 - \overset{(0)}{k}_1)^\mu (k_1 - \overset{(0)}{k}_1)^\nu \left. \frac{\partial^2 I^{i.s.}_{A_2}(p, q; k_1; M)}{\partial k_1^\mu \partial k_1^\nu} \right|_{k_1 = \overset{(0)}{k}_1}. \tag{8.28}$$

Inspection of integral (8.27) shows that it can be calculated term by term if we use the Pauli–Villars regularization with $N = 2$. One can also check that

$$\Pi^{i.s}_{A_2} \overset{df}{=} \int d^4p d^4q \; I^{i.s.}_{A_2}(p, q; k_1; M)$$

depends on the Lorentz scalar k_1^2 when $k_1^2 < 0$, and also on $\overset{(0)}{k}_1 \overset{(0)}{k}_2 = \mu^2/3$ and $(\overset{(0)}{k}_1)^2 = (\overset{(0)}{k}_2)^2 = -\mu^2$. When $k_1^2 \geq 0$, a dependence on sign(k_1^0) may also appear, because the sign of k_1^0 is Lorentz invariant for time- and light-like k_1.

The subtracted terms in formula (8.28) explicitly contain the fixed four-vector $\overset{(0)}{k}_1$. One may worry that this is not compatible with the Lorentz invariance. The point is that, as we shall see in Chap. 10, if the model is Lorentz invariant then $\tilde{G}(k_1)$ introduced by the formula

$$\tilde{G}^{(2)}(k_1, k_2) = \delta(k_1 + k_2)\tilde{G}(k_1), \tag{8.29}$$

has the property

$$\tilde{G}(Lk_1) = \tilde{G}(k_1) \tag{8.30}$$

for arbitrary proper, orthochronous Lorentz transformations ($L \in L_+^\uparrow$). In the perturbative expansion, this property should hold separately for the total contribution in each order (that is for the sum of the contributions from all graphs in the given order). Graph A_2 is the only second order graph contributing to $\tilde{G}^{(2)}$. Therefore, it should give a Lorentz invariant expression. It turns out that this is not the case. Let us assume that $k_1^2 < 0$. Then, $\Pi^{i.s}_{A_2}$ depends only on k_1^2, and

$$\left. \frac{\partial \Pi^{i.s}_{A_2}(k_1^2; M; \mu)}{\partial k_1^\mu} \right|_{k_1 = \overset{(0)}{k}_1} = 2 \overset{(0)}{k}_{1\mu} \left. \frac{\partial \Pi^{i.s}_{A_2}(k_1^2; M; \mu)}{\partial (k_1^2)} \right|_{k_1^2 = -\mu^2},$$

$$\left. \frac{\partial^2 \Pi^{i.s}_{A_2}(k_1^2; M; \mu)}{\partial k_1^\mu \partial k_1^\nu} \right|_{k_1 = \overset{(0)}{k}_1} =$$

$$2\eta_{\mu\nu} \left. \frac{\partial \Pi^{i.s}_{A_2}(k_1^2; M; \mu)}{\partial (k_1^2)} \right|_{k_1^2 = -\mu^2} + 4 \overset{(0)}{k}_{1\mu} \overset{(0)}{k}_{1\nu} \left. \frac{\partial^2 \Pi^{i.s}_{A_2}(k_1^2; M; \mu)}{\partial (k_1^2) \partial (k_1^2)} \right|_{k_1^2 = -\mu^2},$$

where $\eta_{\mu\nu}$ are components of the Minkowski metric tensor. Therefore,

$$\int d^4p d^4q \; R_{A_2}(p, q; k_1; M) = \Pi^{i.s}_{A_2}(k_1^2; M; \mu) - \Pi^{i.s}_{A_2}(-\mu^2; M; \mu)$$

$$-(k_1^2 + \mu^2) \left. \frac{\partial \Pi^{i.s}_{A_2}(k_1^2; M; \mu)}{\partial(k_1^2)} \right|_{k_1^2=-\mu^2} - 2(k_1 \overset{(0)}{k}_1 + \mu^2)^2 \left. \frac{\partial^2 \Pi^{i.s}_{A_2}(k_1^2; M; \mu)}{\partial(k_1^2)\partial(k_1^2)} \right|_{k_1^2=-\mu^2} .$$

The last term is not compatible with Lorentz invariance because in general $(Lk_1)\overset{(0)}{k}_1 \neq k_1 \overset{(0)}{k}_1$.

Fortunately, the harmful term may be omitted, because it is not necessary for the removal of the UV divergences. One can easily see that by taking a particular subtraction point, namely such that $\mu^2 = 0$ and $\overset{(0)}{k}_1 = 0$. Then that term simply vanishes, so the UV divergences are removed despite its absence. One can also give another argument. The two derivatives with respect to k_1^2 lower the dimension by 4, hence that term has a superficial degree of divergence equal to -2. Therefore, it is finite in the limit $M \to \infty$, and in consequence it is irrelevant for the removal of UV divergences.

In order to preserve the Lorentz invariance we have to modify the subtraction procedure for the 2-point Green's function: after introducing the intermediate regularization we first compute $\Pi^{i.s}_{A_2}(k_1^2; M; \mu)$, and next we subtract the first two terms of the Taylor series with respect to k_1^2 at $k_1^2 = -\mu^2$,

$$\Pi^{ren}_{A_2}(k_1^2; \mu) = \lim_{M\to\infty} \Bigg(\Pi^{i.s}_{A_2}(k_1^2; M; \mu)$$

$$-\Pi^{i.s}_{A_2}(-\mu^2; M; \mu) - (k_1^2 + \mu^2) \left. \frac{\partial \Pi^{i.s}_{A_2}(k_1^2; M; \mu)}{\partial(k_1^2)} \right|_{k_1^2=-\mu^2} \Bigg). \tag{8.31}$$

Note that such $\Pi^{ren}_{A_2}(k_1^2; \mu)$ obeys the following identities (the renormalization conditions)

$$\Pi^{ren}_{A_2}(-\mu^2; \mu) = 0, \qquad \left. \frac{\partial \Pi^{ren}_{A_2}(k_1^2; \mu)}{\partial(k_1^2)} \right|_{k_1^2=-\mu^2} = 0. \tag{8.32}$$

Similarly as in the case of graph A_1, the subtractions (8.31) can be interpreted as the result of adding to the interaction $\tilde{V}_{Ig}[\tilde{\beta}]$ the counterterm

$$\delta_2 \tilde{V}_{Ig}[\tilde{\beta}] = \frac{\lambda_0^2}{12(2\pi)^8} \int d^4q_1 d^4q_2 \; \delta(q_1 + q_2) \tilde{\beta}(q_1) \tilde{\beta}(q_2) \left[B_1 + B_2(q_1^2 + \mu^2) \right], \tag{8.33}$$

where

$$B_1 = \Pi^{i.s}_{A_2}(-\mu^2; M; \mu), \quad B_2 = \left. \frac{\partial \Pi^{i.s}_{A_2}(k_1^2; M; \mu)}{\partial (k_1^2)} \right|_{k_1^2=-\mu^2}.$$

This corresponds to adding the term

$$\delta_2 S = -\frac{\lambda_0^2}{12(2\pi)^8} \int d^4x \left[(B_1 + B_2\mu^2)\phi^2(x) + B_2 \partial_\mu \phi(x) \partial^\mu \phi(x) \right] \tag{8.34}$$

to the action functional (8.6). The internal subtractions are implemented by the counterterm (8.21).

The counterterms $\delta_1 \tilde{V}_{Ig}$ and $\delta_2 \tilde{V}_{Ig}$ remove all UV divergences in the order λ^2. In the next order the divergences reappear, and new counterterms have to be included. They also have the general form (8.21), (8.33), because there are no other types of divergent graphs than the ones already considered: quadratically divergent 1PI graphs with 2 legs and logarithmically divergent 1PI graphs with 4 legs. The constants C_1, B_1 and B_2 will have new values. Thus, in spite of the ad hoc modifications (the inclusion of the counterterms), the action functional preserves its original form (8.6). Generally, the action functional in perturbatively renormalizable models may change its form, but after a finite number of such changes it reaches its stable form, in which only coefficients are changed when we go to still higher orders.

Note that as far as the removal of the UV divergences is concerned, the constants B_1, B_2 and C_1 may be changed by adding to them finite constants b_1, b_2 and c_1 of appropriate dimensionality which are independent of M. Then, the renormalization conditions change their form, e.g., instead of (8.32) we have

$$\Pi^{ren}_{A_2}(-\mu^2; \mu) = -b_1, \quad \left. \frac{\partial \Pi^{ren}_{A_2}(k_1^2; \mu)}{\partial (k_1^2)} \right|_{k_1^2=-\mu^2} = -b_2.$$

The new $\Pi^{ren}_{A_2}(k_1^2; \mu)$ differs from the one given by formula (8.31) by the term $-b_1 - (k_1^2 + \mu^2)b_2$. Such freedom in the concrete form of subtracted terms implies that actually the renormalized perturbative expansion contains arbitrary finite constants. Their number is equal to the number of renormalization conditions, and is finite in renormalizable models. On the other hand, in nonrenormalizable models the number of such constants increases indefinitely with the increasing order of the perturbative expansion—this fact greatly diminishes the predictive power of these models.

8.5 The Multiplicative Renormalization

We have seen how to renormalize separate graphs. Now we will look at the effect of renormalization on the whole Green's functions, which are given by an infinite series of graphs.[8] The theory is regularized in order to avoid mathematically meaningless expressions that correspond to the 1PI graphs with loops. We use the version of the Pauli–Villars regularization with $N = 2$, described at the beginning of Sect. 8.2. The existence of the limit $M \to \infty$ is secured by adding the counterterms to the interaction $\tilde{V}_{Ig}[\tilde{\beta}]$ given by formula (7.56). Their general form reads

$$\delta \tilde{V}_{Ig} = \delta_1 \tilde{V}_{Ig} + \delta_2 \tilde{V}_{Ig},$$

where

$$\delta_1 \tilde{V}_{Ig} = (Z_1 - 1)\frac{\lambda_0}{4!}\int \prod_{i=1}^{4} d^4 q_i \; \tilde{g}(q_1, q_2, q_3, q_4)\; \tilde{\beta}(q_1)\tilde{\beta}(q_2)\tilde{\beta}(q_3)\tilde{\beta}(q_4), \quad (8.35)$$

and

$$\delta_2 \tilde{V}_{Ig} = \frac{1}{2}\int d^4 q_1 d^4 q_2 \; \delta(q_1 + q_2)\tilde{\beta}(q_1)\tilde{\beta}(q_2)\left[(1 - Z_3)(q_1^2 - m_0^2) + \delta m^2 Z_3\right]. \quad (8.36)$$

The constants Z_1, Z_3 and δm^2 are adjusted order by order in the perturbative expansion. Because these constants are divergent in the limit $M \to \infty$, they are called infinite renormalization constants.

For example, a comparison with the results of Sects. 8.2, 8.4 for the graphs A_1 and A_2 gives

$$Z_1 = 1 + \frac{3\lambda_0 C_1}{2(2\pi)^4}, \quad Z_3 = 1 - \frac{\lambda_0^2 B_2}{6(2\pi)^8}, \quad \delta m^2 = \frac{\lambda_0^2 B_1 + \lambda_0^2 (m_0^2 + \mu^2) B_2}{6(2\pi)^8},$$

where we have neglected all terms which give higher than second powers of λ_0 in the perturbative expansion for the Green's functions.

Let us denote by $\tilde{G}_s^{(n)}(p_1, p_2, \ldots, p_n; \lambda_0, m_0^2, \mu, M)$ the Fourier transform of the n-point Green's function, calculated by means of the regularized perturbative series with the BPHZ subtractions. Here the subscript s refers to the subtractions, M to the Pauli–Villars regularization with $N = 2$, and μ to the subtraction point. The

[8] This series is likely not convergent. Typically, one expects that perturbative expansions in quantum field theory yield a so called asymptotic series which form a special class of divergent series. In most applications of the perturbative expansions, the series is either cut to a finite sum of graphs (then the problem of convergence disappears), or it is restricted to an infinite subclass of graphs which are distinguished by their particularly simple analytical contributions (and then sometimes one can compute the sum).

subtractions are implemented by the counterterms (8.35), (8.36). The limit $M \to \infty$ of this function exists, and it is called the renormalized Green's function,

$$\tilde{G}^{(n)}_{ren}(p_1, p_2, \ldots, p_n; \lambda_0, m_0^2, \mu) \equiv \lim_{M\to\infty} \tilde{G}^{(n)}_s(p_1, p_2, \ldots, p_n; \lambda_0, m_0^2, \mu, M). \tag{8.37}$$

There exists a certain, very important relation between $\tilde{G}^{(n)}_s(p_i; \lambda_0, m_0^2, \mu, M)$, and the corresponding regularized Green's function without any subtractions denoted by $\tilde{G}^{(n)}(p_1, p_2, \ldots, p_n; \lambda_0, m_0^2, M)$. The relation has the following form

$$\tilde{G}^{(n)}_s(p_1, p_2, \ldots, p_n; \lambda_0, m_0^2, \mu, M) = Z_3^{-\frac{n}{2}} \tilde{G}^{(n)}(p_1, p_2, \ldots, p_n; \lambda_b, m_b^2, M), \tag{8.38}$$

where

$$\lambda_b = \lambda_0 Z_1 Z_3^{-2}, \quad m_b^2 = m_0^2 + \delta m^2. \tag{8.39}$$

Relation (8.38) shows that the subtractions are equivalent to a shift of the mass parameter $m_0^2 \to m_b^2$, and to a rescaling of the Green's function by the factor $Z_3^{-n/2}$ and of the coupling constant λ_0 by the factor Z_1/Z_3^2. It is often called the formula of multiplicative renormalization. The constants λ_b, m_b^2 are called the bare coupling constant and the bare mass parameter, respectively.

In order to prove relation (8.38), we just calculate the effects of the counterterms $\delta_1 \tilde{V}_{Ig}$ and $\delta_2 \tilde{V}_{Ig}$ on a graph Γ constructed in the regularized model without the counterterms. Let us start with $\delta_1 \tilde{V}_{Ig}$. The graphs are generated from formulas (7.59), (7.60), but now $\tilde{V}_{Ig}$ is replaced by

$$\tilde{V}_{Ig} + \delta_1 \tilde{V}_{Ig} = \frac{\lambda_0 Z_1}{4!} \int \prod_{i=1}^{4} d^4 q_i \; \tilde{g}(q_1, q_2, q_3, q_4) \; \tilde{\beta}(q_1)\tilde{\beta}(q_2)\tilde{\beta}(q_3)\tilde{\beta}(q_4).$$

Therefore, the net effect of the counterterm $\delta_1 \tilde{V}_{Ig}$ is that λ_0 is replaced by $\lambda_0 Z_1$ in all internal vertices.

The counterterm $\delta_2 \tilde{V}_{Ig}$ contributes to the functional $\tilde{Z}^{(n)}_I$ the factor

$$\exp(-i \int d^4x \; \delta_2 V_I[\tilde{\beta}]) =$$

$$\exp\left(\frac{i}{2} \int d^4 q_1 d^4 q_2 \left[(Z_3 - 1)(q_1^2 - m_0^2) - \delta m^2 Z_3\right] \delta(q_1 + q_2)\tilde{\beta}(q_1)\tilde{\beta}(q_2)\right).$$

It yields a new internal vertex with two legs and the factor $f_0 \equiv i(Z_3 - 1)(q_1^2 - m_0^2) - i\delta m^2 Z_3$ associated with it.[9] With the new internal vertex available, each line of a given graph Γ can be 'decorated' by putting the new vertex on it arbitrarily many times. We consider here graphs for Green's functions that contain external vertices

[9]The factor 1/2 is canceled by the combinatorial factor 2, which appears because the vertex can be connected to two lines in two ways.

as well as propagators Δ_F on the external lines. Any two neighbouring vertices are connected by a line which represents Δ_F, and the four-leg internal vertices (7.64) contain the regularizing function (8.7). Summing all graphs obtained from Γ by decorating all its lines with the new internal vertex, we effectively obtain the graph Γ again, but each line of it now represents the whole sum

$$\begin{aligned}&\Delta_F(p)+\Delta_F(p)f_0\Delta_F(p)+\Delta_F(p)(f_0\Delta_F(p))^2+\ldots\\&=\Delta_F(p)\frac{1}{1-f_0\Delta_F(p)}=\frac{1}{Z_3}\frac{i}{p^2-(m_0^2+\delta m^2)+i0_+}.\end{aligned}$$

Thus, the inclusion of the counterterm $\delta_2\tilde{V}_{Ig}$ is equivalent to multiplying each line by $1/Z_3$ and shifting the mass parameter m_0^2 by δm^2.

In the last step in the derivation of formula (8.38), we collect the factors $1/Z_3$ from all lines of the graph: this gives an overall factor $(1/Z_3)^{I+n}$. Formula (8.1) implies that $I+n=2l+n/2$. Hence, the overall factor can be written in the form $(Z_3^{-2})^l Z_3^{-n/2}$, which shows that we may ascribe the factor Z_3^{-2} to each internal vertex (there are l of them), and $Z_3^{-1/2}$ to each external vertex. In other words, the factor $1/Z_3$ on each line of the graph is written as $(1/\sqrt{Z_3})^2$ and each $1/\sqrt{Z_3}$ is moved to the internal or external vertices adjacent to the line. In this way $1/Z_3$ disappears from all lines. The net result of the inclusion of the counterterms is that each internal vertex (7.64) is multiplied by $Z_1Z_3^{-2}$, the mass parameter m_0^2 is replaced by $m_0^2+\delta m^2$, and the graph is multiplied by $Z_3^{-n/2}$. This holds for each graph Γ in the perturbative expansion for $\tilde{G}^{(n)}(p_1,p_2,\ldots,p_n;\lambda_0,m_0^2,M)$.

One particle irreducible (1PI) Green's functions $\tilde{\Gamma}^{(n)}$, often also called proper vertices, are obtained from the perturbative expansion for $\tilde{G}^{(n)}$ by throwing away all graphs that are not 1PI and removing the propagators Δ_F from the external lines. Calculation of the effects of the inclusion of the counterterms differs only slightly from the one presented above. Because the external lines are absent now, we miss the factors $1/\sqrt{Z_3}$ needed for obtaining λ_b for the internal vertices adjacent to the external lines. We introduce these factors by multiplying the graph by $1=(\sqrt{Z_3}/\sqrt{Z_3})^n$. Therefore, in the present case we have a factor $\sqrt{Z_3}$ for each external vertex instead of $1/\sqrt{Z_3}$ obtained in the case of Green's functions. Thus, the formula for the proper vertices analogous to (8.38) has the form

$$\tilde{\Gamma}_s^{(n)}(p_1,p_2,\ldots,p_n;\lambda_0,m_0^2,\mu,M)=Z_3^{\frac{n}{2}}\tilde{\Gamma}^{(n)}(p_1,p_2,\ldots,p_n;\lambda_b,m_b^2,M).\tag{8.40}$$

In the limit $M\to\infty$ the l.h.s. of this formula gives the renormalized 1PI Green's function,

$$\tilde{\Gamma}_{ren}^{(n)}(p_1,p_2,\ldots,p_n;\lambda_0,m_0^2,\mu)\equiv\lim_{M\to\infty}\tilde{\Gamma}_s^{(n)}(p_1,p_2,\ldots,p_n;\lambda_0,m_0^2,\mu,M).\tag{8.41}$$

Formulas (8.38), (8.40) lose mathematical meaning in the limit $M\to\infty$ because then Z_3, λ_b, and m_b^2 are divergent. Of course, these divergences cancel each other

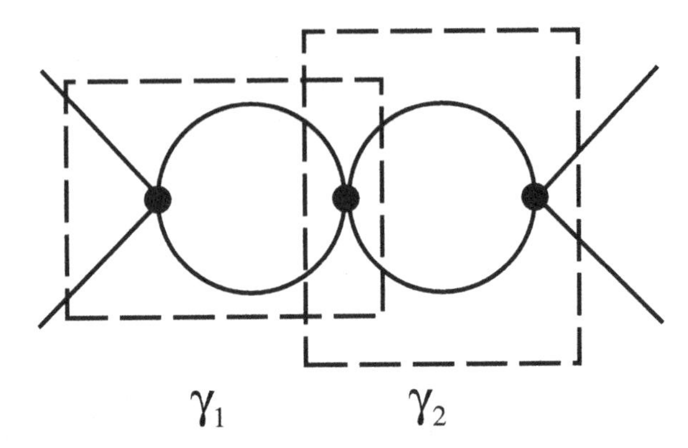

Fig. 8.11 The graph Γ

and produce a finite limit—the renormalized Green's functions or the proper vertices, respectively—but the renormalized functions do not have the forms given on the r.h.s.'s of formulas (8.38), (8.40) with meaningful constants Z_3, λ_b and m_b^2.

Exercises

8.1 Show that a graph equivalent to the one presented in Fig. 8.2 can be drawn on a plane (without the intersections of lines) if the external lines have a finite length.

8.2 Check the renormalizability of the models $\lambda_0 : \phi_d^n :$, where $n > 2$ is a natural number and $d \geq 2$ is the dimension of space-time, by computing:
(a) the dimension of the coupling constant λ_0,
(b) the superficial degree of divergence.

8.3 Construct an explicit example of the symmetric subtraction point.
Hint. Assume that

$$\overset{(0)}{k_1} = (0, 0, 0, \mu), \quad \overset{(0)}{k_2} = (0, 0, \alpha, \beta), \quad \overset{(0)}{k_3} = (0, x, y, z),$$

and choose α, β, x, y, z in order to obey conditions (8.18). Check that $\overset{(0)}{k_4} = -\overset{(0)}{k_1} - \overset{(0)}{k_2} - \overset{(0)}{k_3}$ obeys these conditions automatically.

8.4 Graph Γ has the form presented in Fig. 8.11. The dashed boxes mark its renormalization parts γ_1 and γ_2. They are not disconnected. Prove that

$$R_\Gamma = \left(\frac{\lambda_0}{4!(2\pi)^4}\right)^{-1} R_{\gamma_1} R_{\gamma_2}.$$

8.5 Prove that the superficial degree of divergence of the QED graph Γ in four-dimensional space-time with n_p external photon lines and n_e external fermion (electron or positron) lines is equal to

$$\omega(\Gamma) = 4 - \frac{3}{2} n_e - n_p.$$

What would the formula for $\omega(\Gamma)$ look like had we considered QED in D-dimensional space-time?
The rules for Feynman diagrams in QED can be found in, e.g., [5, 9, 11].

8.6 Derive the general form of Feynman's parametric representation:

$$\prod_{i=1}^{n} \frac{1}{A_i^{\alpha_i}} = \frac{\Gamma(\alpha)}{\prod_{i=1}^{n} \Gamma(\alpha_i)} \int_0^1 \prod_{i=1}^{n} dx_i \, \delta\left(1 - \sum_{i=1}^{n} x_i\right) \frac{\prod_{i=1}^{n} x_i^{\alpha_i - 1}}{\left[\sum_{i=1}^{n} x_i A_i\right]^{\alpha}},$$

where $\alpha = \sum_{i=1}^{n} \alpha_i$.

8.7 Prove the formula:

$$\int \frac{d^D k}{(k^2 + 2k \cdot Q - M^2 + i0_+)^n} = i(-1)^n \pi^{\frac{D}{2}} \frac{\Gamma(n - \frac{D}{2})}{\Gamma(n)} (Q^2 + M^2)^{\frac{D}{2} - n},$$

where for all $D-$vectors the scalar product is $a \cdot b = a^0 b^0 - \vec{a} \cdot \vec{b}$.
Hint. Deform the contour of integration over k^0 onto the imaginary axis in the complex k^0 plane and perform the resulting integral by rewriting it in the spherical coordinates of D-dimensional Euclidean space. The value of the integral over the angular variables can be obtained by comparing the results of calculating the D-dimensional Gaussian integral

$$I_D = \int_{-\infty}^{\infty} \prod_{k=1}^{D} dx_k \, \mathrm{e}^{-\sum_{k=1}^{D} (x_k)^2}$$

in Cartesian and spherical coordinates; the integral over the radial direction can be performed with the help of an integral representation of the Euler beta function and its relation to the gamma function.

8.8 Graph (a) in Fig. 8.12 represents the one-loop, momentum space contribution to the electron self-energy.
(a) Write down its integral representation and denote it by $-i(2\pi)^4 \Sigma_1(\not{p}, m_0)$.
(b) In 4-dimensional space-time the integral appearing in $-i(2\pi)^4 \Sigma_1(\not{p}, m_0)$ is divergent, with a superficial degree of divergence equal to 1 (compare with Problem 8.5). Regularize it by assuming that the number of space-time dimensions D is sufficiently small and then evaluate it using Feynman's parametric representation to combine the denominators and using the formula derived in the problem 8.7.

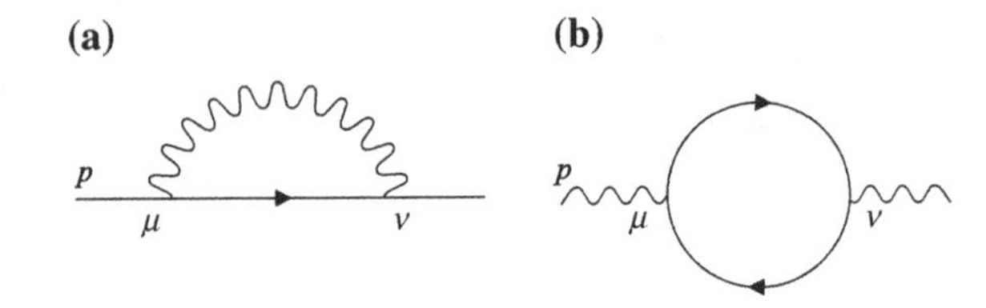

Fig. 8.12 One-loop contributions to the two-point, 1PI, QED Green's functions

Hint. Since the index μ of the Dirac matrices takes D values in D-dimensional space-time, we have the following useful formulae

$$\gamma_\mu \gamma^\mu = D I_4, \qquad \gamma_\mu \not{p} \gamma^\mu = (2-D)\not{p}$$

(derive them).
Answer:

$$\Sigma_1(\not{p}, m_0) = \frac{e_0^2}{8\pi^{2+\varepsilon}}\Gamma(\varepsilon)\int_0^1 dx\, \left((2-\varepsilon)m_0 - (1-\varepsilon)(1-x)\not{p}\right)\left(x m_0^2 - x(1-x)p^2\right)^{-\varepsilon}$$

where $D = 4 - 2\varepsilon$.
(c) The divergence of $\Sigma_1(\not{p}, m_0)$ in four-dimensional space-time reveals itself as a pole of the gamma function for $\varepsilon \to 0$. To obtain the renormalized one-loop contribution to the electron self-energy apply the subtraction procedure (with the subtraction point $p = 0$), i.e. calculate

$$\Sigma_1^{ren}(\not{p}, m_0) = \lim_{\varepsilon \to 0}\left(\Sigma_1(\not{p}, m_0) - \Sigma_1(0, m_0) - p^\mu \frac{\partial \Sigma_1(\not{p}, m_0)}{\partial p^\mu}\bigg|_{p=0}\right).$$

(d) Prove that the one-loop corrected, renormalized electron propagator is given by the formula

$$S_F(\not{p}, m_0) = \frac{i}{(2\pi)^4}\left(\not{p} - m_0 - \Sigma_1^{ren}(\not{p}, m_0) + i0_+\right)^{-1} + \mathcal{O}\left(e_0^4\right).$$

8.9 Denote by $\Pi_{\mu\nu}(p) = -i(2\pi)^4\left(p_\mu p_\nu - p^2 \eta_{\mu\nu}\right)\pi_1(p^2)$ the 1PI contribution to the photon propagator (the so called vacuum polarization tensor) specified by graph (*b*) in Fig. 8.12.
(a) Calculate it using the same strategy as in Exercise 8.8.
Hint. Check that in D-dimensional space-time

$$D\int d^D k\, k_\mu k_\nu f(k^2) = \eta_{\mu\nu}\int d^D k\, k^2 f(k^2)$$

where $f(k^2)$ is some function and, using the result of Exercise 8.7, derive the formula

$$\int d^D k \frac{2k_\mu k_\nu - \eta_{\mu\nu} k^2}{\left(k^2 - m_0^2 + x(1-x)p^2\right)^2} = -i\pi^{2-\varepsilon}\eta_{\mu\nu}\Gamma(\varepsilon)\left(m_0^2 - x(1-x)p^2\right)^{1-\varepsilon}.$$

Answer:

$$\pi_1(p^2) = -\frac{e_0^2}{2\pi^{2+\varepsilon}}\Gamma(\varepsilon)\int_0^1 dx\, x(1-x)\left(m_0^2 - x(1-x)p^2\right)^{-\varepsilon}.$$

(b) Compute the renormalized, one-loop contribution to the photon propagator choosing the subtraction point at $\overset{(0)}{p}$ with $(\overset{(0)}{p})^2 = -\mu^2$.
Answer:

$$\pi_1^{ren}(p^2) = \frac{e_0^2}{2\pi^2}\int_0^1 dx\, x(1-x)\log\frac{m_0^2 - x(1-x)p^2}{m_0^2 + x(1-x)\mu^2}.$$

(c) Choose the free photon propagator in the so called transverse form

$$D_0^{\mu\nu}(p) = \frac{i}{(2\pi)^4}\left(-\eta^{\mu\nu} + \frac{p^\mu p^\nu}{p^2 + i0_+}\right)\frac{1}{p^2 + i0_+}.$$

Prove that the one-loop corrected, renormalized photon propagator is given by the formula

$$D_1^{\mu\nu}(p) = \frac{i}{(2\pi)^4}\left(-\eta^{\mu\nu} + \frac{p^\mu p^\nu}{p^2 + i0_+}\right)\frac{1}{(1 - \pi_1^{ren}(p^2))(p^2 + i0_+)} + \mathcal{O}\left(e_0^4\right).$$

8.10 Start with the QED Lagrangian with the physical quantities (field operators ψ and A_μ, electron mass m_0 and electric charge e_0) replaced by the "bare" quantities ψ_b, $A_{b\,\mu}$, m_b and e_b :

$$\mathcal{L}_b = -\frac{1}{4}F_b^{\mu\nu}F_{b\,\mu\nu} + \overline{\psi_b}\left[i\gamma_\mu\left(\partial^\mu + ie_b A_b^\mu\right) - m_b\right]\psi_b.$$

Relate the bare and physical (renormalized) quantities through the renormalization constants

$$\psi = Z_2^{-1/2}\psi_b, \quad A^\mu = Z_3^{-1/2}A_b^\mu, \quad e_0 = Z_3^{1/2}e_b, \quad m_0 = m_b + \Delta m.$$

The fact that the electromagnetic field renormalization constant is equal to the inverse of the electric charge renormalization constant is a peculiarity of QED related to the gauge invariance of this theory.
(a) Determine the form of

$$\delta\mathcal{L} = \mathcal{L}_b - \mathcal{L}$$

where $\mathcal{L}$ is of the same form as $\mathcal{L}_b$, with the bare quantities replaced by the physical ones, i.e.:

$$\mathcal{L} = -\frac{1}{4} F^{\mu\nu} F_{\mu\nu} + \overline{\psi_b} \left[i\gamma_\mu \left(\partial^\mu + i e_0 A^\mu \right) - m_0 \right] \psi.$$

(b) Treating $-\delta\mathcal{L}$ as an additional contribution to the interaction Hamiltonian (the counterterms) derive the form of the additional Feynman diagram vertices which appear thanks to its presence.
(c) Calculate the values of Z_2, Z_3 and Δm which result in the contribution to the electron and photon propagators equivalent to the subtractions applied in Exercises 8.8 and 8.9.

Answer:

$$Z_2 = 1 - \frac{e_0^2}{8\pi^{2+\varepsilon}} (1-\varepsilon)\Gamma(\varepsilon) \int_0^1 dx\, (1-x) \left(x^2 m_0^2 \right)^{-\varepsilon},$$

$$\Delta m = m_0 \frac{e_0^2}{8\pi^{2+\varepsilon}} \Gamma(\varepsilon) \int_0^1 dx\, \left(1 + (1-\varepsilon)x \right) \left(x^2 m_0^2 \right)^{-\varepsilon},$$

$$Z_3 = 1 - \frac{e_0^2}{2\pi^{2+\varepsilon}} \Gamma(\varepsilon) \int_0^1 dx\, x(1-x) \left(m_0^2 + x(1-x)\mu^2 \right)^{-\varepsilon}.$$

Chapter 9
The Renormalization Group

Abstract The relation between the subtracted Green's functions with different choices of subtraction point in the ϕ_4^4 model. The running coupling constant. Functional equations of the renormalization group. Differential renormalization group equations of the Gell-Mann–Low and the Callan–Symanzik type. The β function. Reliability of the perturbative approximations. The phenomenon of dimensional transmutation in renormalized quantum field theory.

The precise form of the perturbatively calculated and renormalized contributions to the Green's functions depends on the adopted scheme of subtractions. In particular, with the subtraction at the symmetric point (8.18) the dependence on the parameter μ appears. The choice of the subtraction point is not dictated by any concrete physical phenomena. On the contrary, the motivation for introducing it has been purely mathematical: the subtractions secure the existence of the limit $M \to \infty$ (the removal of the regularization), and the mathematical formalism itself does not point to any specific value of μ. Therefore, it is desirable to investigate the dependence of the renormalized Green's functions on μ in more detail. Another arbitrariness, also present in the renormalized perturbative expansion, has the form of the finite constants which can be included in the BPHZ subtractions, as discussed at the end of Sect. 8.4—it should be controlled too.

9.1 Renormalization Group Equations

Renormalization group equations for the renormalized Green's functions (in the $:\phi_4^4:$ model) follow essentially from formula (8.38). That formula implies a relation between the Green's functions $\tilde{G}_s^{(n)}$, obtained with two choices, μ and μ', of the symmetric subtraction point, because on the r.h.s. there is the Green's function in it without any subtractions. Simple calculation shows that

$$\tilde{G}_s^{(n)}(p_1, p_2, \dots, p_n; \lambda_0, m_0^2, \mu, M) = z_3^{\frac{n}{2}} \tilde{G}_s^{(n)}(p_1, p_2, \dots, p_n; \lambda_0 z_1^{-1} z_3^2, m_0^2 - \Delta m^2, \mu', M), \quad (9.1)$$

H. Arodź and L. Hadasz, *Lectures on Classical and Quantum Theory of Fields*, Graduate Texts in Physics, DOI 10.1007/978-3-319-55619-2_9

where

$$z_3 = \frac{Z_3'}{Z_3}, \quad z_1 = \frac{Z_1'}{Z_1}, \quad \Delta m^2 = \delta m'^2 - \delta m^2.$$

In this chapter we regard $p_1, \ldots, p_n, \lambda_0 > 0, m_0^2 > 0, \mu^2 > 0, M^2 > 0$ as variables. Taking the limit $M \to \infty$ on both sides of formula (9.1), we obtain a relation between the renormalized Green's functions with two choices for the subtraction point. Below we prove that the constants $z_1, z_3, \Delta m^2$ remain finite in that limit. Their asymptotic values (at $M \to \infty$) depend on $\mu, \mu', m_0^2, \lambda_0$.

The constants z_1 and z_3 are dimensionless. Therefore, they may be regarded as functions of the following three independent dimensionless variables: μ'/μ, m_0^2/μ^2, λ_0 :

$$z_1 = z_1(\frac{\mu'}{\mu}, \frac{m_0^2}{\mu^2}, \lambda_0), \quad z_3 = z_3(\frac{\mu'}{\mu}, \frac{m_0^2}{\mu^2}, \lambda_0).$$

The asymptotic value of Δm^2 is written in the form

$$\Delta m^2 = m_0^2 \left[1 - \frac{\mu'^2}{\mu^2} \underline{m}(\frac{\mu'}{\mu}, \frac{m_0^2}{\mu^2}, \lambda_0) \right], \tag{9.2}$$

where $\underline{m}$ is a dimensionless function of the indicated variables. Let us also introduce the running coupling constant (often called the effective coupling constant)

$$\underline{\lambda}(\frac{\mu'}{\mu}, \frac{m_0^2}{\mu^2}, \lambda_0) \stackrel{df}{=} \lambda_0 z_1^{-1} z_3^2. \tag{9.3}$$

Of course, for $\mu' = \mu$ the two subtraction points and the corresponding counterterms coincide, hence $z_1 = 1, z_3 = 1, \underline{m} = 1$ and $\underline{\lambda} = \lambda_0$.

The relation between the renormalized Green's functions that follows from (9.1) in the limit $M \to \infty$ can be written in the form

$$\tilde{G}_{ren}^{(n)}(p_1, p_2, \ldots, p_n; \lambda_0, m_0^2, \mu) = z_3^{\frac{n}{2}}(\frac{\mu'}{\mu}, \frac{m_0^2}{\mu^2}, \lambda_0)$$
$$\tilde{G}_{ren}^{(n)}(p_1, p_2, \ldots, p_n; \underline{\lambda}(\frac{\mu'}{\mu}, \frac{m_0^2}{\mu^2}, \lambda_0), m_0^2 \frac{\mu'^2}{\mu^2} \underline{m}(\frac{\mu'}{\mu}, \frac{m_0^2}{\mu^2}, \lambda_0), \mu'). \tag{9.4}$$

Thus, if the change of the subtraction point $\mu \to \mu'$ is accompanied by the substitutions

$$\lambda_0 \to \lambda' = \underline{\lambda}(\frac{\mu'}{\mu}, \frac{m_0^2}{\mu^2}, \lambda_0), \quad m_0^2 \to m'^2 = m_0^2 \frac{\mu'^2}{\mu^2} \underline{m}(\frac{\mu'}{\mu}, \frac{m_0^2}{\mu^2}, \lambda_0), \tag{9.5}$$

and by multiplication by $z_3^{n/2}$, we recover $\tilde{G}_{ren}^{(n)}(p_1, p_2, \ldots, p_n; \lambda_0, m_0^2, \mu)$. Formulas (9.5) can be regarded as a one-parameter family of transformations parameterized by

$t \equiv \mu'/\mu$. This parameter has values in the infinite interval $(0, \infty)$, and $t = 1$ gives the identity transformation. One may consider the pair (λ_0, m_0^2) as coordinates on a plane. Then transformations (9.5), considered for a continuous range of t around $t = 1$, give the curve

$$(\underline{\lambda}(t, \frac{m_0^2}{\mu^2}, \lambda_0),\; m_0^2 t^2 \underline{m}(t, \frac{m_0^2}{\mu^2}, \lambda_0))$$

in that plane. Such a curve is called the renormalization group trajectory (or r.g. flow) passing through the point (λ_0, m_0^2). The family of transformations (9.4) and (9.5) parameterized but $t \in (0, \infty)$ is called the renormalization group.

The formula for the proper vertices analogous to (9.4) has the form

$$\tilde{\Gamma}^{(n)}_{ren}(p_1, p_2, \ldots, p_n; \lambda_0, m_0^2, \mu)$$
$$= z_3^{-\frac{n}{2}} \tilde{\Gamma}^{(n)}_{ren}(p_1, p_2, \ldots, p_n;\; \underline{\lambda}(\frac{\mu'}{\mu}, \frac{m_0^2}{\mu^2}, \lambda_0),\; m_0^2 \frac{\mu'^2}{\mu^2} \underline{m}(\frac{\mu'}{\mu}, \frac{m_0^2}{\mu^2}, \lambda_0),\; \mu'). \quad (9.6)$$

This formula follows, in the limit $M \to \infty$, from a relation between the proper vertices analogous to (9.1).

Note that formulas (9.4) and (9.6) may easily be generalized to cases where the two subtraction schemes differ by much more than merely the concrete values of μ. The renormalization constants $Z_1, Z_3, \delta m^2$ and $Z_1', Z_3', \delta m^{2'}$, which are divergent in the limit $M \to \infty$ may correspond to any two renormalization schemes that can differ by the choice of regularization, as well as by the method of subtracting the divergent terms. Still, we define $z_1 = Z_1'/Z_1$, etc., as above, and obtain formulas that relate the renormalized Green's functions calculated in the two renormalization schemes.

Formulas (9.4) and (9.6) provide a convenient starting point for the calculation of z_1, z_3 and Δm^2, or equivalently z_3, $\underline{\lambda}$ and $\underline{m}$. We shall use the continuity of the renormalized 1PI Feynman graphs contributing to $\tilde{\Gamma}^{(2)}_{ren}$ and $\tilde{\Gamma}^{(4)}_{ren}$ with respect to the parameter μ', see below. The perturbative contributions are generalized functions of the external momenta, and therefore they can be singular at some momenta. For example, $A_1^{ren}(k^2)$ given by formula (8.19) is singular at $k^2 = 0$ in the case $m_0^2 = 0$. However, one can prove[1] that the renormalized contributions to $\tilde{G}^{(n)}_{ren}(p_1, \ldots, p_n)$, after dropping the $\delta(\sum_{i=1}^n p_i)$ factor and eliminating p_n (because $p_n = -\sum_{i=1}^{n-1} p_i$), become smooth functions of the external four-momenta $p_1, \ldots, p_{n-1}$ in the space-like domain defined by the inequalities $p_i^2 < 0,\; p_i p_k < 0$, where $i, k = 1, \ldots n-1$. In fact, this is the reason why the four-momenta of the symmetric subtraction point are space-like.

Let us consider formula (9.6) with $n = 4$ and $p_i = \overset{(0)}{p}_i(\mu')$, where $\overset{(0)}{p}_i(\mu')$ are the four-momenta from the symmetric point (8.18) with μ replaced by μ'. Due to the subtractions, the contributions of all the graphs with one or more loops to

---

[1]The proof can be found in, e.g., [8].

Fig. 9.1 The graph representing the first order contribution to $\tilde{\Gamma}^{(4)}$

$$\tilde{\Gamma}^{(4)}_{ren}(\overset{(0)}{p}_1(\mu'), \ldots, \overset{(0)}{p}_4(\mu');\ \underline{\lambda}(\frac{\mu'}{\mu}, \frac{m_0^2}{\mu^2}, \lambda_0),\ m_0^2\frac{\mu'^2}{\mu^2}\underline{m}(\frac{\mu'}{\mu}, \frac{m_0^2}{\mu^2}, \lambda_0),\ \mu'),$$

vanish, c.f. the renormalization condition (8.17). The only nonvanishing contribution, equal to $-i\underline{\lambda}/(2\pi)^4$, comes from the tree graph shown in Fig. 9.1. Therefore, just for these particular external four-momenta, relation (9.6) can be written in the form

$$\tilde{\Gamma}_4(\overset{(0)}{p}_1(\mu'), \overset{(0)}{p}_2(\mu'), \overset{(0)}{p}_3(\mu'); \lambda_0, m_0^2, \mu) = -i\frac{\underline{\lambda}(\frac{\mu'}{\mu}, \frac{m_0^2}{\mu^2}, \lambda)}{(2\pi)^4 z_3^2} = -\frac{i\lambda_0}{(2\pi)^4 z_1}, \tag{9.7}$$

where $\tilde{\Gamma}_4$ is defined by the formula

$$\tilde{\Gamma}^{(4)}_{ren}(p_1, p_2, p_3, p_4; \lambda_0, m_0^2, \mu) = \delta(\sum_{i=1}^{4} p_i)\ \tilde{\Gamma}_4(p_1, p_2, p_3; \lambda_0, m_0^2, \mu).$$

The l.h.s. of formula (9.7) is a nontrivial sum of 1PI graphs unless $\mu' = \mu$. In the latter case it is equal to $-i\lambda_0/(2\pi)^4 \neq 0$. Also, because of continuity with respect to μ', it does not vanish for $\mu' \neq \mu$, at least when μ' is sufficiently close to μ. Therefore, formula (9.7) implies that for such μ'

$$z_1 \neq 0,\ z_1 < \infty.$$

The functions z_3 and $\underline{m}$ can be determined from formula (9.4), in which we put $n = 2$ and $p_i = \overset{(0)}{p}_i(\mu')$. The Green's function on the r.h.s. is given by the zeroth order contribution. Therefore formula (9.4) can be written in the form

$$\tilde{G}(-\mu'^2; \lambda_0, m_0^2, \mu) = -\frac{i\mu^2 z_3}{\mu'^2\left(\mu^2 + m_0^2\,\underline{m}(\frac{\mu'}{\mu}, \frac{m_0^2}{\mu^2}, \lambda_0)\right)}, \tag{9.8}$$

where the function $\tilde{G}$ is defined by (8.29) and is represented by a nontrivial sum of graphs. The r.h.s. is given by the tree graph (7.68). Contributions from all other graphs vanish due to the subtractions, c.f. the renormalization conditions (8.32). The subtractions for the 1PI graphs with two external lines contain two terms, see formula (8.31). The presence of the term with the derivative implies the second renormalization condition (8.32), which leads to the formula

$$\tilde{G}'(-\mu'^2; \lambda_0, m_0^2, \mu) = -\frac{i\mu^4 z_3}{\mu'^4 \left(\mu^2 + m_0^2\, \underline{m}(\frac{\mu'}{\mu}, \frac{m_0^2}{\mu^2}, \lambda_0)\right)^2}, \tag{9.9}$$

where

$$\tilde{G}'(-\mu'^2; \lambda_0, m_0^2, \mu) = \left.\frac{\partial \tilde{G}(p^2; \lambda_0, m_0^2, \mu)}{\partial (p^2)}\right|_{p^2=-\mu'^2}.$$

The notation in (9.8) and (9.9) takes into account the fact that $\tilde{G}(p)$ is actually a function of p^2 when the four-momentum p_1 is space-like. This function of p^2 is denoted by $\tilde{G}(p^2)$. Relations (9.8) and (9.9) give

$$z_3(\frac{\mu'}{\mu}, \frac{m_0^2}{\mu^2}, \lambda_0) = \frac{i\tilde{G}^2}{\tilde{G}'}, \quad \underline{m}(\frac{\mu'}{\mu}, \frac{m_0^2}{\mu^2}, \lambda_0) = \frac{\mu^2}{m_0^2}\left(\frac{\tilde{G}}{\mu'^2 \tilde{G}'} - 1\right), \tag{9.10}$$

where for brevity we have omitted the arguments of $\tilde{G}$ and $\tilde{G}'$. We conclude that z_3 and $\underline{m}$ are finite, at least for μ' sufficiently close to μ.

Apart from proving the finiteness of z_1, z_3 and $\underline{m}$, formulas (9.7) and (9.10) show also that these functions are uniquely determined by the renormalized Green's functions. Thus, we may say that formula (9.4) determines $\underline{\lambda}$, z_3 and $\underline{m}$ uniquely at least for μ' close enough to μ. This fact is crucial for the derivation of the renormalization group equations presented below.

In the first step we derive a set of functional equations for $\underline{\lambda}$, z_3 and $\underline{m}$. Let us add a third subtraction point μ''. Formula (9.4) relates the corresponding renormalized Green's functions

$$\begin{aligned}&\tilde{G}^{(n)}_{ren}(p_i; \lambda_0, m_0^2, \mu) \\ &= z_3^{\frac{n}{2}}(\frac{\mu''}{\mu}, \frac{m_0^2}{\mu^2}, \lambda_0)\, \tilde{G}^{(n)}_{ren}(p_i; \underline{\lambda}(\frac{\mu''}{\mu}, \frac{m_0^2}{\mu^2}, \lambda_0), m_0^2 \frac{\mu''^2}{\mu^2} \underline{m}(\frac{\mu''}{\mu}, \frac{m_0^2}{\mu^2}, \lambda_0), \mu'').\end{aligned} \tag{9.11}$$

On the other hand,

$$\begin{aligned}&\tilde{G}^{(n)}_{ren}(p_i; \lambda', m'^2, \mu') \\ &= z_3^{\frac{n}{2}}(\frac{\mu''}{\mu'}, \frac{m'^2}{\mu'^2}, \lambda')\, \tilde{G}^{(n)}_{ren}(p_i; \underline{\lambda}(\frac{\mu''}{\mu'}, \frac{m'^2}{\mu'^2}, \lambda'), m'^2 \frac{\mu''^2}{\mu'^2} \underline{m}(\frac{\mu''}{\mu'}, \frac{m'^2}{\mu'^2}, \lambda'), \mu''),\end{aligned} \tag{9.12}$$

where λ' and m'^2 are defined in formulas (9.5). Inserting formula (9.12) on the r.h.s. of (9.4), we obtain the relation

$$\tilde{G}^{(n)}_{ren}(p_i;\lambda_0,m_0^2,\mu)=z_3^{\frac{n}{2}}(\frac{\mu'}{\mu},\frac{m_0^2}{\mu^2},\lambda_0)\,z_3^{\frac{n}{2}}(\frac{\mu''}{\mu'},\frac{m'^2}{\mu'^2},\lambda')$$

$$\tilde{G}^{(n)}_{ren}(p_i;\underline{\lambda}(\frac{\mu''}{\mu'},\frac{m'^2}{\mu'^2},\lambda'),m'^2\frac{\mu''^2}{\mu'^2}\underline{m}(\frac{\mu''}{\mu'},\frac{m'^2}{\mu'^2},\lambda'),\mu''). \quad (9.13)$$

Because of the above mentioned uniqueness of $\underline{\lambda}$, z_3 and $\underline{m}$, and by comparing relations (9.11) and (9.13), we conclude that $\underline{\lambda}$, z_3 and $\underline{m}$ obey the following functional equations[2]

$$\underline{\lambda}(\frac{\mu''}{\mu},\frac{m_0^2}{\mu^2},\lambda_0)=\underline{\lambda}\left(\frac{\mu''}{\mu'},\frac{m_0^2}{\mu^2}\underline{m}(\frac{\mu'}{\mu},\frac{m_0^2}{\mu^2},\lambda_0),\underline{\lambda}(\frac{\mu'}{\mu},\frac{m_0^2}{\mu^2},\lambda_0)\right), \quad (9.14)$$

$$\underline{m}(\frac{\mu''}{\mu},\frac{m_0^2}{\mu^2},\lambda_0)=\underline{m}(\frac{\mu'}{\mu},\frac{m_0^2}{\mu^2},\lambda_0)\,\underline{m}\left(\frac{\mu''}{\mu'},\frac{m_0^2}{\mu^2}\underline{m}(\frac{\mu'}{\mu},\frac{m_0^2}{\mu^2},\lambda_0),\underline{\lambda}(\frac{\mu'}{\mu},\frac{m_0^2}{\mu^2},\lambda_0)\right), \quad (9.15)$$

and

$$z_3(\frac{\mu''}{\mu},\frac{m_0^2}{\mu^2},\lambda_0)=z_3(\frac{\mu'}{\mu},\frac{m_0^2}{\mu^2},\lambda_0)\,z_3\left(\frac{\mu''}{\mu'},\frac{m_0^2}{\mu^2}\underline{m}(\frac{\mu'}{\mu},\frac{m_0^2}{\mu^2},\lambda_0),\underline{\lambda}(\frac{\mu'}{\mu},\frac{m_0^2}{\mu^2},\lambda_0)\right). \quad (9.16)$$

The functional equations obtained above can be used in order to generate various differential equations (or rather identities). Equations of the Gell-Mann–Low type are obtained by differentiating both sides of the functional equations with respect to μ'', and putting $\mu''=\mu'$ afterwards. In the case of (9.14) we obtain

$$t\frac{\partial\underline{\lambda}(t,m_0^2/\mu^2,\lambda_0)}{\partial t}=\beta(\frac{m_0^2}{\mu^2}\underline{m}(t,\frac{m_0^2}{\mu^2},\lambda_0),\underline{\lambda}(t,\frac{m_0^2}{\mu^2},\lambda_0)), \quad (9.17)$$

where

$$t=\mu'/\mu,$$

and the Gell-Mann–Low function $\beta(m^2/\mu^2,\lambda)$ is defined as follows:

$$\beta(\frac{m^2}{\mu^2},\lambda)\stackrel{df}{=}\left.\frac{\partial\underline{\lambda}(x,m^2/\mu^2,\lambda)}{\partial x}\right|_{x=1}. \quad (9.18)$$

Here x just denotes the first argument of the function $\underline{\lambda}$. Here it does not matter which letter we use, because we finally put $x=1$. From (9.15), we analogously obtain

$$t\frac{\partial\ln\underline{m}(t,m_0^2/\mu^2,\lambda_0)}{\partial t}=\gamma_m\left(\frac{m_0^2}{\mu^2}\underline{m}(t,\frac{m_0^2}{\mu^2},\lambda_0),\ \underline{\lambda}(t,\frac{m_0^2}{\mu^2},\lambda_0)\right), \quad (9.19)$$

[2]In the presented approach to the renormalization group they are just identities which follow from the definitions of $\underline{\lambda}$, $\underline{m}$ and z_3. Nevertheless, we shall call them equations as in most textbooks.

where

$$\gamma_m(\frac{m^2}{\mu^2}, \lambda) \stackrel{df}{=} \left.\frac{\partial \underline{m}(x, m^2/\mu^2, \lambda)}{\partial x}\right|_{x=1}. \tag{9.20}$$

Finally, (9.16) gives

$$t\frac{\partial \ln z_3(t, m_0^2/\mu^2, \lambda_0)}{\partial t} = \gamma\left(\frac{m_0^2}{\mu^2}\underline{m}(t, \frac{m_0^2}{\mu^2}, \lambda_0),\ \underline{\lambda}(t, \frac{m_0^2}{\mu^2}, \lambda_0)\right), \tag{9.21}$$

where

$$\gamma(\frac{m^2}{\mu^2}, \lambda) \stackrel{df}{=} \left.\frac{\partial z_3(x, m^2/\mu^2, \lambda)}{\partial x}\right|_{x=1}. \tag{9.22}$$

These differential equations are supplemented with the 'initial conditions'

$$\underline{\lambda}(1, \frac{m_0^2}{\mu^2}, \lambda_0) = \lambda_0,\ \underline{m}(1, \frac{m_0^2}{\mu^2}, \lambda_0) = 1,\ z_3(1, \frac{m_0^2}{\mu^2}, \lambda_0) = 1. \tag{9.23}$$

Note that in order to calculate the functions β, γ_m and γ it is sufficient to know the functions $\underline{\lambda}$, $\underline{m}$ and z_3 for all t from an arbitrarily small open interval containing $t = 1$. The differential equations (9.17), (9.19) and (9.21) can be used in order to calculate these functions for t outside that arbitrarily small interval.

Another set of differential equations, called the Callan–Symanzik equations, is obtained by differentiation of the functional equations (9.14)–(9.16) with respect to μ', next putting $\mu' = \mu$, and finally changing the notation $\mu'' \to \mu'$. The derivatives of the l.h.s.'s of the functional equations vanish, while on the r.h.s.'s we obtain derivatives with respect to all three arguments. For example, (9.14) gives the Callan–Symanzik equation for the running coupling constant:

$$t\frac{\partial\underline{\lambda}(t, m_0^2/\mu^2, \lambda_0)}{\partial t} - \frac{m_0^2}{\mu^2}\gamma_m(\frac{m_0^2}{\mu^2}, \lambda_0)\frac{\partial\underline{\lambda}(t, m_0^2/\mu^2, \lambda_0)}{\partial(m_0^2/\mu^2)} - \beta(\frac{m_0^2}{\mu^2}, \lambda_0)\frac{\partial\underline{\lambda}(t, m_0^2/\mu^2, \lambda_0)}{\partial\lambda_0} = 0. \tag{9.24}$$

The differentiation with respect to μ' and subsequent substitution $\mu' = \mu$ applied to formula (9.4) gives the Callan–Symanzik equation for $\tilde{G}^{(n)}_{ren}$:

$$\mu\frac{\partial\tilde{G}^{(n)}_{ren}}{\partial\mu} + \beta(\frac{m_0^2}{\mu^2}, \lambda_0)\frac{\partial\tilde{G}^{(n)}_{ren}}{\partial\lambda_0} + m_0^2\big[\gamma_m(\frac{m_0^2}{\mu^2}, \lambda_0) + 2\big]\frac{\partial\tilde{G}^{(n)}_{ren}}{\partial m_0^2} + \frac{n}{2}\gamma(\frac{m_0^2}{\mu^2}, \lambda_0)\tilde{G}^{(n)}_{ren} = 0. \tag{9.25}$$

The Callan–Symanzik equation for $\tilde{\Gamma}^{(n)}_{ren}$ has a similar form. It can readily be obtained from relation (9.6).

9.2 The Running Coupling Constant

The running coupling constant plays an important role in assessing the reliability of the perturbative approximation. Let us introduce a dimensionless function $\tilde{g}^{(n)}$ (which should not be confused with the regularizing function $\tilde{g}$ considered in the previous chapters) such that

$$\tilde{G}^{(n)}_{ren}(p_i; \lambda_0, m_0^2, \mu) = m_0^{d_0}\, \tilde{g}^{(n)}(\frac{p_i}{\mu}; \lambda_0, \frac{m_0^2}{\mu^2}), \tag{9.26}$$

where $d_0 = -3n$ is the dimension of $\tilde{G}^{(n)}_{ren}$ in units of mass.[3] We have assumed that $m_0 \neq 0$. The perturbative contributions to the renormalized Green's function can always be written in the form (9.26). One can see this from the formulas for the BPHZ subtractions and for the free propagator $\Delta_F(k)$: all external and internal four-momenta k_j are written as $k_j = \mu\, k_j/\mu$, where $\mu > 0$, and the factors μ are extracted, e.g.,

$$\frac{i}{k^2 - m_0^2 + i0_+} = \mu^{-2} \frac{i}{k^2/\mu^2 - m_0^2/\mu^2 + i0_+}.$$

The factors μ can also be extracted from the four-momenta $\overset{(0)}{p}_i$ which appear in the symmetric subtraction point (8.18), from the cutoff parameter M ($M^2 = \mu^2 M^2/\mu^2$), as well as from the four-momenta in $\delta(\sum p_i)$. Finally, we write $\mu = m_0\, \mu/m_0$ and collect all factors m_0. This gives the overall factor $m_0^{d_0}$. The definition (9.26), used on both sides of relation (9.4), gives

$$\tilde{g}^{(n)}(\frac{p_i}{\mu}; \lambda_0, \frac{m_0^2}{\mu^2}) = \left(\frac{\mu'}{\mu}\right)^{d_0} \underline{m}^{d_0/2}(\frac{\mu'}{\mu}, \frac{m_0^2}{\mu^2}, \lambda_0)$$
$$z_3^{n/2}(\frac{\mu'}{\mu}, \frac{m_0^2}{\mu^2}, \lambda_0)\; \tilde{g}^{(n)}\left(\frac{p_i}{\mu'}; \underline{\lambda}(\frac{\mu'}{\mu}, \frac{m_0^2}{\mu^2}, \lambda_0), \frac{m_0^2}{\mu^2}\underline{m}(\frac{\mu'}{\mu}, \frac{m_0^2}{\mu^2}, \lambda_0)\right). \tag{9.27}$$

Let us take particular four-momenta p_i,

$$p_i = \frac{\mu'}{\mu} \underline{p}_i, \tag{9.28}$$

where the momenta $\underline{p}_i$ are fixed. Then, the four-vectors $p_i/\mu,\ p_i/\mu'$ present in formula (9.27) can be written as

[3] In the natural units ($\hbar = 1,\ c = 1$) the field $\phi(x)$ has the dimension cm^{-1}, and the vacuum state vector $|0\rangle$ is dimensionless, hence $[G^{(n)}] = \mathrm{cm}^{-n}$. The Fourier transform changes the dimension by $+4n$. Therefore, $[\tilde{G}^{(n)}] = \mathrm{cm}^{+3n} = [m_0]^{-3n}$.

$$\frac{p_i}{\mu} = t\frac{\underline{p}_i}{\mu}, \quad \frac{p_i}{\mu'} = \frac{\underline{p}_i}{\mu},$$

where $t = \mu'/\mu$. Therefore,

$$\tilde{g}^{(n)}(t\frac{\underline{p}_i}{\mu}; \lambda_0, \frac{m_0^2}{\mu^2}) = t^{d_0}\underline{m}^{d_0/2} z_3^{n/2}\, \tilde{g}^{(n)}\left(\frac{\underline{p}_i}{\mu}; \underline{\lambda}, \frac{m_0^2}{\mu^2}\underline{m}\right), \tag{9.29}$$

where $\underline{m}$, z_3 and $\underline{\lambda}$ are functions of t, m_0^2/μ^2 and λ_0 as shown in (9.27). Using the definition (9.26) again, we see that

$$\tilde{G}^{(n)}_{ren}(t\underline{p}_i; \lambda_0, m_0^2, \mu) = t^{d_0} z_3^{n/2} \tilde{G}^{(n)}_{ren}(\underline{p}_i; \underline{\lambda}, m_0^2\, \underline{m}, \mu). \tag{9.30}$$

On both sides of this relation we have the renormalized Green's functions with the same subtraction point μ.

Relation (9.30) shows that the renormalized Green's functions calculated at the four-momenta $t\underline{p}_i$ in the model with coupling constant λ_0, are related to the Green's functions calculated at the four-momenta $\underline{p}_i$ in the model with the coupling constant $\underline{\lambda}$. It is essentially a consequence of straightforward dimensional analysis applied to relation (9.4).

Suppose that there exists t_0 such that $\underline{\lambda}(t, m_0^2/\mu^2, \lambda_0) \to 0$ when $t \to t_0$. Because $\underline{\lambda}$ is the actual coupling constant on the r.h.s. of formula (9.30), one may hope that for $p_i \approx t_0\underline{p}_i$ one can obtain a good approximation to $\tilde{G}^{(n)}_{ren}(t\underline{p}_i; \lambda_0, m_0^2, \mu)$ by taking into account only the first few terms in the perturbative expansion for $\tilde{G}^{(n)}_{ren}$ on the r.h.s. of formula (9.30). On the other hand, if $\underline{\lambda}$ diverges at a certain $t = t_\infty$, i.e., $\underline{\lambda}(t, m_0^2/\mu^2, \lambda_0) \to \infty$ when $t \to t_\infty$, the perturbative approximation is not trustworthy at the four-momenta $p_i \approx t_\infty\underline{p}_i$. A more precise meaning of these statements is as follows. Suppose that we have calculated the perturbative approximation for $\tilde{G}^{(n)}_{ren}(\underline{p}_i; \lambda_0, m_0^2, \mu)$ up to a certain finite order in λ_0 at certain four-momenta $\underline{p}_i$. Relation (9.30) says that when we use such a perturbative formula with the rescaled four-momenta $t\underline{p}_i$ instead of $\underline{p}_i$, we may take as the four-momenta again $\underline{p}_i$, but the coupling constant λ_0 should then be replaced by $\underline{\lambda}$ (of course one should also include the prefactors $t^{d_0} z_3^{n/2}$, and $\underline{m}$). It is clear that we can trust the perturbative formula with the four-momenta equal to $t\underline{p}_i$ when $\underline{\lambda} < \lambda_0$, and we should be concerned about its usefulness if $\underline{\lambda} \gg \lambda_0$. At t_∞ our approximation completely breaks down. In practice, such considerations yield information about the reliability of the perturbative approximation only for very small or very large four-momenta, because working with the perturbative approximations for $\underline{\lambda}$ one usually finds that t_0 and t_∞ are either equal to 0 or very large. Also note that all components of all four-momenta in formula (9.30) are rescaled by the same factor t. It remains an open question what happens if we keep finite, e.g., the momenta $\vec{p}_i$ and rescale only the components p_i^0 (the energies).

We have just seen that the behavior of the running coupling constant as a function of t is crucial for checking in which asymptotic region we may trust the perturbative approximation. In order to investigate the behavior of $\underline{\lambda}$ we use the Gell-Mann–Low equation (9.17), which has to be considered together with (9.19) for $\underline{m}$. The functions β and γ_m can be computed from their definitions (9.18) and (9.20). To this end, we only need to know $\underline{\lambda}$ and $\underline{m}$ for $t \approx 1$. For this we may use formulas (9.7) and (9.10) in which $\tilde{\Gamma}_4$ and $\tilde{G}$ are calculated perturbatively.

As an example, let us find the form of equations (9.17) and (9.19) in the 1-loop approximation, in which only graph A_1 (Fig. 8.7) is present, apart from the zeroth order graphs. The self-energy graph A_2, Fig. 8.9, has two independent loops, therefore it is discarded. Thus,

$$Z_1 = 1 + \frac{3\lambda_0 C_1}{2(2\pi)^4}, \;\; Z_3 = 1, \;\; \delta m^2 = 0.$$

It follows that

$$z_3^{(1)} = 1, \quad \underline{m}^{(1)}(\frac{\mu'}{\mu}, \frac{m_0^2}{\mu^2}, \lambda_0) = \frac{\mu^2}{\mu'^2} = \frac{1}{t^2},$$

where the superscript $^{(1)}$ denotes the 1-loop approximation. Definition (9.20) gives $\gamma_m^{(1)} = -2$. In order to find $\beta^{(1)}$ we use the one-loop approximation for $\tilde{\Gamma}_4$,

$$\tilde{\Gamma}_4 = -\frac{i\lambda_0}{(2\pi)^4} - \frac{3i\lambda_0^2}{8(2\pi)^6}\int_0^1 dz \, \ln\frac{3m_0^2/\mu^2 + 4t^2 z(1-z)}{3m_0^2/\mu^2 + 4z(1-z)} + \mathcal{O}(\lambda_0^3), \tag{9.31}$$

where we have used result (8.19) together with the combinatorial factor $(4!)^2$ shown in the first line of Fig. 7.6. The factor 3 in front of the integral appears because there are three 1PI graphs (the first line in Fig. 7.6) which contribute to $\tilde{\Gamma}_4$—because the subtraction point is the symmetric one they give identical contributions. Since $z_3^{(1)} = 1$, formulas (9.3), (9.7), (9.18), and (9.31) give

$$\beta^{(1)}(\frac{m_0^2}{\mu^2}, \lambda_0) = \frac{3\lambda_0^2}{16\pi^2}\int_0^1 dz \frac{4z(1-z)}{4z(1-z) + 3m_0^2/\mu^2}. \tag{9.32}$$

Therefore, the Gell-Mann–Low equation (9.17) in the 1-loop order has the form

$$t\frac{\partial\underline{\lambda}}{\partial t} = \frac{3\underline{\lambda}^2}{16\pi^2}\int_0^1 dz \, \frac{4z(1-z)}{4z(1-z) + 3m_0^2/(\mu^2 t^2)}. \tag{9.33}$$

The integral over z is elementary, but it gives a rather complicated function of t. Result (9.33) holds in the renormalization scheme used in Chap. 8.

One can simplify the approximate form of the β function by adopting a special renormalization scheme. Especially attractive in this respect is the so called mass independent (MI) renormalization scheme in which one puts $m_0 = 0$ in the constants Z_1 and Z_3, but of course not in the original Feynman graphs. It turns out that such subtractions are sufficient for the removal of the UV divergences.

For example, let us reconsider the graph A_1 from Sect. 8.2. In the MI scheme we replace definition (8.16) of the renormalized contribution by

$$A_1^{MI}(k^2) \stackrel{df}{=} \lim_{M\to\infty} \left(A_1(k^2; M) - A_1((\overset{(0)}{k})^2; M)|_{m_0=0} \right), \tag{9.34}$$

where in the first term on the r.h.s. we still keep the original value $m_0 > 0$. With this new definition, formula (8.19) for A_1^{ren} is replaced by

$$A_1^{MI} = i\pi^2 \frac{\lambda_0^2}{(4!)^2(2\pi)^8} \int_0^1 dz \ln \frac{3m_0^2 - 3k^2 z(1-z)}{4\mu^2 z(1-z)}.$$

It is clear that A_1^{MI} does not obey the renormalization condition (8.17).

The MI renormalization scheme has the same types of counterterms as discussed in Sect. 8.5, only the concrete values of the constants Z_1, Z_3 and δm^2 are different. Therefore, the multiplicative renormalization formulas (8.38) and (8.40) are still valid.

Let us calculate the β function in the MI scheme in the 1-loop approximation. Because Z_1 and Z_3 do not depend on m_0^2, the same is true for z_1, z_3 and, in consequence, for $\underline{\lambda}$ defined by formula (9.3). Therefore, the dimensional analysis applied to $\underline{\lambda}$ in the MI scheme implies that it is a function of $t = \mu'/\mu$ and λ_0,

$$\underline{\lambda} = \underline{\lambda}^{MI}(t, \lambda_0).$$

Definition (9.18) implies that β in (9.17) depends only on $\underline{\lambda}$. Hence, in the MI scheme, the Gell-Mann–Low equation (9.17) decouples from the equation for $\underline{m}$,

$$t \frac{\partial \underline{\lambda}^{MI}(t, \lambda_0)}{\partial t} = \beta^{MI}(\underline{\lambda}^{MI}(t, \lambda_0)). \tag{9.35}$$

In order to calculate β^{MI} we need $\underline{\lambda}^{MI}$ for $t \approx 1$. The perturbative approximation for it can be directly found from definition (9.3), and the definitions of z_1 and z_3 given at the beginning of Sect. 9.1. For example, if we calculate β^{MI} in the 1-loop order, we may put $z_3 = 1$ as before. The counterterm giving Z_1 is essentially defined by formula (9.34). Including appropriate numerical factors and taking the limit $M \to \infty$ we find that

$$z_1^{MI} = \lim_{M\to\infty} \frac{Z_1'}{Z_1} = 1 - \frac{3\lambda_0}{16\pi^2} \ln t + \mathcal{O}(\lambda_0^2).$$

Therefore,

$$\underline{\lambda}^{MI} = \lambda_0 + \frac{3\lambda_0^2}{16\pi^2} \ln t + \mathcal{O}(\lambda_0^3), \tag{9.36}$$

and finally in the 1-loop approximation

$$\beta^{MI}(\lambda_0) = \frac{3\lambda_0^2}{16\pi^2}. \tag{9.37}$$

Let us insert formula (9.37) on the r.h.s. of (9.35),

$$t\frac{\partial \underline{\lambda}^{MI}(t, \lambda_0)}{\partial t} = \frac{3}{16\pi^2}(\underline{\lambda}^{MI}(t, \lambda_0))^2.$$

The solution of this equation with the 'initial condition'

$$\underline{\lambda}^{MI}(1, \lambda_0) = \lambda_0$$

has the form

$$\underline{\lambda}^{MI}(t, \lambda_0) = \frac{\lambda_0}{1 - \frac{3}{16\pi^2}\lambda_0 \ln t}. \tag{9.38}$$

Comparing (9.38) with formula (9.36), we see that the first two terms in the expansions in powers of λ_0 coincide, but (9.38) contains terms of an arbitrarily high order. Formula (9.38) is often called the renormalization group improved version of (9.36). Of course, formula (9.38) is not the exact formula for the running coupling constant, because we have used the approximate form of the β function.

Formula (9.38) implies that

$$t_0 = 0, \quad t_\infty = \exp(\frac{16\pi^2}{3\lambda_0}) \approx (7.25 \times 10^{22})^{1/\lambda_0}.$$

Thus, we may expect that when the four-momenta become large, the quality of the perturbative approximation will worsen.

In the model $\lambda_0 :\phi_6^3:$, which involves the real scalar field in a six-dimensional space-time with the (self)interaction $\lambda_0 :\phi^3:$, one finds that the first non-vanishing contribution to the β function has the form

$$\beta^{MI}(\lambda_0) = -a_1\lambda_0^3,$$

where a_1 is a positive constant. In this case, the Gell-Mann–Low equation (9.35) gives

$$\underline{\lambda}^{MI}(t, \lambda_0) = \frac{\lambda_0^2}{1 + 2a_1\lambda_0^2 \ln t}.$$

Hence, in this model

$$t_0 = \infty, \quad t_\infty = \exp(-\frac{1}{2a_1\lambda_0^2}) < 1.$$

Now the accuracy of the perturbative approximation is better at the large four-momenta $t\underline{p}_i$, $t \gg 1$, and worse at the four-momenta $t\underline{p}_i$, $t < 1$. Models in which $\underline{\lambda} \to 0$ as $t \to \infty$ are called asymptotically free. The $:\phi_6^3:$ model is renormalizable, but it is not very interesting because of the large dimensionality of the space-time, and also because the corresponding quantum Hamiltonian is likely not bounded from below. A much more interesting asymptotically free theory is provided by the quantum Yang–Mills fields. This theory is the main ingredient of modern theories of interactions of particles. It is discussed in Chap. 12.

9.3 Dimensional Transmutation

Dimensional transmutation is the phenomenon of an emerging physical mass scale in superficially massless quantum field models. For example, the classical theory of the Yang–Mills fields contains a dimensionless coupling constant g and no explicit mass parameter ($m_0 = 0$). On the other hand, there are many indications of particles called glueballs with a non-zero rest mass in the quantum version of that theory. It is a puzzle as to how a non-zero rest mass can be obtained in a theory, in which no dimensional parameter is available. In fact, it cannot be obtained in the classical theory, but the quantum theory of the Yang–Mills field actually contains a dimensional parameter, namely the subtraction parameter μ, or equivalent parameters in other renormalization schemes. This answer is not fully satisfactory because μ has no physical meaning—it can have arbitrary positive values. However, it turns out that by using μ one can construct a parameter of the dimension of mass which is constant on the renormalization group trajectory, hence that parameter belongs to the set of physical characteristics of the model.

A physically meaningful quantity $F(\lambda_0, m_0^2, \mu)$, defined within the perturbative approach, should be constant on the trajectories (9.5) of the renormalization group transformations—in other words, F should be invariant under the renormalization group transformations,

$$F(\lambda_0, m_0^2, \mu) = F\left(\underline{\lambda}(\frac{\mu'}{\mu}, \frac{m_0^2}{\mu^2}, \lambda_0), m_0^2\frac{\mu'^2}{\mu^2}\underline{m}(\frac{\mu'}{\mu}, \frac{m_0^2}{\mu^2}, \lambda_0), \mu'\right). \tag{9.39}$$

The differential form of this condition is obtained by differentiation with respect to μ' and putting $\mu' = \mu$,

$$\mu\frac{\partial F(\lambda_0, m_0^2, \mu)}{\partial \mu} + \beta(\frac{m_0^2}{\mu^2}, \lambda_0)\frac{\partial F(\lambda_0, m_0^2, \mu)}{\partial \lambda_0} + m_0^2\left[2 + \gamma_m(\frac{m_0^2}{\mu^2}, \lambda_0)\right]\frac{\partial F(\lambda_0, m_0^2, \mu)}{\partial m_0^2} = 0. \tag{9.40}$$

It is the Callan–Symanzik equation for F. Note that the renormalized Green's functions $\tilde{G}^{(n)}_{ren}$ do not obey condition (9.39)—they are not invariant with respect to the renormalization group transformations (9.5).

In the massless case ($m_0 = 0$) (9.40) reduces to

$$\mu \frac{\partial F(\lambda_0, \mu)}{\partial \mu} + \beta(\lambda_0) \frac{\partial F(\lambda_0, \mu)}{\partial \lambda_0} = 0. \tag{9.41}$$

It is clear that $F(\lambda_0, \mu) = \mu$ does not obey this condition. On the other hand, let us take

$$F(\lambda_0, \mu) = \Lambda(\lambda_0, \mu) \stackrel{df}{=} \mu \exp\left(-\int_a^{\lambda_0} \frac{d\lambda'}{\beta(\lambda')}\right), \tag{9.42}$$

where a is a constant. Simple calculation shows that $\Lambda(\lambda_0, \mu)$ obeys condition (9.40), hence it is a renormalization group invariant. It provides the physically meaningful mass scale. Of course, λ_0 and the constant a should be chosen in such a way that the integral in the exponent exists.

Exercises

9.1 Compute $\Lambda(\lambda_0, \mu)$ for the massless $:\phi_4^4:$ and $:\phi_6^3:$ models using the results of Sect. 9.2. Analyze the behavior of Λ when $\lambda_0 \to 0_+$.

9.2 Using the value of the Z_3 renormalization constant, calculated for the subtraction point $\overset{(0)}{p}$ such that $(\overset{(0)}{p})^2 = -\mu^2$ in Exercise 8.10,

$$Z_3 = Z_3(\mu) = 1 - \frac{e_0^2}{2\pi^{2+\varepsilon}} \Gamma(\varepsilon) \int_0^1 dx\, x(1-x) \left(m_0^2 + x(1-x)\mu^2\right)^{-\varepsilon},$$

and the relation $e(\mu) = \sqrt{Z_3(\mu)}\, e_b$, find the form of the one-loop beta function in QED in the case $\mu \gg m_0$.

Answer:

$$\beta(e(\mu)) = \frac{e^3(\mu)}{12\pi^2} + \mathcal{O}\left(e^5(\mu)\right).$$

Chapter 10
Relativistic Invariance and the Spectral Decomposition of $G^{(2)}$

Abstract The requirements for a relativistically invariant quantum field theory. Generators of the unitary representations of the universal covering group of the Poincaré group, and their commutation relations. The spectral decomposition of the two-point function $G^{(2)}$ in the quantum theory of the real scalar field. The contribution of the single particle states. The pole of $\tilde{G}^{(2)}$ at the physical value of p^2 of the single particle. Finite mass corrections to the renormalized two-point function.

We have seen how one can perturbatively compute Green's functions in the $:\phi_4^4:$ model. For purely mathematical reasons, we have had to introduce the regularizing function g, which does not have any physical meaning. Next, we have shown that one can redefine the model (by including the subtractions) in such a way that the regularizing function can be removed. This is done graph by graph, and the sum of all such renormalized graphs up to a certain finite order defines the renormalized, perturbative Green's functions. Computations of infinite sums of graphs are possible only in rather special cases, because calculations of contributions represented by graphs with a large number of loops in general are prohibitively complicated.

In the presence of the regularizing function, the model, and in particular the interaction Hamiltonian $\hat{V}_{Ig}$, is well-defined in the Fock space spanned by the basis states $|0_I\rangle$, $\hat{a}_I^\dagger(\vec{k})|0_I\rangle, \ldots$. We expect that the perturbatively calculated renormalized Green's functions are approximations of Green's functions of a certain relativistic model which can be called the exact $:\phi_4^4:$ model. We have already mentioned in Chap. 7 that we do not know how to construct such an exact model. Nevertheless, accepting a number of reasonable assumptions about its properties, we can derive a certain formula for the exact Green's function $\tilde{G}^{(2)}$, known as the spectral decomposition. The assumptions include the relativistic invariance and the particle interpretation. Next, by comparing the spectral decomposition with the perturbative, renormalized Green's function $\tilde{G}^{(2)}_{ren}$, we shall see that if the latter is to be an approximation to the exact Green's function, the mass parameter m_0^2 of the initial Lagrangian (7.1) has to be chosen in a special way. Only in the zeroth order is this parameter

H. Arodź and L. Hadasz, *Lectures on Classical and Quantum Theory of Fields*,
Graduate Texts in Physics, DOI 10.1007/978-3-319-55619-2_10

equal to m^2, that is to the square of the rest mass of the scalar particle associated with the field $\hat{\phi}(x)$. In general, $m_0^2 = m^2\,(1 + a_2\lambda_0^2 + a_3\lambda_0^3 + \ldots)$, where $a_2, a_3, \ldots$ are dimensionless functions of m^2/μ^2. They can be calculated within the framework of the renormalized perturbative expansion.

10.1 Relativistic Invariance in QFT

Similarly as in Sect. 3.1, we consider the proper orthochronous Lorentz transformations, which form the group $L_+^{\uparrow}$, and the translations in Minkowski space-time (the group T_4). Together they form the Poincaré group $\mathcal{P}$,

$$\mathcal{P} = \{(\hat{L}, a) : \hat{L} \in L_+^{\uparrow}, a \in T_4\}.$$

In the present context of relativistic invariance we adopt the so called passive interpretation of the Poincaré transformations

$$x^{'\mu} = L^{\mu}{}_{\nu} x^{\nu} + a^{\mu}, \tag{10.1}$$

where $\hat{L} = (L^{\mu}{}_{\nu})$, $a = (a^{\mu})$ do not depend on x^{μ}. Namely, formula (10.1) is regarded as a change of Cartesian coordinates in Minkowski space-time M. Thus, x^{μ} and $x^{'\mu}$ are coordinates of the same point in M. The alternative (so called active) interpretation assumes that we use one Cartesian coordinate system in M and (10.1) defines a transformation of the points in M: the point x with the coordinates x^{μ} is moved to the point x' with the coordinates $x^{'\mu}$.

Thus, the Poincaré transformation (10.1) now represents a change of inertial reference frame in which we investigate the fields. The fundamental assumption is that such frames are equivalent in the sense that all physical laws, which in particular say which phenomena are possible and which are not, are identical in all of them.[1]

The group multiplication in $\mathcal{P}$ has the form

$$(\hat{L}_2, a_2)(\hat{L}_1, a_1) = (\hat{L}_2\hat{L}_1, \hat{L}_2 a_1 + a_2). \tag{10.2}$$

The Poincaré group is the most important group of symmetries of Minkowski space-time. Its unitary irreducible representations (UIR's) appear in the definition of the relativistically invariant quantum field theory (QFT) given below.

[1] This does not have to be true if one generalizes Poincaré transformations (10.1). For example, often one performs a Lorentz transformation to the rest frame of an accelerated particle. Such a Lorentz transformation is time-dependent, because the particle changes its velocity. The rest frame is non-inertial, and the transformation is not a symmetry. The physics in the rest frame is different from that of the inertial laboratory frame, because in the former case any physical object (particles or fields) is affected by special forces like the centrifugal one. In the rest frame they are real forces, which in quantum field theory may lead, e.g., to creation of particle-antiparticle pairs. Such forces are absent in the inertial laboratory frame.

The theory of the symmetry of quantum systems was developed mainly by E.P. Wigner. It belongs among the most beautiful pieces of theoretical physics. Below we briefly outline the main points of that theory.

Let us begin from the observation that physical states[2] of the quantum system are represented by (that is, they are in one-to-one correspondence with) rays in a Hilbert space $\mathcal{H}$, and not with vectors in $\mathcal{H}$. The ray $[\psi]$ is the set of all vectors in $\mathcal{H}$ obtained from a single vector $|\psi\rangle$ by multiplying it by an arbitrary complex number different from 0. Thus,

$$[\psi] = \{c|\psi\rangle : c \in C, c \neq 0\}. \tag{10.3}$$

Any concrete vector belonging to the given ray is called a representative of that ray. Actually, it is sufficient to consider normalized rays, obtained by adding the restrictions

$$|c| = 1, \quad \langle\psi|\psi\rangle = 1.$$

In the following we use only the normalized rays. The space of physical pure states can be identified with the space of all normalized rays in $\mathcal{H}$. We will denote it by $R_{\mathcal{H}}$.

As we know from quantum mechanics, physical predictions are obtained by calculating scalar products of vectors from $\mathcal{H}$. More precisely, the physically relevant quantity is

$$([\psi] \mid [\chi]) \overset{df}{=} |\langle\psi|\chi\rangle|.$$

It does not depend on the choice of the representatives of the rays, as opposed to the scalar product. Expectation values of an observable $\hat{A}$, given by the formula $\langle\psi|\hat{A}\psi\rangle$, also do not depend on the choice of representative $c|\psi\rangle$ of the normalized ray $[\psi]$.

Let us consider a certain Poincaré transformation of the states of a quantum system. It is represented by an operator U_R in the space $R_{\mathcal{H}}$. Thus, U_R transforms each normalized ray into a normalized ray. Both $[\psi]$ and $U_R[\psi]$ represent states of the field with respect to the reference frame (x^μ). We may look at the field in the state $[\psi]$ also from a reference frame (x'^μ) defined by (10.1). Then we shall see the field in a state represented by $[\psi']$. The operator U_R is defined by the formula $[\psi'] = U_R[\psi]$. Thus, the state $U_R[\psi]$ of the field seen from reference frame (x^μ), and the state $[\psi]$ seen from reference frame (x'^μ), look the same. If the transformation is to be a symmetry of the system, it should leave invariant both the space of states and the product $([\psi] \mid [\chi])$, that is

$$(i) \qquad U_R R_{\mathcal{H}} = R_{\mathcal{H}}, \qquad (ii) \qquad (U_R[\psi] \mid U_R[\chi]) = ([\psi] \mid [\chi])$$

[2]For brevity, we discuss here only pure states. The most general space of states includes mixed states, represented by density operators. However, such mixed states can be regarded as being composed of several pure states, therefore one may introduce the notion of symmetry using only the pure states.

for all $[\psi], [\chi] \in R_{\mathcal{H}}$. The meaning of condition (i) is that the full space $R_{\mathcal{H}}$ of states of the quantum field in reference frame (x^{μ}) coincides with the full space of states $U_R R_{\mathcal{H}}$ of that field in reference frame $(x^{'\mu})$. Condition (ii) says that the probability of finding the state $[\psi]$ in the state $[\chi]$, both states given with respect to reference frame (x^{μ}), does not change if we look at these states from reference frame $(x^{'\mu})$. Conditions (i) and (ii) give the precise formulation of the equivalence of the two inertial reference frames.

E.P. Wigner has shown that every symmetry transformation U_R can be represented in the Hilbert space $\mathcal{H}$ by an operator U such that:

$$(Ia) \qquad U\mathcal{H} = \mathcal{H}$$

and

$$(Ib) \qquad \langle U\psi|U\psi\rangle = \langle\psi|\psi\rangle$$

for all $|\psi\rangle$ from $\mathcal{H}$. Moreover,

$$(Ic) \qquad U(|\psi\rangle + |\chi\rangle) = U|\psi\rangle + U|\chi\rangle$$

for all $|\psi\rangle$ and $|\chi\rangle$ from $\mathcal{H}$, and either

$$(Id) \qquad U(c|\psi\rangle) = c\, U|\psi\rangle$$

or

$$(Ie) \qquad U(c|\psi\rangle) = c^*\, U|\psi\rangle,$$

where c is an arbitrary complex number with c^* its complex conjugation. It is clear that U transforms rays into rays, and precisely this transformation of rays coincides with U_R.

In the case (Id) the operator U is unitary, while in the case (Ie) it is called antiunitary. Properties $(Ib - Ie)$ allow us to compute $\langle U\psi|U\chi\rangle$ also when $\psi \neq \chi$ because

$$\langle\psi_1|\psi_2\rangle = \frac{1}{4}\langle\psi_1+\psi_2|\psi_1+\psi_2\rangle - \frac{1}{4}\langle\psi_1-\psi_2|\psi_1-\psi_2\rangle - \frac{i}{4}\langle\psi_1+i\psi_2|\psi_1+i\psi_2\rangle + \frac{i}{4}\langle\psi_1-i\psi_2|\psi_1-i\psi_2\rangle.$$

The r.h.s. of this formula contains only the norms of vectors $|\psi_1\rangle \pm |\psi_2\rangle$ and $|\psi_1\rangle \pm i|\psi_2\rangle$, to which we may apply (Ib). Using that formula for $|\psi_1\rangle = U|\psi\rangle$ and $|\psi_2\rangle = U|\chi\rangle$ we find that in the unitary case (Id)

$$\langle U\psi|U\chi\rangle = \langle\psi|\chi\rangle.$$

In the antiunitary case $U|\psi\rangle \pm iU|\chi\rangle = U(|\psi\rangle \mp i|\chi\rangle)$ and therefore

$$\langle U\psi|U\chi\rangle = \langle\chi|\psi\rangle.$$

In the relativistically invariant QFT we demand that each Poincaré transformation be a symmetry. Hence, for each element $(\hat{L}, a)$ from $\mathcal{P}$ we have an operator $U(\hat{L}, a)$ in $\mathcal{H}$, which has the properties (Ia)–(Ic), and also (Id).

We choose (Id) and not (Ie) for the following reason. The Poincaré group includes the trivial transformation $(I_4, 0)$. It is natural to demand that it be represented by the unit operator I in $\mathcal{H}$

$$U(I_4, 0) = I, \tag{10.4}$$

and this operator is of course unitary. Another natural assumption is that the operator $U(\hat{L}, a)$ depends on $\hat{L}$ and a in a continuous manner. In particular, $U(\hat{L}_n, a_n) \to I$ if the sequence $(\hat{L}_n, a_n)$ is convergent to the trivial element $(I_4, 0)$ when $n \to \infty$. Now, suppose that the operators $U(\hat{L}_n, a_n)$ are antiunitary. Then, for any $|\psi\rangle \in \mathcal{H}$

$$U(\hat{L}_n, a_n)(i|\psi\rangle) = -iU(\hat{L}_n, a_n)|\psi\rangle.$$

Because of the continuity the l.h.s. is convergent to $i|\psi\rangle$ while the r.h.s. to $-i|\psi\rangle$, and we obtain a contradiction. Thus, all operators $U(\hat{L}, a)$ have to be unitary.

Yet another requirement imposed on the operators $U(\hat{L}, a)$ stems from the fact that two consecutive Poincaré transformations, first $g_1 = (\hat{L}_1, a_1)$ and then $g_2 = (\hat{L}_2, a_2)$, are equivalent to the product transformation $g_2 g_1 = (\hat{L}_2\hat{L}_1, \hat{L}_2 a_1 + a_2)$. It is natural to demand that the same holds for the corresponding transformations of the rays in $\mathcal{H}$, that is that

$$U_R(\hat{L}_2, a_2)U_R(\hat{L}_1, a_1) = U_R(\hat{L}_2\hat{L}_1, \hat{L}_2 a_1 + a_2). \tag{10.5}$$

On the level of the operators in the Hilbert space $\mathcal{H}$ property (10.5) is represented by the formula

$$U(g_2)U(g_1) = \exp(i\omega(g_2, g_1))\, U(g_2 g_1), \tag{10.6}$$

where $\omega(g_2, g_1)$ is a real-valued function of the indicated variables. The phase factor $\exp(i\omega)$ is called the cocycle. We assume that it is a continuous function of g_1 and g_2. The set of all unitary operators $U(g)$ in the given Hilbert space $\mathcal{H}$, where $g \in \mathcal{P}$, is called a unitary, projective representation of the Poincaré group if all $U(g)$ obey conditions (10.4) and (10.6), and also the condition of continuity with respect to g. 'Projective' refers to the presence of the cocycle—in the case $\omega(g_2, g_1) = 1$ for all $g_1, g_2 \in \mathcal{P}$ we just say 'unitary representation'.

The presence of the cocycle is a characteristic feature of symmetries of quantum systems. For many groups, e.g., $SU(N)$ groups, it can be removed just by redefining the representation operators $U(g)$. In the case of rotations (the $SO(3)$ group), as well as for the Lorentz and Poincaré groups which contain $SO(3)$ as a subgroup, the cocycle cannot be completely removed. Wigner has proved that all unitary projective representations of $\mathcal{P}$ can be divided into two classes. In the first class, relevant for bosonic fields and integer spin particles, the cocycle can be completely removed just by redefining $U(g)$. In the second class, related to fermionic fields and particles of half-integer spin, the cocycle can be removed only if we introduce a double-valued unitary representation: for any given $g \in \mathcal{P}$ we have two operators $\pm U(g)$. It is a well-known fact that in the theory of continuous multivalued complex functions of a complex variable $z \in C$ one can remove the multivaluedness by extending the domain of the z variable from C to an appropriate Riemann surface. In the case of representations of the Poincaré group, there exists an analogous construction: each double-valued unitary representation U of $\mathcal{P}$ in $\mathcal{H}$ is equivalent to a single-valued unitary representation (also in $\mathcal{H}$) of a group $\tilde{\mathcal{P}}$ larger than $\mathcal{P}$. That new group is called the universal covering group. It consists of all pairs of the form (Λ, a), where a is an arbitrary translation as before, while Λ is an arbitrary element of the $SL(2, C)$ group. Let us recall that the $SL(2, C)$ group consists of all the 2 by 2 complex matrices with the determinant equal to $+1$. Such a set of matrices forms the group with respect to the matrix product. The relation between $SL(2, C)$ and $L_+^\uparrow$ was discussed in Sect. 5.1, see formulas (5.22) and (5.23). We recall it in Sect. 10.3 below. Because Λ and $-\Lambda$ give the same $\hat{L} \in L_+^\uparrow$, $SL(2, C)$ covers $\mathcal{P}$ twice. The group product in $\tilde{\mathcal{P}}$ has the form $(\Lambda_1, a_1)(\Lambda_2, a_2) = (\Lambda_1\Lambda_2, \hat{L}(\Lambda_1)a_2 + a_1)$, where $\hat{L}(\Lambda_1)$ is the Lorentz transformation corresponding to Λ_1. The unit element has the form $(\sigma_0, 0)$. Thus,

$$\tilde{U}(\Lambda, a)\mathcal{H} = \mathcal{H},$$

$$\langle \tilde{U}(\Lambda, a)\psi | \tilde{U}(\Lambda, a)\chi\rangle = \langle \psi | \chi \rangle$$

for all $|\psi\rangle$, $|\chi\rangle \in \mathcal{H}$, and

$$\tilde{U}(\Lambda_1, a_1)\tilde{U}(\Lambda_2, a_2) = \tilde{U}(\Lambda_1\Lambda_2, \hat{L}(\Lambda_1)a_2 + a_1).$$

The correspondence between $SL(2, C)$ and $L_+^\uparrow$ becomes an isomorphism if we take Λ's from a certain not-too-large vicinity of the 2 by 2 unit matrix σ_0. Such Λ's can be smoothly parameterized by the 6 real parameters known from Sect. 3.1: $\omega^{12}, \omega^{23}, \omega^{31}, \omega^{01}, \omega^{02}, \omega^{03}$. Thus, in that vicinity of the unit element of $\tilde{\mathcal{P}}$, we may use ω and a as the parameters: $g = (\Lambda(\omega), a)$, where ω denotes the six real parameters specified above. It is clear that $\Lambda(\omega = 0) = \sigma_0$.

It turns out that in the case of continuous unitary representations of $\tilde{\mathcal{P}}$ which appear in QFT, the operators $\tilde{U}(\Lambda(\omega), a)$ can be written as an infinite series with respect to $\omega^{\mu\nu}$'s and a^μ's:

$$\tilde{U}(\Lambda(\omega), a) = I + ia^\mu \hat{P}_\mu + \frac{i}{2}\omega^{\mu\nu}\hat{M}_{\mu\nu} + \dots, \tag{10.7}$$

where by definition

$$\hat{M}_{\mu\nu} = -\hat{M}_{\nu\mu}.$$

This last condition is related to the fact that $\omega^{\mu\nu} = -\omega^{\nu\mu}$. The factor 1/2 is introduced in order to cancel the factor 2 from

$$\omega^{\mu\nu}\hat{M}_{\mu\nu} = 2\Big(\sum_{i=1}^{3}\omega^{0i}\hat{M}_{0i} + \omega^{12}\hat{M}_{12} + \omega^{23}\hat{M}_{23} + \omega^{31}\hat{M}_{31}\Big).$$

Because $\tilde{U}(g)$ are unitary operators, the operators $\hat{P}_\mu$ and $\hat{M}_{\mu\nu}$ are Hermitian—this is the reason for extracting the factors i in the second and third term on the r.h.s. of formula (10.7). $\hat{P}_\mu$ and $\hat{M}_{\mu\nu}$ are called the generators of the representation $\tilde{U}$ in the chosen parametrization (ω, a).

The group structure of $\tilde{\mathcal{P}}$ implies commutation relations for $\hat{P}_\mu$ and $\hat{M}_{\mu\nu}$. In order to derive them we first notice that

$$\hat{P}_\mu = -i\frac{\partial\tilde{U}(\Lambda(\omega), a)}{\partial a^\mu}\bigg|_{\omega=0,\, a=0}, \qquad \hat{M}_{\mu\nu} = -i\frac{\partial\tilde{U}(\Lambda(\omega), a)}{\partial\omega^{\mu\nu}}\bigg|_{\omega=0,\, a=0}. \tag{10.8}$$

Now, consider the following identity

$$\tilde{U}(\sigma_0, a_1)\,\tilde{U}(\sigma_0, a_2) = \tilde{U}(\sigma_0, a_1 + a_2) = \tilde{U}(\sigma_0, a_2)\,\tilde{U}(\sigma_0, a_1).$$

The derivative with respect to a_1^μ taken at $a_1 = 0$ gives

$$i\hat{P}_\mu\tilde{U}(\sigma_0, a_2) = \frac{\partial\tilde{U}(\sigma_0, a_2)}{\partial a_2^\mu} = i\tilde{U}(\sigma_0, a_2)\hat{P}_\mu. \tag{10.9}$$

This formula implies that the operators $\hat{P}_\mu$ are invariant with respect to translations, that is that

$$\tilde{U}^{-1}(\sigma_0, a_2)\hat{P}_\mu\tilde{U}(\sigma_0, a_2) = \hat{P}_\mu. \tag{10.10}$$

The l.h.s. of this formula is, by definition, the transformation of the operator $\hat{P}_\mu$ corresponding to the symmetry represented by $\tilde{U}$. It is a general postulate of quantum theory that the action of a unitary symmetry transformation $\tilde{U}(g)$ on an operator $\hat{Q}$ in the Hilbert space $\mathcal{H}$ has the form

$$\hat{Q} \to \hat{Q}' \stackrel{df}{=} \tilde{U}^{-1}(g)\hat{Q}\tilde{U}(g). \tag{10.11}$$

Let us take the derivative of both sides of formula (10.9) with respect to a_2^ν at $a_2 = 0$. The result can be written as

$$[\hat{P}_\mu,\ \hat{P}_\nu] = 0. \tag{10.12}$$

We see that the generators of space-time translations commute with each other. In a relativistically invariant theory, the generator of time translations $\hat{P}^0$ coincides with the quantum Hamiltonian of the considered field, and $\hat{P}^i$ coincide with components of the operator of the total momentum of the field.

The first part of formula (10.9), namely

$$i\hat{P}_\mu \tilde{U}(\sigma_0, a_2) = \frac{\partial \tilde{U}(\sigma_0, a_2)}{\partial a_2^\mu},$$

can actually be regarded as a set of differential equations for $\tilde{U}(\sigma_0, a_2)$. Because $\hat{P}_\mu$ commute, the solution which obeys the condition $\tilde{U}(\sigma_0, 0) = I$ has the form

$$\tilde{U}(\sigma_0, a_2) = \exp(ia_2^\mu \hat{P}_\mu). \tag{10.13}$$

Acting with $-i\partial/\partial a^\mu$ on both sides of another identity, namely

$$\tilde{U}(\Lambda, 0)\, \tilde{U}(\sigma_0, a) = \tilde{U}(\sigma_0, L(\Lambda)a)\, \tilde{U}(\Lambda, 0), \tag{10.14}$$

and setting $a = 0$, we obtain

$$\tilde{U}(\Lambda, 0)\hat{P}_\mu = \hat{P}_\nu L(\Lambda)^\nu{}_\mu \tilde{U}(\Lambda, 0). \tag{10.15}$$

This formula can be written in the form

$$\tilde{U}^{-1}(\Lambda, 0)\hat{P}^\mu \tilde{U}(\Lambda, 0) = L(\Lambda)^\mu{}_\nu \hat{P}^\nu, \tag{10.16}$$

(as always, we raise the indices using $\eta^{\mu\nu}$: $\hat{P}^\nu = \eta^{\nu\mu}\hat{P}_\mu$). Formula (10.16) says that the operators $\hat{P}^\mu$ transform under Lorentz transformations as components of a four-vector. Formula (10.15) implies the commutation relation between $\hat{P}_\mu$ and $\hat{M}_{\rho\lambda}$: we take the derivative of both sides of it with respect to $\omega^{\rho\lambda}$ and we put $\omega = 0$. Because $L(\Lambda)^\nu{}_\mu = \delta^\nu_\mu + \omega^\nu{}_\mu + \mathcal{O}(\omega^2)$, we have

$$\left.\frac{\partial L(\Lambda)^\nu{}_\mu}{\partial \omega^{\rho\lambda}}\right|_{\omega=0} = \delta^\nu_\rho \eta_{\mu\lambda} - \delta^\nu_\lambda \eta_{\mu\rho},$$

and therefore

$$\hat{M}_{\rho\lambda}\hat{P}_\mu = -i(\hat{P}_\rho \eta_{\mu\lambda} - \hat{P}_\lambda \eta_{\mu\rho}) + \hat{P}_\mu \hat{M}_{\rho\lambda},$$

or

$$[\hat{M}_{\rho\lambda}, \hat{P}_\mu] = i(\eta_{\mu\rho}\hat{P}_\lambda - \eta_{\mu\lambda}\hat{P}_\rho). \tag{10.17}$$

Finally, let us consider the identity

$$\tilde{U}^{-1}(\Lambda(\omega), 0)\, \tilde{U}(\Lambda(\omega_1), 0)\, \tilde{U}(\Lambda(\omega), 0) = \tilde{U}(\Lambda^{-1}(\omega)\Lambda(\omega_1)\Lambda(\omega), 0),$$

where $\Lambda^{-1}(\omega) \equiv (\Lambda(\omega))^{-1}$. Its derivative with respect to $\omega_{1\mu\nu}$ at $\omega_1 = 0$ gives

$$\tilde{U}^{-1}(\Lambda, 0)\hat{M}^{\mu\nu}\tilde{U}(\Lambda, 0) = L(\Lambda)^{\mu}{}_{\sigma}L(\Lambda)^{\nu}{}_{\rho}\hat{M}^{\sigma\rho}, \tag{10.18}$$

where $\Lambda = \Lambda(\omega)$ (Exercise 10.1). This formula shows that the operators $\hat{M}^{\mu\nu}$ transform as components of a second rank tensor.

Note that formula (10.14) implies that the operators $M_{\sigma\rho}$ also have a nontrivial transformation law with respect to translations:

$$\tilde{U}^{-1}(\sigma_0, a)\hat{M}_{\sigma\rho}\tilde{U}(\sigma_0, a) = \hat{M}_{\sigma\rho} + \tilde{U}^{-1}(\sigma_0, a)(a_\rho\hat{P}_\sigma - a_\sigma\hat{P}_\rho),$$

(Exercise 10.2).

Taking the derivative of both sides of formula (10.18) with respect to $\omega_{\alpha\beta}$ at $\omega = 0$, and lowering the indices, we find that

$$[\hat{M}_{\alpha\beta}, \hat{M}_{\mu\nu}] = i(\eta_{\alpha\mu}\hat{M}_{\beta\nu} - \eta_{\alpha\nu}\hat{M}_{\beta\mu} + \eta_{\beta\nu}\hat{M}_{\alpha\mu} - \eta_{\beta\mu}\hat{M}_{\alpha\nu}). \tag{10.19}$$

We have emphasized in Chap. 2 that a symmetry transformation in classical field theory transforms solutions of the pertinent field equations into solutions of the same equations. Similarly as in Chap. 2, we will use the general notation $u_i(x)$ for the classical fields. Their relativistic transformation law can be written in the general form as

$$u'_i(x) = V_{ik}(\hat{L})u_k(\hat{L}^{-1}(x-a)). \tag{10.20}$$

In particular, $V_{ik}(\hat{L}) = \delta_{ik}$ when $u_i(x)$ is a set of scalar fields, $V_{ik}(\hat{L}) = L^{\mu}{}_{\nu}$ for a vector field $u_i(x) = W^{\nu}(x)$, or $V(\hat{L}) = S(\hat{L})$ if $\{u_i\} = \psi$ is the Dirac field. The corresponding quantum fields in the Heisenberg picture are denoted by $\hat{u}_i(x)$. By definition, their transformation law has the form (10.11), that is

$$\hat{u}'_i(x) \stackrel{df}{=} \tilde{U}^{-1}(\Lambda, a)\, \hat{u}_i(x)\, \tilde{U}(\Lambda, a). \tag{10.21}$$

The quantum field is called a scalar, vector, bispinor, etc., if the definition (10.21) implies that

$$(II) \qquad \hat{u}'_i(x) = V_{ik}(\hat{L}(\Lambda))\, \hat{u}_k(\hat{L}^{-1}(\Lambda)(x-a)), \tag{10.22}$$

where $V_{ik}(\hat{L}(\Lambda))$ has the same form as in the classical case (10.20).

In particular, the quantum field $\hat{\phi}(x)$ is called a relativistic scalar field if it obeys the condition

$$(II') \qquad \tilde{U}^{-1}(\Lambda, a)\hat{\phi}(x)\tilde{U}(\Lambda, a) = \hat{\phi}(\hat{L}^{-1}(\Lambda)(x-a)) \tag{10.23}$$

for all $(\Lambda, a) \in \tilde{\mathcal{P}}$. Differentiation of this formula with respect to a^μ and $\omega^{\mu\nu}$ gives, after setting $a = 0$ and $\omega = 0$, the conditions

$$[\hat{P}_\mu, \hat{\phi}(x)] = -i\partial_\mu\hat{\phi}(x), \tag{10.24}$$

$$[\hat{M}_{\mu\nu}, \hat{\phi}(x)] = -i(x_\nu\partial_\mu - x_\mu\partial_\nu)\hat{\phi}(x). \tag{10.25}$$

Actually, one can prove that they are equivalent to (10.23).

To summarize, the first requirement for a relativistically invariant quantum field theory is that in the Hilbert space of the model there exists a unitary representation $\tilde{U}$ of the group $\tilde{\mathcal{P}}$. The second requirement concerns the quantum fields: we demand that the transformed quantum field $\hat{u}'_i(x)$, which is defined by formula (10.21), be a solution of the Heisenberg equation of motion together with $\hat{u}_i(x)$, and that the quantum field $\hat{u}_i(x)$ obeys condition (10.22).

It turns out that in order to obtain the representation $\tilde{U}$ it is sufficient to know the operators $\hat{P}_\mu$ and $\hat{M}_{\mu\nu}$ obeying commutation relations (10.12), (10.17), (10.19), (10.24) and (10.25). The proof of this theorem is based on the fact that any element of the group $\tilde{\mathcal{P}}$ can be written as a product of sufficiently many elements from a small vicinity of the unit element $(\sigma_0, 0)$. The same is true for representation operators $\tilde{U}$. For each factor in that product we may use expansion (10.7), in which the terms denoted by dots may be neglected. Therefore, in practice one rarely explicitly introduces the unitary operators $\tilde{U}$—it is sufficient to consider the generators $\hat{P}_\mu$ and $\hat{M}_{\nu\sigma}$.

The third group of requirements for a relativistically invariant quantum field theory is related to its particle interpretation. Such an interpretation means that in the Hilbert space of the model there exists a basis which consists of states with definite numbers of particles, including a single state without any particles[3]: the vacuum state $|0\rangle$. A generic state is a superposition of these basis states—it can have components with various numbers of particles. In general, such basis states are not eigenstates of the Hamiltonian of the quantum field, because of interactions between particles which can lead to the creation or annihilation of them, while the eigenstates can change in time only by a phase factor, hence their particle content is constant in time. In the free field models discussed in Chap. 6 such interactions are absent, and in consequence the basis states in the Fock space can be chosen in such a way that they are eigenstates of the pertinent Hamiltonians and particle number operators.

In the theory with the particle interpretation, physical characteristics of a given state of the field can be regarded as contributions from the particles present in that state. For example, the total energy of the field in a certain state with a definite number of particles, that is the expectation value of the Hamiltonian in that state, has the form of the sum of the kinetic energies of the particles and energies of interactions between

[3]If in the classical system spontaneous symmetry breaking is present, one has to pick one of the several classical ground states in order to construct the corresponding quantum model, and then the quantum vacuum state corresponds to that chosen classical ground state. The remaining classical ground states are not incorporated into such quantum theory.

them, weighted by appropriate probability densities. The vacuum state $|0\rangle$ does not contain any particles. Hence, there is no kinetic or interaction energy involved, and such a state should be the eigenstate of the quantum Hamiltonian of the field with vanishing eigenvalue, $E = 0$:

$$\hat{H}|0\rangle = 0. \tag{10.26}$$

For the same reason, it is assumed that $E = 0$ is the smallest eigenvalue of $\hat{H}$—when particles are present the energy is larger because of the relativistic kinetic energies $\sqrt{\vec{p}^{\,2} + m^2}$, where m is the rest mass of the particle.[4] Furthermore, the state without any particles should have vanishing total momentum,

$$\hat{P}^i|0\rangle = 0. \tag{10.27}$$

Now we are ready to state the third group of requirements for relativistic invariance in QFT. In accordance with conditions (10.26) and (10.27), we demand that the vacuum be invariant under space-time translations:

$$(IIIa) \qquad \tilde{U}(\sigma_0, a)|0\rangle = |0\rangle. \tag{10.28}$$

One more requirement is that the vacuum state should look identical to all observers related to each other by the Lorentz transformations:

$$\tilde{U}(\Lambda, 0)|0\rangle = e^{i\chi(\Lambda)}|0\rangle,$$

where $e^{i\chi(\Lambda)}$ is a phase factor, which can depend on $\Lambda \in SL(2, C)$. This phase factor has the property

$$e^{i\chi(\Lambda_1)}e^{i\chi(\Lambda_2)} = e^{i\chi(\Lambda_1\Lambda_2)}$$

for all $\Lambda_1, \Lambda_2 \in SL(2, C)$, which is obtained by applying both sides of the identity $\tilde{U}(\Lambda_1, 0)\tilde{U}(\Lambda_2, 0) = \tilde{U}(\Lambda_1\Lambda_2, 0)$ to the vacuum state. One can show that the mapping $SL(2, C) \ni \Lambda \to e^{i\chi(\Lambda)}$ is a one dimensional unitary representation of the $SL(2, C)$ group. On the other hand, it is known that all unitary representations of this group are infinite dimensional, except for the trivial one for which $e^{i\chi(\Lambda)} = 1$. Thus, the phase factors are equal to 1, and

$$(IIIb) \qquad \tilde{U}(\Lambda, 0)|0\rangle = |0\rangle \tag{10.29}$$

for all $\Lambda \in SL(2, C)$.

In the classical theory the energy can always be shifted by a constant. In the relativistically invariant quantum field theory this is no longer true. The structure of

[4] Notice that this means that we hope that the particles and the vacuum state can be defined in such a manner that the interaction energies cannot render the total energy of the states with particles negative.

such a theory is so tight, that such freedom is not allowed. To see this, let us suppose that the vacuum state has non vanishing energy or momentum,

$$\hat{P}^{\mu}|0\rangle = p^{\mu}_{(0)}|0\rangle.$$

Applying both sides of formula (10.16) to the vacuum state and using (10.29) we find that

$$p^{\nu}_{(0)} = L(\Lambda)^{\nu}{}_{\mu}p^{\mu}_{(0)}$$

for all $\hat{L}$ from the $L_{+}^{\uparrow}$ group. This is possible only if $p^{\nu}_{(0)} = 0$. Thus, the vacuum has to have vanishing energy and momentum if the quantum model is to be relativistically invariant. Another consequence of the lack of freedom of adding a constant to the energy is that the energy of a single free particle of momentum $\vec{p}$, which is equal to $\sqrt{\vec{p}^{\,2} + m^2}$ as we have found when discussing the free quantum field models, also cannot be shifted by a constant.

Apart from the presence of the vacuum state, we also assume that there are states of the quantum field which contain just a single stable particle. In general, a quantum field theoretic model can predict the existence of several species of stable particles. We shall label them with the index $K = a, b, \ldots$. In order to simplify the discussion we assume that all these particles are massive, that is that their rest masses m_K are strictly positive. States of K-th particle are represented by rays in a subspace $\mathcal{H}^{(1)}_K$ of the full Hilbert space $\mathcal{H}$. Such single particle states have the special property that they evolve in time as states of a free relativistic particle, because, by assumption, in these states there are no other particles with which the given particle could interact. As a basis in $\mathcal{H}^{(1)}_K$ we may take the normalized eigenstates of the total momentum $\vec{\hat{P}}$, and of a certain component of the spin operator. In particular,

$$\hat{P}^{i}|\vec{p}, \lambda, K\rangle = p^{i}|\vec{p}, \lambda, K\rangle,$$

where λ stands for the projection of spin of the K-th particle on, e.g., the x^3-axis. In a single particle subspace, $\vec{p}$ is of course equal to the momentum of the particle. The energy eigenvalue is a function of the momentum,

$$\hat{P}^{0}|\vec{p}, \lambda, K\rangle = E_K(\vec{p}\,)|\vec{p}, \lambda, K\rangle, \tag{10.30}$$

where

$$E_K(\vec{p}\,) = \sqrt{\vec{p}^{\,2} + m_K^2}. \tag{10.31}$$

Note that formula (10.31) contains the square of m_K, which is insensitive to the sign of m_K. It is merely a convention that non-vanishing masses of particles in relativistically invariant theories are positive.

Let us stress that the masses m_K should not be confused with the mass parameters present in classical Lagrangians, e.g., with m_0 present in Lagrangian (7.1) in the case of the $:\phi_4^4:$ model, see Sect. 10.4 for a detailed discussion. Only in special cases, like the free quantum fields, or models with special symmetries, is the rest mass of the particle (m_K) equal to the corresponding mass parameter (m_0) in the pertinent classical Lagrangian.

A general vector $|\psi\rangle$ from the Hilbert space $\mathcal{H}_K^{(1)}$ has the form

$$|\psi\rangle = \sum_\lambda \int d^3p\, \psi_\lambda(\vec{p}\,)|\vec{p}, \lambda, K\rangle, \tag{10.32}$$

where

$$\langle\psi|\psi\rangle = \sum_\lambda \int d^3p\, \overline{\psi_\lambda}(\vec{p}\,)\psi_\lambda(\vec{p}\,) < \infty.$$

All vectors of the form (10.32) are eigenvectors of the operator $\hat{P}^\mu\hat{P}_\mu = (\hat{P}_0)^2 - (\vec{\hat{P}})^2$. In fact,

$$\begin{aligned}\hat{P}^\mu\hat{P}_\mu|\psi\rangle &= \sum_\lambda \int d^3p\, \psi_\lambda(\vec{p}\,)\hat{P}^\mu\hat{P}_\mu|\vec{p}, \lambda, K\rangle \\ &= \sum_\lambda \int d^3p\, \psi_\lambda(\vec{p}\,)(E_K^2(\vec{p}\,) - \vec{p}^{\,2})|\vec{p}, \lambda, K\rangle = m_K^2|\psi\rangle.\end{aligned}$$

Because they have finite norm, they are true eigenvectors of the operator $\hat{P}^\mu\hat{P}_\mu$. For a comparison consider the two-particle sector of the free scalar field, see Sect. 6.1. The vectors $|\vec{k}_1, \vec{k}_2\rangle$ are eigenvectors of $\hat{P}^\mu\hat{P}_\mu$ in the sense that

$$\hat{P}^\mu\hat{P}_\mu|\vec{k}_1, \vec{k}_2\rangle = M^2(\vec{k}_1, \vec{k}_2)|\vec{k}_1, \vec{k}_2\rangle,$$

where

$$\begin{aligned}M^2(\vec{k}_1, \vec{k}_2) &= \left(\sqrt{\vec{k}_1^{\,2} + m_0^2} + \sqrt{\vec{k}_2^{\,2} + m_0^2}\right)^2 - (\vec{k}_1 + \vec{k}_2)^2 \\ &= 2(m_0^2 + \sqrt{\vec{k}_1^{\,2} + m_0^2}\sqrt{\vec{k}_2^{\,2} + m_0^2} - \vec{k}_1\vec{k}_2),\end{aligned}$$

but there is the crucial difference that in the latter case the eigenvalues of $\hat{P}^\mu\hat{P}_\mu$ form a continuous set, hence the corresponding eigenvectors do not have a finite norm. Therefore, they do not belong to the Hilbert space $\mathcal{H}$. The vectors (10.32) have finite norm, they belong to the Hilbert space, and the eigenvalues m_K^2 are a part of the discrete spectrum of $\hat{P}^\mu\hat{P}_\mu$.

Each space $\mathcal{H}_K^{(1)}$, $K = a, b, \ldots,$ is invariant under the representation $\tilde{U}$ of $\tilde{P}$, that is

$$\tilde{U}(\Lambda, a)|\psi\rangle \in \mathcal{H}_K^{(1)} \quad \text{if} \quad |\psi\rangle \in \mathcal{H}_K^{(1)}.$$

This mathematical fact has an obvious physical meaning—the type of the particle does not change if we look at the particle from another inertial reference frame. In particular, all states $\tilde{U}(\Lambda, a)|\psi\rangle$ belong to the same eigenspace of the operator $\hat{P}^{\mu}\hat{P}_{\mu}$. This follows from formulas (10.10) and (10.16):

$$\hat{P}^{\mu}\hat{P}_{\mu}\tilde{U}(\Lambda, a)|\psi\rangle = \tilde{U}(\Lambda, a)\tilde{U}^{-1}(\Lambda, a)\hat{P}^{\mu}\tilde{U}(\Lambda, a)\tilde{U}^{-1}(\Lambda, a)\hat{P}_{\mu}\tilde{U}(\Lambda, a)|\psi\rangle$$
$$= \tilde{U}(\Lambda, a)L(\Lambda)_{\mu}{}^{\rho}\hat{P}_{\rho}L(\Lambda)^{\mu}{}_{\sigma}\hat{P}^{\sigma}|\psi\rangle = \tilde{U}(\Lambda, a)\hat{P}_{\rho}\hat{P}^{\rho}|\psi\rangle = m_K^2\tilde{U}(\Lambda, a)|\psi\rangle.$$

Moreover, if the space $\mathcal{H}_K^{(1)}$ could be split into two or more nontrivial[5] subspaces $\mathcal{H}_K^{(1a)}$ and $\mathcal{H}_K^{(1b)}$, etc., each of them being invariant under the representation $\tilde{U}$, we would rather regard the states from these subspaces as states of different particles, K_a and K_b, etc. In such a case, we accordingly redefine the particle label K in (10.30), so that finally the spaces $\mathcal{H}_K^{(1)}$ do not contain any nontrivial invariant subspaces. In mathematical language, the unitary representation $\tilde{U}$ restricted to such a subspace is irreducible. Mathematical investigations of unitary irreducible representations of the group $\tilde{P}$ have shown that in the case $m_K^2 > 0$ the basis states in $\mathcal{H}_K^{(1)}$ are labelled by the momentum $\vec{p}$, and the projection of spin $\lambda = -s, -s+1, \ldots, s-1, s$, where the spin s has one value chosen from the set of numbers $0, 1/2, 1, \ldots$. The value s of the spin is included in the particle label K. It does not change when we look at the particle from various inertial reference frames, i.e., it is invariant with respect to the Poincaré transformations, in contrary to the spin projection which can be changed, for example, by a rotation.

The particle label K also includes the rest mass m_K, as well as other characteristics of the particle such as its electric charge, various parities, strangeness, etc. Each of them is invariant with respect to the Poincaré transformations.

Let us summarize:

(*IIIc*) The pure states of the quantum field that contain only a single particle of type K are represented by rays in the subspace $\mathcal{H}_K^{(1)}$ of the full Hilbert space $\mathcal{H}$. The representation $\tilde{U}$ restricted to this subspace is irreducible. In particular, $\hat{P}^{\mu}\hat{P}_{\mu}|\psi\rangle = m_K^2|\psi\rangle$ for all $|\psi\rangle \in \mathcal{H}_K^{(1)}$. Such $|\psi\rangle$ are normalizable.

Note that with such a definition of a relativistic quantum particle—as a subclass of the states of the quantum field—a stable bound state of two or more particles is a particle too. Of course, such a particle should not be called an elementary one.

Let us return to the real scalar quantum field. In the case of the free field, the operators $\hat{P}_0 \equiv \hat{H}, \hat{P}^i, \hat{M}_{ik}$ and $\hat{M}_{0i}$ constructed in Sect. 6.1 obey the commutation relations (10.12), (10.17) and (10.19). Therefore, we have the representation $\tilde{U}$ of the group $\tilde{P}$ in the Fock space $\mathcal{H}_F$. Also the commutation relations (10.24) and (10.25) are satisfied. Hence, this field is indeed a relativistic scalar quantum field. The vacuum state $|0\rangle$ has the properties (10.28) and (10.29). The vectors $|\vec{k}\rangle$, which form a basis of the single particle subspace do not have any additional label λ. This suggests

[5]That is, larger than the trivial space consisting of the single zero vector.

that the particle is spinless, $s = 0$. In order to check that, one should rotate the basis vector with momentum equal to zero. This actually means acting with $\tilde{U}(u, 0)$, where $u \in SU(2) \subset SL(2, C)$, on the vector $|\vec{0}\,\rangle$.[6] In the case of a spinless particle, this state should be invariant with respect to all rotations. It is sufficient to check this for infinitesimal rotations, when we may use formula (10.7) with $a = 0$, $\omega^{0i} = 0$, and with omission of the terms denoted by dots. Using formula (6.47), we find that $\hat{M}_{ik}|\vec{0}\,\rangle = 0$, hence indeed $\tilde{U}(u, 0)|\vec{0}\,\rangle = |\vec{0}\,\rangle$. The rest mass of the particle coincides with the mass parameter m in the Lagrangian (6.1).

In the case of the $:\phi_4^4:$ model,[7] we are not able to provide even the Hilbert space $\mathcal{H}$, not to mention the representation $\tilde{U}$. We hope that at least for small λ_0, such a quantum model exists, and that its properties do not differ drastically from those of the free real scalar field (which is obtained when $\lambda_0 = 0$). In particular, we expect that there exists a single vacuum state $|0\rangle$, which is invariant under the Poincaré group, and a sector $\mathcal{H}^{(1)}$ describing a single spinless particle with rest mass $m > 0$. Such expectations are to some extent supported by the fact that using the renormalized perturbative expansion in λ_0, one can construct approximate generators $\hat{P}_\mu$ and $\hat{M}_{\mu\nu}$ in the interaction picture Fock space, introduced in Sect. 7.1. They obey the required commutation relations up to the considered order of the perturbative expansion. The problem with the renormalized perturbative expansion is that we do not know whether it really approximates (in the sense of the theory of asymptotic series) that hypothetical exact theory.

10.2 The Spectral Decomposition of $G^{(2)}$

In this Section we derive a very important formula for the Green's function $G^{(2)}$, known as the spectral decomposition. It follows from the postulates of relativistic invariance, and from the assumptions about the particle interpretation of the quantum field. For the sake of simplicity we will again discuss the real scalar quantum field only.

Let us first introduce the 2-point Wightman's function $W^{(2)}$. It is defined as follows

$$W^{(2)}(x_1, x_2) = \langle 0|\hat{\phi}(x_1)\hat{\phi}(x_2)|0\rangle, \tag{10.33}$$

where $\hat{\phi}(x)$ is the quantum field operator in the Heisenberg picture, and x_1 and x_2 are points in Minkowski space-time. $W^{(2)}(x_1, x_2)$ is a generalized function of x_1 and x_2. The Green's function $G^{(2)}$ is defined by the formula

[6] The subgroup $SU(2)$ of $SL(2, C)$ consists of all 2 by 2 matrices which are unitary ($u^\dagger = u^{-1}$) and unimodular ($\det u = 1$). It is the universal covering group of the $SO(3)$ subgroup of $L_+^\uparrow$.

[7] We mean here a model without the regularizing function g.

$$
\begin{aligned}
G^{(2)}(x_1, x_2) &= \langle 0|T(\hat{\phi}(x_1)\hat{\phi}(x_2))|0\rangle \\
&= \Theta(x_1^0 - x_2^0)\langle 0|\hat{\phi}(x_1)\hat{\phi}(x_2)|0\rangle + \Theta(x_2^0 - x_1^0)\langle 0|\hat{\phi}(x_2)\hat{\phi}(x_1)|0\rangle. \qquad (10.34)
\end{aligned}
$$

Therefore,

$$G^{(2)}(x_1, x_2) = \Theta(x_1^0 - x_2^0)W^{(2)}(x_1, x_2) + \Theta(x_2^0 - x_1^0)W^{(2)}(x_2, x_1). \qquad (10.35)$$

Formula (10.23) with $\Lambda = \sigma_0$, $a = x$, and formula (10.13) give

$$\hat{\phi}(x) = \exp(i\hat{P}_\mu x^\mu)\hat{\phi}(0)\exp(-i\hat{P}_\nu x^\nu). \qquad (10.36)$$

Using this formula and the property (10.28) of the vacuum state, we obtain the following expression for Wightman's function

$$W^{(2)}(x_1, x_2) = \langle 0|\hat{\phi}(0)\exp[-i\hat{P}_\mu(x_1 - x_2)^\mu]\hat{\phi}(0)|0\rangle. \qquad (10.37)$$

Thus, the translational invariance of the quantum field theory implies that $W^{(2)}$ depends only on $x_1 - x_2$. In consequence, also $G^{(2)}(x_1, x_2)$ is a generalized function of $x_1 - x_2$ only.

Invariance with respect to Lorentz transformations implies that

$$\hat{\phi}(x) = \tilde{U}^{-1}(\Lambda, 0)\hat{\phi}(\hat{L}(\Lambda)x)\tilde{U}(\Lambda, 0), \quad \tilde{U}(\Lambda, 0)|0\rangle = |0\rangle. \qquad (10.38)$$

Therefore,

$$W^{(2)}(\hat{L}x_1, \hat{L}x_2) = W^{(2)}(x_1, x_2) \qquad (10.39)$$

for all $\hat{L} \in L_+^\uparrow$.

In the next step we use the completeness relation in the full Hilbert space $\mathcal{H}$ of the model

$$|0\rangle\langle 0| + \int d^3p\, |\vec{p}\,\rangle\langle\vec{p}\,| + \int\!\!\!\!\sum_\alpha |\alpha\rangle\langle\alpha| = I, \qquad (10.40)$$

where $\{|0\rangle, |\vec{p}\,\rangle, |\alpha\rangle\}$ is a basis in $\mathcal{H}$. The vectors $|\alpha\rangle$ form a basis in the part of the Hilbert space orthogonal to the vacuum and the single particle subspaces—these vectors are enumerated by a set of quantum numbers denoted here by α. The symbol $\int\!\sum_\alpha$ is used in order to denote that among these quantum numbers there can be continuous as well as discrete ones. The basis is chosen in such a way that each vector $|\alpha\rangle$ is an eigenstate of the total four-momentum of the field,

$$\hat{P}^\mu|\alpha\rangle = p_\alpha^\mu|\alpha\rangle. \qquad (10.41)$$

Of course,

$$\hat{P}^\mu|0\rangle = 0, \quad \hat{P}^\mu|\vec{p}\,\rangle = p^\mu|\vec{p}\,\rangle, \qquad (10.42)$$

where

$$p^0 = E(\vec{p}\,) = \sqrt{\vec{p}^{\,2} + m^2},$$

and m is the rest mass of the particle. Inserting (10.40) on the r.h.s. of formula (10.37), and using (10.41) and (10.42), we obtain

$$W^{(2)}(x_1, x_2) = |\langle 0|\hat{\phi}(0)|0\rangle|^2 + \int d^3p\, |\langle 0|\hat{\phi}(0)|\vec{p}\,\rangle|^2 \exp(-ip(x_1 - x_2)) + \int \sum_{\alpha} |\langle 0|\hat{\phi}(0)|\alpha\rangle|^2 \exp(-ip_{\alpha}(x_1 - x_2)). \quad (10.43)$$

In the contribution from the single particle sector, given by the last term in the first line, we have $p = (p^0, \vec{p}\,)$, where $p^0 = E(\vec{p}\,) = \sqrt{\vec{p}^{\,2} + m^2}$.

In the next section we prove that

$$|\langle 0|\hat{\phi}(0)|\vec{p}\,\rangle|^2 = \frac{m}{E(\vec{p}\,)} |\langle 0|\hat{\phi}(0)|\vec{0}\,\rangle|^2, \quad (10.44)$$

where $|\vec{0}\,\rangle$ is the basis vector in the single particle sector with momentum equal to zero. Therefore, the contribution of the single particle states can be written in the form

$$\int d^3p\, |\langle 0|\hat{\phi}(0)|\vec{p}\,\rangle|^2 \exp(-ip(x_1 - x_2)) = c_0 W_m^{(2)}(x_1, x_2), \quad (10.45)$$

where

$$W_m^{(2)}(x_1, x_2) = \frac{1}{2(2\pi)^3} \int \frac{d^3p}{E(\vec{p}\,)} \exp(-ip(x_1 - x_2)), \quad (10.46)$$

and

$$c_0 = 2(2\pi)^3 m |\langle 0|\hat{\phi}(0)|\vec{0}\,\rangle|^2. \quad (10.47)$$

Note that $c_0 \geq 0$.

$W_m^{(2)}(x_1, x_2)$ is the 2-point Wightman's function for the free scalar field with mass parameter equal to m. This fact can easily be checked with the help of formula (6.16) for the free scalar field. The $W_m^{(2)}(x_1, x_2)$ function is of course Lorentz invariant: for all $\hat{L} \in L_+^{\uparrow}$

$$W_m^{(2)}(\hat{L}x_1, \hat{L}x_2) = W_m^{(2)}(x_1, x_2), \quad (10.48)$$

because the theory of the free scalar field constructed in Sect. 6.1 is Lorentz invariant. Formula (10.48) can also be obtained directly by rewriting the integral in formula (10.46) in the Lorentz invariant form,

$$W_m^{(2)}(x_1, x_2) = \frac{1}{(2\pi)^3} \int d^4p\, \Theta(p_0)\delta(p^2 - m^2) \exp(-ip(x_1 - x_2)). \quad (10.49)$$

In the case of the free scalar field, formula (6.16), simple calculation gives $c_0 = 1$. Therefore, we expect that c_0 is also strictly positive, $c_0 > 0$, for a sufficiently small $\lambda_0 > 0$.

Now let us consider the contribution of the multi-particle states. It is given by the last term on the r.h.s. of formula (10.43). It is convenient to introduce the generalized function

$$\rho(q) = (2\pi)^3 \int\!\!\!\!\!\!\sum_{\alpha} |\langle 0|\hat{\phi}(0)|\alpha\rangle|^2 \delta^4(q - p_\alpha). \tag{10.50}$$

Then, formula (10.43) can be rewritten in the form

$$W^{(2)}(x_1, x_2) = (\langle 0|\hat{\phi}(0)|0\rangle)^2 + c_0 W_m^{(2)}(x_1, x_2) + (2\pi)^{-3} \int d^4q\ \rho(q) \exp(-i(x_1 - x_2)q). \tag{10.51}$$

The function $\rho(q)$ is positive, $\rho(q) \geq 0$, in the sense that

$$\int d^4q\ \rho(q)\chi(q) \geq 0$$

for any non-negative test function $\chi(q)$. This property of $\rho(q)$ follows directly from its definition:

$$\int d^4q\ \rho(q)\chi(q) = (2\pi)^3 \int\!\!\!\!\!\!\sum_{\alpha} |\langle 0|\hat{\phi}(0)|\alpha\rangle|^2 \chi(p_\alpha) \geq 0.$$

Comparing formulas (10.39), (10.48) and (10.51) we obtain the equality

$$\int d^4q\ \rho(q) \exp(-iq(x_1 - x_2)) = \int d^4q\ \rho(q) \exp(-iq(\hat{L}x_1 - \hat{L}x_2))$$

for any $\hat{L} \in L_+^\uparrow$. The r.h.s. of this formula is equal to

$$\int d^4q\ \rho(\hat{L}q) \exp(-iq(x_1 - x_2)),$$

because the scalar product in the exponent as well as the four-dimensional volume element d^4q are Lorentz invariant. The Fourier transformation in the space of generalized functions is invertible. Therefore,

$$\rho(\hat{L}q) = \rho(q) \quad \text{for all} \quad \hat{L} \in L_+^\uparrow, \tag{10.52}$$

that is, $\rho(q)$ is Lorentz invariant.

Another important property of $\rho(q)$ is that it vanishes when $q^0 < 0$. The reason for this is that the energies p_α^0 of the multiparticle states are positive because, for the assumed small value of the coupling constant λ_0, attractive interactions between particles are not strong enough to form bound states with negative total energy. Taking into account property (10.52), we may write $\rho(q)$ in the standard form

$$\rho(q) = \Theta(q^0)\underline{\sigma}(q^2), \tag{10.53}$$

where $\underline{\sigma}(q^2)$ is called the multiparticle spectral function. Furthermore, we expect that if λ_0 is small enough, so that no bound states of particles can be formed, then

$$\underline{\sigma}(q^2) = 0 \quad \text{for} \quad q^2 < 4m^2,$$

because the smallest value of p_α^2 is obtained for two particles with total momentum $\vec{q} = \vec{0}$ (then $q^0 = 2m$, and $q^2 = 4m^2$). In the case of the free scalar field $\underline{\sigma}(q^2) = 0$, because the states $|\alpha\rangle$ contain at least two particles, while in the free field operator there is only one annihilation operator.

Formula (10.51) can now be written in the form

$$\begin{aligned} W^{(2)}(x_1, x_2) &= (\langle 0|\hat{\phi}(0)|0\rangle)^2 + c_0 W_m^{(2)}(x_1, x_2) \\ &+ \frac{1}{(2\pi)^3} \int_{4m^2}^{\infty} d\, M^2\, \underline{\sigma}(M^2) \int d^4q\, \Theta(q^0)\delta(q^2 - M^2) \exp(-iq(x_1 - x_2)) \\ &= (\langle 0|\hat{\phi}(0)|0\rangle)^2 + c_0 W_m^{(2)}(x_1, x_2) + \int_{4m^2}^{\infty} dM^2\, \underline{\sigma}(M^2) W_M^{(2)}(x_1, x_2), \end{aligned} \tag{10.54}$$

where $W_M^{(2)}(x_1, x_2)$ denotes the 2-point Wightman's function of the free scalar field with mass parameter M. The integration variable is M^2. Formula (10.54) is called the spectral decomposition of Wightman's function.

The spectral decomposition for $G^{(2)}$ is obtained by inserting (10.54) on the r.h.s. of formula (10.35):

$$G^{(2)}(x_1, x_2) = |\langle 0|\hat{\phi}(0)|0\rangle|^2 + c_0 G_m^{(2)}(x_1, x_2) + \int_{4m^2}^{\infty} dM^2\, \underline{\sigma}(M^2) G_M^{(2)}(x_1, x_2). \tag{10.55}$$

Here $G_m^{(2)}$ and $G_M^{(2)}$ denote Green's functions of the free scalar field with mass parameters equal to m and $M \geq 2m$, respectively.

The Fourier transform of formula (10.55), (see the definition (7.49) with $n = 2$) has the form $\tilde{G}^{(2)}(k_1, k_2) = \delta(k_1 + k_2)\tilde{G}(k_1)$, where

$$\begin{aligned} \tilde{G}(k_1) &= (2\pi)^4 |\langle 0|\hat{\phi}(0)|0\rangle|^2 \delta^4(k_1) \\ &+ \frac{ic_0}{k_1^2 - m^2 + i0_+} + \int_{4m^2}^{\infty} dM^2\, \underline{\sigma}(M^2) \frac{i}{k_1^2 - M^2 + i0_+}. \end{aligned} \tag{10.56}$$

It is clear that $\tilde{G}(k_1)$ obeys property (8.30).

The spectral decomposition (10.56) shows that $\tilde{G}(k_1)$ has a simple pole at $k_1^2 = m^2$ with residue ic_0 where $c_0 > 0$. The perturbative results for $\tilde{G}(k_1)$, discussed in Chaps. 7 and 8, have to be reconsidered in this respect. This will be done in Sect. 10.4.

10.3 The Contribution of the Single Particle Sector

This section is devoted to the derivation of formula (10.44). We shall see how powerful the requirement of relativistic invariance is: it implies that all basis states $|\vec{p}\,\rangle$ can be obtained from, e.g., the state $|\vec{0}\,\rangle$, by applying the representation operators $\tilde{U}$.

We shall use so called Hermitian boosts: the Hermitian, positive definite matrices $H_p \in SL(2, C)$ determined from the condition

$$m H_p^2 = p^0 \sigma_0 + p^i \sigma_i, \tag{10.57}$$

where σ_i are Pauli matrices, $p = (p^0, p^i)$ is a given four-momentum such that $p^\mu p_\mu = m^2$ and $p^0 > 0,\ m > 0$. Simple calculation shows that

$$H_p = \frac{(p^0 + m)\sigma_0 + p^k \sigma_k}{\sqrt{2m(p^0 + m)}}.$$

We know from Chap. 5 that

$$\Lambda^{-1} \sigma^\mu (\Lambda^\dagger)^{-1} = L(\Lambda)^\mu{}_\nu \sigma^\nu$$

for any $\Lambda \in SL(2, C)$, or equivalently

$$\Lambda \sigma^\mu \Lambda^\dagger = L(\Lambda^{-1})^\mu{}_\nu \sigma^\nu = \sigma^\nu L(\Lambda)_\nu{}^\mu. \tag{10.58}$$

It is convenient to introduce the matrix

$$\hat{a} \stackrel{df}{=} a_\mu \sigma^\mu = a^\mu \sigma_\mu.$$

Multiplying both sides of formula (10.58) by a_μ and summing over μ we obtain

$$\Lambda \hat{a} \Lambda^\dagger = a'^\mu \sigma_\mu, \tag{10.59}$$

where

$$a'^\mu = L(\Lambda)^\mu{}_\nu a^\nu. \tag{10.60}$$

Comparing (10.57) with (10.59) and (10.60) we see that $\hat{L}(H_p)$ is a Lorentz transformation which transforms the 4-vector $(m, 0, 0, 0)$ into $(p^0, \vec{p})$.

Note that instead of H_p we may take $H'_p = H_p u$, with arbitrary $u \in SU(2)$. The corresponding Lorentz transformation $\hat{L}(H'_p) = \hat{L}(H_p)\hat{L}(u)$ contains $\hat{L}(u)$, which is a spatial rotation because it does not change the 4-vector $(m, 0, 0, 0)$: $u\, m\sigma_0\, u^\dagger = m\sigma_0$. The boost H'_p is not Hermitian in general. One can prove that an arbitrary matrix $\Lambda \in SL(2, C)$ can be written in the form $\Lambda = H_p u$, where H_p is the Hermitian boost and $u \in SU(2)$.

Now, let us consider the vector $\tilde{U}(\Lambda, 0)|\vec{q}\,\rangle$ from the space $\mathcal{H}^{(1)}$. Formula (10.16) implies that it is an eigenvector of $\hat{P}^\mu$:

$$\hat{P}^\mu \tilde{U}(\Lambda, 0)|\vec{q}\,\rangle = \tilde{U}(\Lambda, 0)\tilde{U}^{-1}(\Lambda, 0)\hat{P}^\mu \tilde{U}(\Lambda, 0)|\vec{q}\,\rangle$$
$$= \tilde{U}(\Lambda, 0)L(\Lambda)^\mu{}_\nu \hat{P}^\nu|\vec{q}\,\rangle = L(\Lambda)^\mu{}_\nu q^\nu \tilde{U}(\Lambda, 0)|\vec{q}\,\rangle.$$

We see that the eigenvalues are equal to $L(\Lambda)^\mu{}_\nu q^\nu$, where $q^0 = E(\vec{q}\,) = \sqrt{\vec{q}^{\,2} + m^2}$. Because the operators $\hat{P}^i$, $i = 1, 2, 3$, form the complete set of commuting observables in $\mathcal{H}^{(1)}$, the vector $\tilde{U}(\Lambda, 0)|\vec{q}\rangle$ has to be proportional to $|\vec{Lq}\rangle$, where $\vec{Lq}$ denotes the spatial part of the 4-vector $\hat{L}(\Lambda)q$, i.e., $(Lq)^i = L^i{}_0 E(\vec{q}\,) + L^i{}_k q^k$. Thus,

$$\tilde{U}(\Lambda, 0)|\vec{q}\,\rangle = N(\Lambda, \vec{q}\,)|\vec{Lq}\rangle, \tag{10.61}$$

where the coefficient N can depend on Λ and $\vec{q}$.

In order to calculate the coefficient N, we use the normalization condition for the basis vectors,

$$\langle \vec{q}\,|\vec{q}\,'\rangle = \delta^3(\vec{q} - \vec{q}\,').$$

Because

$$\langle \vec{q}\,|\vec{q}\,'\rangle = \langle \vec{q}\,|\tilde{U}^\dagger(\Lambda, 0)\tilde{U}(\Lambda, 0)|\vec{q}\,'\rangle = \overline{N(\Lambda, \vec{q}\,)}N(\Lambda, \vec{q}\,')\langle \vec{Lq}|\vec{Lq}'\rangle,$$

we have the condition

$$\delta^3(\vec{q} - \vec{q}\,') = \overline{N(\Lambda, \vec{q})}N(\Lambda, \vec{q}\,')\delta^3(\vec{Lq} - \vec{Lq}').$$

Next, on the r.h.s. of this condition we use the formula

$$\delta^3(\vec{Lq} - \vec{Lq}') = \frac{E(\vec{q}\,)}{E(\vec{Lq})}\delta^3(\vec{q} - \vec{q}\,') \tag{10.62}$$

which is proved at the end of this section. It follows that

$$|N(\Lambda, \vec{q}\,)|^2 = \frac{E(\vec{Lq})}{E(\vec{q}\,)}.$$

Thus,

$$N(\Lambda, \vec{q}\,) = \sqrt{\frac{E(\vec{Lq})}{E(\vec{q}\,)}} \exp(i\chi(\Lambda, \vec{q})),$$

where $\exp(i\chi)$ is a phase factor.

Let us now take $\vec{q} = 0$ and $\Lambda = H_p$. Then

$$N(H_p, 0) = \sqrt{\frac{E(\vec{p}\,)}{m}} \exp(i\chi(H_p, 0)),$$

and formula (10.61) says that

$$\tilde{U}(H_p, 0)|\vec{0}\,\rangle = \sqrt{\frac{E(\vec{p})}{m}} \exp(i\chi(\Lambda, \vec{q}))|\vec{p}\,\rangle,$$

or

$$|\vec{p}\,\rangle = \sqrt{\frac{m}{E(\vec{p}\,)}} \exp(-i\chi(\Lambda, \vec{q}\,))\tilde{U}(H_p, 0)|\vec{0}\,\rangle. \tag{10.63}$$

Formula (10.44) follows immediately from (10.63), (10.23) and (10.29):

$$|\langle 0|\hat{\phi}(0)|\vec{p}\,\rangle|^2 = \frac{m}{E(\vec{p}\,)}|\langle 0|\hat{\phi}(0)\tilde{U}(H_p, 0)|\vec{0}\,\rangle|^2$$
$$= \frac{m}{E(\vec{p}\,)}|\langle 0|\tilde{U}(H_p, 0)\tilde{U}^{-1}(H_p, 0)\hat{\phi}(0)\tilde{U}(H_p, 0)|\vec{0}\,\rangle|^2 = \frac{m}{E(\vec{p}\,)}|\langle 0|\hat{\phi}(0)|\vec{0}\,\rangle|^2.$$

It remains to prove formula (10.62). Let us regard $\vec{q}\,'$ as a fixed vector and $\vec{q}$ as a variable. We shall use the general formula

$$\delta^3(\vec{F}(\vec{q}\,)) = \frac{1}{|\det\hat{M}(\vec{q}_0)|}\delta^3(\vec{q} - \vec{q}_0), \tag{10.64}$$

where $\vec{q}_0$ is the vector such that $\vec{F}(\vec{q}_0) = \vec{0}$, and the Jacobi matrix

$$\hat{M} = \left[\frac{\partial F^i}{\partial q^j}\right]\Bigg|_{\vec{q}=\vec{q}_0}$$

is nonsingular. It is assumed that apart from $\vec{q}_0$ there are no other vectors $\vec{q}$ for which $\vec{F}(\vec{q}\,) = \vec{0}$. In our case

$$\vec{F}(\vec{q}\,) = \vec{Lq} - \vec{Lq}',$$

that is

$$F^i(\vec{q}\,) = L^i{}_0 E(\vec{q}\,) + L^i{}_s q^s - L^i{}_0 E(\vec{q}\,') - L^i{}_s q'^s,$$

where

$$E(\vec{q}\,) = \sqrt{\vec{q}^{\,2} + m^2}, \quad E(\vec{q}\,') = \sqrt{\vec{q}^{\,\prime 2} + m^2}. \tag{10.65}$$

Let us first prove that $\vec{F}(\vec{q}\,) = \vec{0}$ only for $\vec{q} = \vec{q}\,'$. These two vectors are momenta of the particle of rest mass m. The corresponding energies have the form (10.65). The energies corresponding to the momenta $\vec{Lq}$ and $\vec{Lq'}$ are given by formulas

$$E(\vec{Lq}) = \sqrt{(\vec{Lq})^2 + m^2} = L^0{}_0 E(\vec{q}\,) + L^0{}_i q^i,$$

$$E(\vec{Lq'}) = \sqrt{(\vec{Lq'})^2 + m^2} = L^0{}_0 E(\vec{q}\,') + L^0{}_i q'^i.$$

Therefore, the equation $\vec{F}(\vec{q}\,) = \vec{0}$ is equivalent to the equality of the 4-momenta

$$\begin{pmatrix} E(\vec{Lq}) \\ \vec{Lq} \end{pmatrix} = \begin{pmatrix} E(\vec{Lq'}) \\ \vec{Lq'} \end{pmatrix}.$$

Acting on both sides of this equality with the inverse Lorentz transformation $\hat{L}^{-1}$ we obtain the equivalent equation

$$\begin{pmatrix} E(\vec{q}\,) \\ \vec{q} \end{pmatrix} = \begin{pmatrix} E(\vec{q}\,') \\ \vec{q}\,' \end{pmatrix},$$

which has $\vec{q} = \vec{q}\,'$ as the only solution.

The elements $M^i{}_k$ of the Jacobi matrix $\hat{M}$ at the point $\vec{q}_0 = \vec{q}\,'$ have the form

$$M^i{}_k(\vec{q}\,') = L^i{}_k + \frac{L^i{}_0 q'^k}{E(\vec{q}\,')}.$$

In order to compute $\det\hat{M}$ we use the following trick. Let us introduce another matrix $\hat{A} = [A^k{}_s]$, where

$$A^k{}_s = L^k{}_s - \frac{L^k{}_0 L^0{}_s}{L^0{}_0},$$

and consider the matrix $\hat{B} = \hat{M}\hat{A}^T$, where T denotes matrix transposition. Using the following properties of the Lorentz transformations

$$L^i{}_s L^r{}_s = L^i{}_0 L^r{}_0 + \delta_{ir}, \tag{10.66}$$

$$L^i{}_s L^0{}_s = L^i{}_0 L^0{}_0, \tag{10.67}$$

we find that

$$B^i{}_r = M^i{}_s(\hat{A}^T)^s{}_r = M^i{}_s A^r{}_s = \delta_{ir} + c_i d_r,$$

where

$$c_i = L^i{}_0, \quad d_r = \frac{1}{E(\vec{q}\,')}\left(L^r{}_s q'^s - \frac{L^r{}_0 L^0{}_s q'^s}{L^0{}_0}\right).$$

Straightforward calculation gives

$$\det\hat{B} = 1 + \vec{c}\,\vec{d} = 1 + \frac{L^i{}_0 L^i{}_s q'^s}{E(\vec{q}\,')} - \frac{L^i{}_0 L^i{}_0 L^0{}_s q'^s}{E(\vec{q}\,')L^0{}_0}.$$

The r.h.s. of this formula can be simplified with the help of another identity satisfied by the Lorentz matrices, namely

$$L^i{}_0 L^i{}_0 = L^0{}_0 L^0{}_0 - 1.$$

On the other hand,

$$\det\hat{B} = \det\hat{M}\,\det\hat{A}.$$

Because, as we show below,

$$\det\hat{A} = \frac{1}{L^0{}_0}, \tag{10.68}$$

we obtain

$$\det\hat{M} = L^0{}_0 + \frac{L^0{}_s q'^s}{E(\vec{q}\,')} = \frac{E(\vec{Lq}\,')}{E(\vec{q}\,')}.$$

Thus, indeed formula (10.64) gives (10.62).

In order to compute $\det\hat{A}$ we use the fact that $\det\hat{L} = 1$. Because

$$1 = \det\hat{L} = \det\left(\begin{array}{c|c} L^0{}_0 & L^0{}_k \\ \hline L^i{}_0 & L^i{}_k \end{array}\right) = L^0{}_0 \det\left(\begin{array}{c|c} 1 & \frac{L^0{}_k}{L^0{}_0} \\ \hline L^i{}_0 & L^i{}_k \end{array}\right)$$

$$= L^0{}_0 \det\left(\begin{array}{c|c} 1 & \frac{L^0{}_k}{L^0{}_0} \\ \hline 0 & L^i{}_k - \frac{L^i{}_0 L^0{}_k}{L^0{}_0} \end{array}\right) = L^0{}_0 \det\hat{A},$$

we see that formula (10.68) is indeed true.

10.4 The Pole of the Perturbative $\tilde{G}^{(2)}$

The perturbative approach to the Green's functions in the $:\phi_4^4:$ model has been discussed in Chap. 7. We have seen that $G^{(1)} \equiv 0$, that is that $\langle 0|\hat{\phi}(x)|0\rangle = 0$. For this reason, the first term in the spectral decompositions (10.54) and (10.56) vanishes.

The renormalized perturbative contribution to the $\tilde{G}(k)$ function is schematically depicted in Fig. 10.1. The lines represent

$$\Delta_F(k) = \frac{i}{k^2 - m_0^2 + i0_+},$$

while the dark circles, denoted by Π^{ren}, stand for the sum of all 1-particle irreducible renormalized graphs contributing to the 2-point function. Analytically, $\tilde{G}(k)$ is given by the geometric series

$$\tilde{G}(k) = \Delta_F(k) + \Delta_F(k)\,\Pi^{ren}\,\Delta_F(k) + \Delta_F(k)(\Pi^{ren}\,\Delta_F(k))^2 + \ldots = \frac{\Delta_F(k)}{1 - \Pi^{ren}\Delta_F(k)}.$$

Therefore, in the perturbative approach

$$\tilde{G}(k) = \frac{i}{k^2 - m_0^2 - i\Pi^{ren} + i0_+}. \tag{10.69}$$

On the other hand, formula (10.56) shows that $\tilde{G}(k)$ is a regular function of k^2 for $k^2 < 4m^2$, apart from the simple pole at $k^2 = m^2$ (remember that $\langle 0|\hat{\phi}(0)|0\rangle = 0$):

$$\lim_{k^2 \to m^2} (k^2 - m^2)\,\tilde{G}(k) = ic_0. \tag{10.70}$$

Moreover, $1/\tilde{G}(k)$ is a smooth function of k^2 in a vicinity of $k^2 = m^2$. Therefore, for $k^2 < 4m^2$ the perturbatively calculated Π^{ren} should also be only a function of k^2, which is smooth in a vicinity of $k^2 = m^2$. Renormalization schemes have to respect these conditions.

Inserting (10.69) on the l.h.s. of formula (10.70), we obtain the condition

$$\lim_{k^2 \to m^2} \frac{k^2 - m^2}{k^2 - m_0^2 - i\Pi^{ren}(k^2) + i0_+} = c_0, \tag{10.71}$$

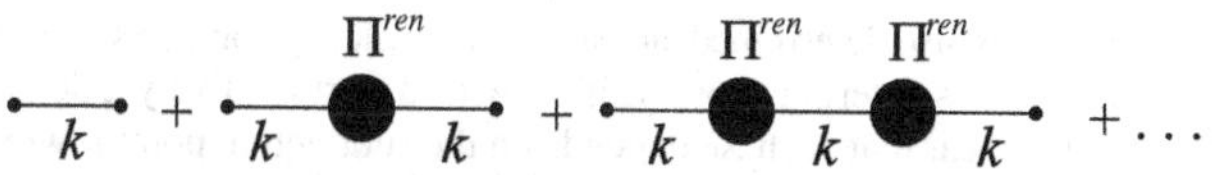

Fig. 10.1 The schematic picture of the perturbative contributions to $\tilde{G}(k)$

where $c_0 > 0$. Therefore, the denominator has to vanish at $k^2 = m^2$:

$$m^2 - m_0^2 - i\Pi^{ren}(k^2 = m^2) = 0. \tag{10.72}$$

This condition determines the mass parameter m_0^2 present in the Lagrangian (7.1). The value of m^2 is provided by measuring the rest mass of the particle.[8]

Let us analyze condition (10.72) order by order. In the lowest order, $\sim \lambda_0^0$, there are no 1-particle irreducible graphs contributing to $\tilde{G}^{(2)}$. Hence, $\Pi^{ren}_{(0)}(k^2) = 0$ and

$$m_0^2 = m^2. \tag{10.73}$$

Thus, in the zeroth order, the mass parameter m_0^2 is equal to the rest mass squared of the scalar particle. Comparing (10.69) with (10.56) we also find that in the zeroth order

$$\underline{\sigma}^{(0)} = 0, \quad c_0^{(0)} = 1.$$

The first non vanishing contribution to $\Pi^{ren}(k^2)$ appears in the λ_0^2 order. It is represented by the graph from Fig. 8.9. Let us denote it by $\Pi^{ren}_{(2)}(k^2)$. Now formula (10.72) has the form

$$m_0^2 = m^2 - i\Pi^{ren}_{(2)}(k^2 = m^2). \tag{10.74}$$

$\Pi^{ren}_{(2)}(m^2)$ contains m_0^2 in the free propagators $\Delta_F(k)$, hence (10.74) is actually an equation for m_0^2. However, because $\Pi^{ren}_{(2)}$ is already proportional to λ_0^2, we may replace m_0^2 by m^2 in the free propagators—this does not change the term proportional to λ_0^2. Thus, in the second order

$$m_0^2 = m^2 - i\, \Pi^{ren}_{(2)}(k^2 = m^2)\Big|_{m_0^2 = m^2}. \tag{10.75}$$

The mass parameter m_0^2, which in the zeroth order was equal to m^2, now has to be corrected in accordance with formula (10.75). The term $-m_0^2\phi^2/2$ in the Lagrangian can be written in the form

$$-\frac{1}{2}m^2\phi^2 + \frac{i}{2}\, \Pi^{ren}_{(2)}(k^2 = m^2)\Big|_{m_0^2 = m^2}\phi^2.$$

The term $\frac{i}{2}\, \Pi^{ren}_{(2)}(k^2 = m^2)\Big|_{m_0^2 = m^2}\phi^2$ is called the finite mass counterterm. It is finite because it is calculated from the renormalized $\Pi^{ren}_{(2)}(k^2)$. Also in higher orders, finite

[8]The coupling constant λ_0 is also determined, at least in principle, by comparison with the the results of measurements of, e.g., a scattering cross section with a perturbatively calculated theoretical prediction. However, it is clear that such λ_0 depends on the subtraction point μ which is present in the perturbative formulas. Hence, in fact it should be regarded as the running coupling constant at that value of μ.

counterterms of this type are necessary. Without them, the perturbative $:\phi_4^4:$ model would not be compatible with relativistic invariance and the particle interpretation.

Exercises

10.1 Derive formula (10.18).
Hint: $\Lambda^{-1}(\omega)\Lambda(\omega_1)\Lambda(\omega) = \Lambda(\tilde{\omega})$, where $\tilde{\omega}$ is determined from the formula $\hat{L}^{-1}(\omega)\hat{L}(\omega_1)\hat{L}(\omega) = \hat{L}(\tilde{\omega})$.

10.2 Obtain the transformation law of $\hat{M}_{\mu\nu}$ with respect to translations in space-time.
Hint: Compute derivatives of both sides of formula (10.14) with respect to $\omega^{\mu\nu}$ assuming that $\Lambda = \Lambda(\omega)$ and next put $\omega = 0$.

10.3 Check that the free real scalar field obeys the relation (10.24).

10.4 Starting from formula (10.71) prove that

$$c_0 = \frac{1}{1 - i\Pi^{ren\,\prime}(k^2 = m^2)},$$

where $'$ denotes the derivative with respect to k^2.
Hint: Apply l'Hospital's rule known from calculus.

Chapter 11
Path Integrals in QFT

Abstract The path integral formulas for the evolution operator in quantum mechanics. The path integral formula for the generating functional $Z[j]$ in the quantum theory of the real scalar field. Rederivation of the perturbative expansion for the ϕ_4^4 model. Integration over Grassmann variables. Path integral formula for the generating functional in the theory of the quantum Dirac field.

The time evolution of the states of an isolated quantum system is described by a unitary operator U in a Hilbert space. Path integrals are used in order to write matrix elements of U in a form which makes the connection with a certain classical theory explicit, hence path integrals facilitate the study of the classical limit of the quantum theory. In many cases in field theory we are not able to construct the quantum theory explicitly. Then path integrals can be used as a heuristic tool, with which we can guess many features of the sought after quantum theory. An outstanding example of such 'reversed' use of path integrals is provided by non-Abelian gauge fields, to be discussed in the next chapter.

We start our introduction to the formalism of path integrals with a very simple example, namely that of a single, spinless, one-dimensional particle where the quantum theory is well-known. Next, we pass to the relativistic quantum scalar field for which we already know the perturbative expansion for the Green's functions. Finally, we introduce path integrals for fermionic fields—in this case anticommuting classical fields appear.

11.1 Path Integrals in Quantum Mechanics

In this section we show how the path integrals are derived in the framework of the operator formalism of quantum mechanics. We consider a spinless, nonrelativistic particle of mass m. It can move only along a straight line, which we call the x axis, and it is subject to forces described by a smooth classical potential $V(x)$. The quantum Hamiltonian for such a particle has the form

H. Arodź and L. Hadasz, *Lectures on Classical and Quantum Theory of Fields*, Graduate Texts in Physics, DOI 10.1007/978-3-319-55619-2_11

$$\hat{H} = T(\hat{p}) + V(\hat{x}), \tag{11.1}$$

where $\hat{x}$ and $\hat{p} = -i\hbar d/dx$ are the position and momentum operators in the Schroedinger picture, and

$$T(\hat{p}) = \frac{\hat{p}^2}{2m}$$

is the kinetic energy operator. In the present section we write the Planck constant $\hbar$ explicitely because natural units are very rarely used in quantum mechanics. The Hamiltonian $\hat{H}$ does not depend on time, therefore the evolution operator is given by the formula

$$U(t'', t') = \exp\left[-\frac{i}{\hbar}\hat{H}(t'' - t')\right]. \tag{11.2}$$

This operator is fully described by its matrix elements $\langle x''|U(t'', t')|x'\rangle$ in the basis of eigenstates $|x\rangle$ of the position operator $\hat{x}$

$$\hat{x}|x\rangle = x\,|x\rangle.$$

The matrix elements $\langle x''|U(t'', t')|x'\rangle$ can be expressed by an integral over a certain set of trajectories in the phase space of the particle. Let us divide the interval $[t'', t']$ into N subintervals $[t_{i-1}, t_i]$, where

$$t_i = t' + \epsilon\, i, \quad i = 0, \ldots, N, \quad \epsilon = (t'' - t')/N,$$

with $t_0 \equiv t',\ t_N \equiv t''$. Then

$$\langle x''|U(t'', t')|x'\rangle = \langle x''|U(t'', t_{N-1})\, U(t_{N-1}, t_{N-2}) \ldots U(t_1, t')|x'\rangle. \tag{11.3}$$

Next, we insert N identity operators of the form

$$I = \int_{-\infty}^{+\infty} dp\, |p\rangle\langle p|,$$

where $|p\rangle$ is the eigenstate of the momentum operator

$$\hat{p}|p\rangle = p|p\rangle,$$

and also $N - 1$ identity operators of the form

$$I = \int_{-\infty}^{+\infty} dx\, |x\rangle\langle x|.$$

For the sake of clarity, the integration variables x and p in all identity operators are appropriately numbered. We obtain the following formula

$$\langle x''|U(t'',t')|x'\rangle = \int_{-\infty}^{+\infty} \dots \int_{-\infty}^{+\infty} dp_N dp_{N-1} dx_{N-1} \dots dp_1 dx_1 \langle x''|p_N\rangle$$
$$\langle p_N|e^{-\frac{i}{\hbar}\epsilon\hat{H}}|x_{N-1}\rangle\langle x_{N-1}|p_{N-1}\rangle\langle p_{N-1}|e^{-\frac{i}{\hbar}\epsilon\hat{H}}|x_{N-2}\rangle\langle x_{N-2}|p_{N-2}\rangle$$
$$\langle p_{N-2}|e^{-\frac{i}{\hbar}\epsilon\hat{H}}|x_{N-3}\rangle\langle x_{N-3}|p_{N-3}\rangle \dots \langle p_1|e^{-\frac{i}{\hbar}\epsilon\hat{H}}|x'\rangle. \quad (11.4)$$

The scalar products of the form $\langle x|p\rangle$ are normalized plane waves

$$\langle x|p\rangle = \frac{1}{\sqrt{2\pi\hbar}} e^{\frac{i}{\hbar}xp}. \quad (11.5)$$

We are interested in the limit $\epsilon \to 0$. Therefore, ϵ is small and the matrix elements of the exponentials in formula (11.4) can be rewritten as follows

$$\langle p_{k+1}|e^{-\frac{i}{\hbar}\epsilon\hat{H}}|x_k\rangle$$
$$= \left[1 - \frac{i}{\hbar}\epsilon\left(T(p_{k+1}) + V(\frac{x_k + x_{k+1}}{2})\right) + \mathcal{O}(\epsilon^2)\right] e^{-\frac{i}{\hbar}p_{k+1}x_k}$$
$$= e^{-\frac{i}{\hbar}\epsilon(T(p_{k+1}) + V(\frac{x_k + x_{k+1}}{2}))} e^{-\frac{i}{\hbar}p_{k+1}x_k} + \mathcal{O}(\epsilon^2). \quad (11.6)$$

Using formulas (11.5), (11.6) we transform (11.4) into the following form

$$\langle x''|U(t'',t')|x'\rangle = \int_{-\infty}^{+\infty} \frac{dp_N}{2\pi\hbar} \int_{-\infty}^{+\infty} \dots \int_{-\infty}^{+\infty} \prod_{l=1}^{N-1} \frac{dp_l dx_l}{2\pi\hbar}$$
$$\exp\left(\frac{i}{\hbar}\epsilon \sum_{k=1}^{N-1} \left[p_{k+1}\frac{x_{k+1} - x_k}{\epsilon} - T(p_{k+1}) - V(\frac{x_k + x_{k+1}}{2})\right]\right)(1 + \mathcal{O}(\epsilon^2)). \quad (11.7)$$

Note that the number of integrals over the momenta is larger by 1 than over the positions. We expect that in the limit $\epsilon \to 0$ the terms marked as $\mathcal{O}(\epsilon^2)$ can be neglected. Unfortunately, precise control of these terms turns out to be very difficult. It is a major obstacle in obtaining a mathematically rigorous definition of the path integrals.

The action functional for the path $(x(t), p(t))$ in the phase space of the particle has the form

$$S[x(t), p(t)] = \int_{t'}^{t''} dt \; [\dot{x}(t)p(t) - H(p(t), x(t))] .$$

Let us take a path $(p_{(N)}(t), x_{(N)}(t))$ in the phase space such that $p_{(N)}(t)$ is constant in each interval $(t_k, t_{k+1}]$ introduced above—the value of $p_{(N)}(t)$ in that interval is denoted as p_{k+1} (here $k = 0, 1, N-1$), see Fig. 11.1. Moreover, the function $x_{(N)}(t)$ is linear in each time interval, namely

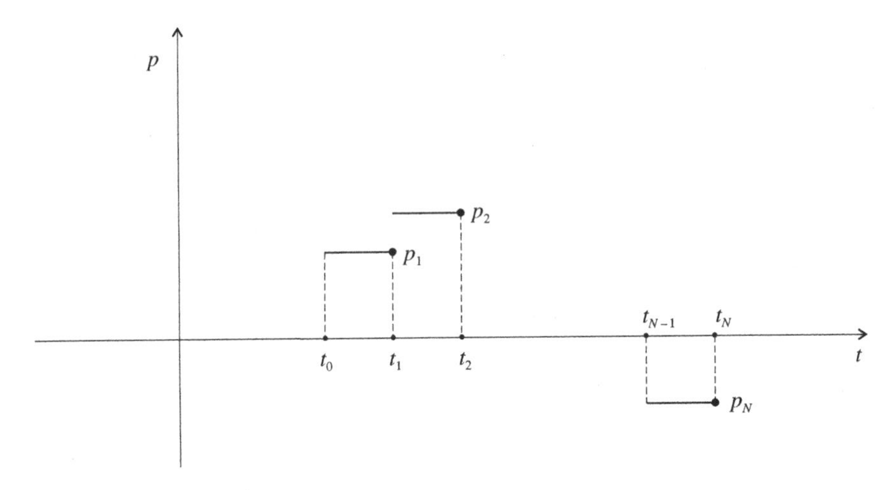

Fig. 11.1 The function $p_{(N)}(t)$

$$x_{(N)}(t) = x_k + (t - t_k)\frac{x_{k+1} - x_k}{\epsilon} \quad \text{if} \quad t \in [t_k, t_{k+1}],$$

see Fig. 11.2. Note that the momentum part of the phase space path is not continuous in general, while the position part is always continuous. The velocity $\dot{x}(t)$ is constant during the introduced time intervals and equal to $(x_{k+1} - x_k)/\epsilon$. It is not correlated at all with the values p_{k+1} of the momentum in these time intervals. In particular, the relation $p_{k+1}/m = (x_{k+1} - x_k)/\epsilon$, which would correspond to $p(t)/m = \dot{x}(t)$, is not true in general—this relation holds only for the paths which are the physical trajectories of the particle, that is for solutions of the classical Hamilton equations, while here we consider arbitrary paths. In the limit $\epsilon \to 0$, equivalent to the limit $N \to \infty$, the functions $x_{(N)}(t)$ remain continuous, but in general they are not differentiable on the whole interval (t', t'').

The value of the action functional S for the path $(p_{(N)}(t), x_{(N)}(t))$, denoted by S_N, is calculated as follows:

$$S_N = \sum_{k=0}^{N-1} \int_{t_k}^{t_{k+1}} dt\ (p\dot{x} - H)$$
$$= \sum_{k=0}^{N-1} \epsilon \left[p_{k+1}\frac{x_{k+1} - x_k}{\epsilon} - T(p_{k+1}) - V(\frac{x_k + x_{k+1}}{2}) \right] \left(1 + \mathcal{O}(\epsilon^2)\right),$$

where we have used the following approximation

$$\int_{t_k}^{t_{k+1}} dt\ V(x_{(N)}) = \epsilon V(\frac{x_k + x_{k+1}}{2}) + \mathcal{O}(\epsilon^2).$$

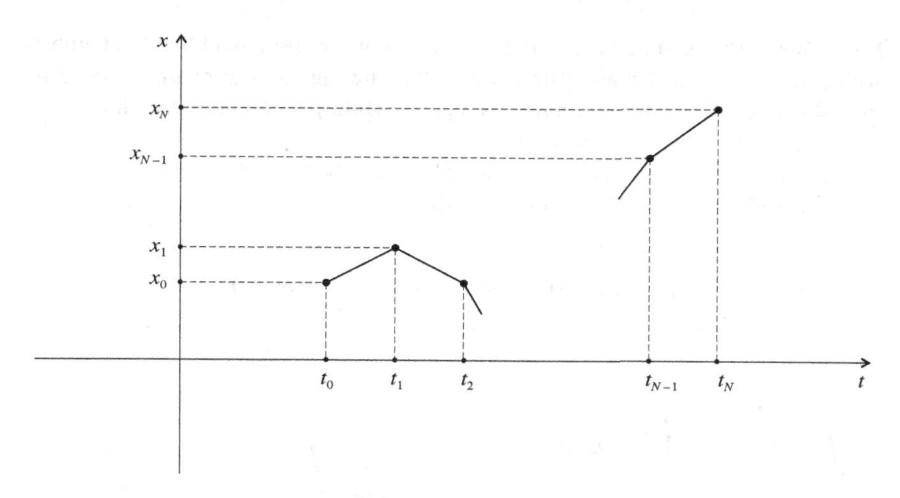

Fig. 11.2 The function $x_{(N)}(t)$

Therefore, formula (11.7) can be written in the form

$$\langle x''|U(t'',t')|x'\rangle$$
$$= \int_{-\infty}^{+\infty} \frac{dp_N}{2\pi\hbar} \int_{-\infty}^{+\infty} \cdots \int_{-\infty}^{+\infty} \prod_{l=1}^{N-1} \frac{dp_l dx_l}{2\pi\hbar} \exp\left(\frac{i}{\hbar} S_N\right) \left(1+\mathcal{O}(\epsilon^2)\right). \tag{11.8}$$

In the cases where the $\mathcal{O}(\epsilon^2)$ terms do not give any contribution to the limit $N \to \infty$ we may write

$$\langle x''|U(t'',t')|x'\rangle = \lim_{N\to\infty} \int_{-\infty}^{+\infty} \frac{dp_N}{2\pi\hbar} \int_{-\infty}^{+\infty} \cdots \int_{-\infty}^{+\infty} \prod_{l=1}^{N-1} \frac{dp_l dx_l}{2\pi\hbar} \exp\left(\frac{i}{\hbar} S_N\right) \tag{11.9}$$

Formula (11.9) gives a representation of the matrix elements $\langle x''|U(t'',t')|x'\rangle$ in terms of integration over the set of paths in the phase space—for each concrete choice of values of the integration variables $x_1, \ldots, x_{N-1}, p_1, \ldots, p_N$ we have the paths $(x_N(t), p_N(t))$ in the phase space. That formula is often written in a concise form as

$$\langle x''|U(t'',t')|x'\rangle = \int_{\substack{x(t')=x' \\ x(t'')=x''}} \prod_{t\in(t',t'')} \frac{dp(t)dx(t)}{2\pi\hbar} \exp\left(\frac{i}{\hbar} S[p,x]\right), \tag{11.10}$$

or, in an even more concise form,

$$\langle x''|U(t'',t')|x'\rangle = \int [\frac{dpdx}{2\pi\hbar}] \exp\left(\frac{i}{\hbar} S[p,x]\right). \tag{11.11}$$

These short forms can be misleading: one does not see from them that the numbers of integrals over p and x are different, and that the functions $p(t)$ are not continuous. Moreover, the paths $x(t)$ have fixed ends, while $p(t)$ do not. One should also remember that $\dot{x}(t)$ is not related to $p(t)$.

The integrals over momenta can be calculated, because $T(p) = p^2/(2m)$ and these integrals have the Gaussian form. Using

$$\int_{-\infty}^{+\infty} dp \, \exp(-ap^2 + bp) = \sqrt{\frac{\pi}{a}} \exp(\frac{b^2}{4a}),$$

we obtain

$$\int_{-\infty}^{+\infty} dp_{k+1} \exp\left(\frac{i}{\hbar}\left[p_{k+1}(x_{k+1} - x_k) - \epsilon \frac{p_{k+1}^2}{2m}\right]\right)$$
$$= \sqrt{\frac{2\pi\hbar m}{i\epsilon}} \exp\left(\frac{im}{2\hbar\epsilon}(x_{k+1} - x_k)^2\right).$$

In consequence,

$$\langle x''|U(t'', t')|x'\rangle = \left(\frac{m}{2\pi i\hbar\epsilon}\right)^{\frac{N}{2}} \int_{-\infty}^{+\infty} \cdots \int_{-\infty}^{+\infty} \prod_{l=1}^{N-1} dx_l$$
$$\exp\left(\sum_{k=0}^{N-1}\left[\frac{im}{2\hbar\epsilon}(x_{k+1} - x_k)^2 - \frac{i\epsilon}{\hbar} V(\frac{x_{k+1} + x_k}{2})\right]\right)\left(1 + \mathcal{O}(\epsilon^2)\right). \qquad (11.12)$$

On the other hand, the action functional for a path $x(t)$ in the configuration space of the particle has the form

$$S[x(t)] = \int_{t'}^{t''} dt \; L(x(t), \dot{x}(t)),$$

where

$$L = \frac{m}{2}\dot{x}^2 - V(x(t)).$$

Therefore,

$$S[x_N(t)] = \sum_{k=0}^{N-1} \int_{t_k}^{t_{k+1}} dt \; L(x_N(t), \dot{x}_N(t))$$
$$= \sum_{k=0}^{N-1}\left[\frac{m(x_{k+1} - x_k)^2}{2\epsilon} - \epsilon \, V(\frac{x_{k+1} + x_k}{2})\right]\left(1 + \mathcal{O}(\epsilon^2)\right). \qquad (11.13)$$

and

$$\langle x''|U(t'',t')|x'\rangle$$
$$= \lim_{N\to\infty}\left(\frac{m}{2\pi i\hbar\epsilon}\right)^{\frac{N}{2}}\int_{-\infty}^{+\infty}\cdots\int_{-\infty}^{+\infty}\left(\prod_{l=1}^{N-1}dx_l\right)\exp\left(\frac{i}{\hbar}S[x_N(t)]\right), \quad (11.14)$$

if the $\mathcal{O}(\epsilon^2)$ terms do not give any contribution in the $N \to \infty$ limit. This formula is written in a concise form as

$$\langle x''|U(t'',t')|x'\rangle = \mathcal{N}\int_{\substack{x(t')=x' \\ x(t'')=x''}}[dx(t)]\exp\left(\frac{i}{\hbar}S[x(t)]\right). \quad (11.15)$$

Formula (11.14) gives the matrix elements of the time evolution operator in terms of the integral over a set of paths $x(t)$ in the classical configuration space of the particle. The paths have fixed ends, they are continuous, but in general not differentiable. Note that the paths do not go back in time—it is clear from Fig. 11.2 that for such paths there would be three or more integration variables at given time t_k, while in our derivation we have introduced just one.

Quantum mechanical Green's functions have the form of matrix elements of time-ordered products of the position operator $\hat{x}_H(t)$ in the Heisenberg picture,

$$\hat{x}_H(t) = \exp(\frac{i}{\hbar}t\hat{H})\,\hat{x}\,\exp(-\frac{i}{\hbar}t\hat{H}), \quad (11.16)$$

namely

$$G^{(n)}(t_1,t_2,\ldots,t_n) = \langle b|T\left(\hat{x}_H(t_1)\hat{x}_H(t_2)\ldots\hat{x}_H(t_n)\right)|a\rangle, \quad (11.17)$$

where $|a\rangle$ and $|b\rangle$ are certain states. Using formula (11.16) and performing the time ordering we obtain

$$G^{(n)}(t_1,t_2,\ldots,t_n)$$
$$= \langle b|\exp\left(\frac{i}{\hbar}t_{i_n}\hat{H}\right)\,\hat{x}\,U(t_{i_n},t_{i_{n-1}})\,\hat{x}\,\ldots U(t_{i_2},t_{i_1})\,\hat{x}\,\exp\left(\frac{i}{\hbar}t_{i_1}\hat{H}\right)|a\rangle, \quad (11.18)$$

where $t_{i_n} \geq t_{i_{n-1}} \geq \ldots t_{i_2} \geq t_{i_1}$ is the time ordered sequence obtained by permuting $t_1, t_2, \ldots, t_n$. In order to obtain the path integral formula for the Green's functions we substitute for each operator $\hat{x}$ in formula (11.18) its spectral representation, namely

$$\hat{x} = \int_{-\infty}^{\infty}dx\,|x\rangle\,x\,\langle x|. \quad (11.19)$$

We distinguish the integration variables in formula (11.19) for n operators $\hat{x}$ in (11.18) by denoting them as $x(t_{i_k})$ with $k = 1, 2, \ldots, n$, namely $x(t_{i_k})$ is used in the spectral representation of that operator $\hat{x}$ in formula (11.18) which has t_{i_k} on both sides. Moreover, we insert two identity operators of the form

$$I = \int_{-\infty}^{\infty} dx_f \; |x_f\rangle\langle x_f|, \quad I = \int_{-\infty}^{\infty} dx_i \; |x_i\rangle\langle x_i|,$$

and the exponentials $\exp(\pm\frac{i}{\hbar}T_f \hat{H})$ and $\exp(\pm\frac{i}{\hbar}T_i \hat{H})$, where $T_f > t_{i_n} \geq t_{i_1} > T_i$. After all these steps, the r.h.s. of formula (11.18) has the following form

$$\int_{-\infty}^{\infty} \cdots \int_{-\infty}^{\infty} dx_f dx_i dx(t_{i_1}) \ldots dx(t_{i_n}) \; \langle b| \exp\left(\frac{i}{\hbar}T_f \hat{H}\right) |x_f\rangle$$
$$\langle x_f|U(T_f, t_{i_n})|x(t_{i_n})\rangle \; x(t_{i_n}) \; \langle x(t_{i_n})|U(t_{i_n}, t_{i_{n-1}})|x(t_{i_{n-1}})\rangle \; x(t_{i_{n-1}}) \ldots$$
$$\langle x(t_{i_2})|U(t_{i_2}, t_{i_2})|x(t_{i_1})\rangle \; x(t_{i_1}) \langle x(t_{i_1})|U(t_{i_1}, T_i)|x_i\rangle \; \langle x_i| \exp\left(-\frac{i}{\hbar}T_i \hat{H}\right) |a\rangle.$$

For each matrix element $\langle x(t_{i_k})|U(t_{i_k}, t_{i_{k-1}})|x(t_{i_{k-1}})\rangle$ we use formula (11.15), which involves paths connecting the points $x(t_{i_k})$ and $x(t_{i_{k-1}})$. These paths from consecutive time intervals are combined to form long paths connecting the points x_f and x_i. Therefore, the path integral representation of the Green's function has the form

$$\begin{aligned} G^{(n)}(t_1, t_2, \ldots, t_n) &= \mathcal{N} \int_{-\infty}^{\infty} dx_f dx_i \; \langle b| \exp\left(\frac{i}{\hbar}T_f \hat{H}\right) |x_f\rangle \; \langle x_i| \exp\left(-\frac{i}{\hbar}T_i \hat{H}\right) |a\rangle \\ &\int_{\substack{x(T_i) = x_i \\ x(T_f) = x_f}} [dx(t)] \; x(t_1)x(t_2) \ldots x(t_n) \; \exp\left(\frac{i}{\hbar}S[x(t)]\right). \end{aligned} \quad (11.20)$$

In the particular case of $|a\rangle$ and $|b\rangle$ being eigenstates of $\hat{H}$ with eigenvalues E_a and E_b, respectively,

$$\begin{aligned} G^{(n)}(t_1, t_2, \ldots, t_n) &= \mathcal{N} \int_{-\infty}^{\infty} dx_f dx_i \; \psi_b^*(x_f)\psi_a(x_i) \; \exp\left(\frac{i}{\hbar}[T_f E_b - T_i E_a]\right) \\ &\int_{\substack{x(T_i) = x_i \\ x(T_f) = x_f}} [dx(t)] \; x(t_1)x(t_2) \ldots x(t_n) \; \exp\left(\frac{i}{\hbar}S[x(t)]\right), \end{aligned} \quad (11.21)$$

where $\psi_a(x_i) = \langle x_i|a\rangle$ and $\psi_b(x_f) = \langle x_f|a\rangle$ are the wave functions corresponding to the states $|a\rangle$ and $|b\rangle$.

The main attractive feature of the path integral representation of time evolution in the quantum theory is the explicit appearance of the classical action, see for example formula (11.15). This fact facilitates a derivation of the classical limit of the quantum theory. The topic of the classical limit of quantum theory lies outside the scope of our considerations, but it is so important that we cannot leave it without a comment. Note that the classical action has appeared in formula (11.15), which has been obtained as a result of a computation in which we have assumed that we know the quantum Hamiltonian (11.1). Thus, the form of the classical action is dictated by the quantum theory, and not *vice versa*. Furthermore, the path integral formulation of quantum mechanics gives a rather simple explanation of the otherwise rather strange fact, that equations of motion for a classical particle often have the form of the Euler–Lagrange equations obtained from the stationary action principle: this principle follows from a certain quantum theory in the path integral formulation by taking the limit $\hbar \to 0$. One may say that the existence of the Lagrangian form of the classical equation of motion points to the fact that the classical theory is just a classical limit of a certain underlying quantum theory.

The path integral representation can also be used as an heuristic tool to help us construct a quantum theory which would correspond to a previously known classical theory. An example of such a use of the path integral is presented in the next chapter, where we construct a renormalizable perturbative expansion for quantized non-Abelian gauge fields. Let us give here another example.

It is a well-known fact in classical mechanics that the Lagrange functions $L(x, \dot{x})$ and $L' = L + \dot{x} f'(x)$, where f is a differentiable function and $f' = df/dx$, are equivalent in the sense that they give the same Euler–Lagrange equation. For simplicity we consider a particle in the one-dimensional space R^1. Let us insert the action

$$S' = \int_{T_i}^{T_f} dt \, \big(L + \frac{df(x(t))}{dt}\big) = S + f(x(T_f)) - f(x(T_i))$$

in formula (11.21) instead of S. Because $x(T_i) = x_i$ and $x(T_f) = x_f$, the net result of such a change of the action is equivalent to changing the wave functions ψ_a and ψ_b by a phase factor $\exp(-if/\hbar)$,

$$\psi_{a,b}(x) \to \exp\big(-\frac{i}{\hbar} f(x)\big)\, \psi_{a,b}(x).$$

Thus, we see that the two quantum theories obtained from the actions S and S', respectively, are equivalent in the sense that there exists a (unitary) transformation from one to the other—it consists in the multiplication of all wave functions by the same x-dependent phase factor $\exp(-if/\hbar)$. The field theoretic version of this fact was used in Sect. 6.2 in order to facilitate the quantization of the Dirac field.

Yet another application of path integrals is based on the fact that various matrix elements, originally given in terms of states and operators in the Hilbert space, can be expressed by path integrals, which subsequently can be computed with the help of efficient numerical approximation techniques.

11.2 The Path Integral for Bosonic Fields

The path integral formula for Green's functions in the case of bosonic fields is obtained essentially by repeating the steps described in the previous section. For brevity, we will discuss just one real, scalar field with the Lagrangian

$$\mathcal{L} = \frac{1}{2}\partial_\mu\phi\partial^\mu\phi - \frac{1}{2}m_0^2\phi^2 - V(\phi), \tag{11.22}$$

and the canonical momentum and Hamiltonian

$$\pi = \partial_0\phi, \quad H = \frac{1}{2}\pi^2 + \frac{1}{2}\partial_i\phi\partial_i\phi + \frac{1}{2}m_0^2\phi^2 + V(\phi). \tag{11.23}$$

We again use the natural units. The time variable is denoted by x^0 or t, as convenient.

The counterpart of the position operator $\hat{x}$ in the Schroedinger representation is the time-independent field operator $\hat{\phi}_S(\vec{x})$. Because

$$\hat{\phi}_S(\vec{x})\hat{\phi}_S(\vec{y}) - \hat{\phi}_S(\vec{y})\hat{\phi}_S(\vec{x}) = 0 \quad \text{for all} \quad \vec{x}, \vec{y} \in R^3,$$

there exist eigenstates of the field operator, denoted as $|\phi\rangle$:

$$\hat{\phi}_S(\vec{x})|\phi\rangle = \phi(\vec{x})|\phi\rangle \quad \text{for all} \quad \vec{x} \in R^3.$$

Thus, the eigenstates are labeled by the functions $\phi(\vec{x})$ defined on the space R^3. The identity operator and the spectral representation of $\hat{\phi}_S$ have the following form

$$I = \int (d\phi)\ |\phi\rangle\langle\phi|, \quad \hat{\phi}_S(\vec{x}) = \int (d\phi)\ |\phi\rangle\phi(\vec{x})\langle\phi|, \tag{11.24}$$

where

$$(d\phi) = \prod_{\vec{y}\in R^3} d\phi(\vec{y}).$$

Of course, this last formula for the integration measure $(d\phi)$ should not be taken literally—rather it is to be understood as a limit in which a discrete and finite set of points $\vec{x}$ from the space R^3 is becoming larger and denser, asymptotically approaching the whole R^3. A mathematically rigorous discussion of such a limit is not necessary for our purposes.

The operators $\hat{\pi}_S(\vec{x})$ and $\hat{\pi}_S(\vec{y})$ also commute with each other, therefore there exist eigenstates $|\pi\rangle$ such that

$$I = \int (d\pi)\ |\pi\rangle\langle\pi|, \quad \hat{\pi}_S(\vec{x}) = \int (d\pi)\ |\pi\rangle\pi(\vec{x})\langle\pi|, \tag{11.25}$$

where

$$(d\pi) = \prod_{\vec{y}\in R^3} d\pi(\vec{y}).$$

The evolution operator has the form (11.2), where now

$$\hat{H} = \frac{1}{2}\hat{\pi}^2 + \frac{1}{2}\partial_i\hat{\phi}\partial_i\hat{\phi} + \frac{1}{2}m_0^2\hat{\phi}^2 + V(\hat{\phi}).$$

Here we assume that the operator expressions are suitably regularized if necessary. Repeating the steps leading to formula (11.10), we obtain

$$\langle\phi''|U(t'',t')|\phi'\rangle = \int_{\substack{\phi(t',\vec{x}) = \phi'(\vec{x}) \\ \phi(t'',\vec{x}) = \phi''(\vec{x})}} [d\pi d\phi]\, e^{iS[\pi,\phi]}, \tag{11.26}$$

where

$$[d\pi d\phi] = \prod_{x^0\in(t',t'')}\prod_{\vec{x}\in R^3} \frac{d\pi(x^0,\vec{x})\, d\phi(x^0,\vec{x})}{2\pi}$$

and

$$S[\pi,\phi] = \int_{R^3} d^3x \int_{t'}^{t''} dx^0 \left[\pi(x^0,\vec{x})\partial_0\phi(x^0,\vec{x}) - \mathcal{L}\right].$$

The integration in formula (11.26) is over paths in the phase space of the field. Because Hamiltonian (11.23) is quadratic in the canonical momentum, we can integrate over it. This gives the analog of formula (11.15),

$$\langle\phi''|U(t'',t')|\phi'\rangle = \mathcal{N}\int_{\substack{\phi(t',\vec{x}) = \phi'(\vec{x}) \\ \phi(t'',\vec{x}) = \phi''(\vec{x})}} [d\phi] e^{iS[\phi]}, \tag{11.27}$$

where

$$S[\phi] = \int_{R^3} d^3x \int_{t'}^{t''} dx^0\, \mathcal{L}(\phi(x^0,\vec{x}), \partial_\mu\phi(x^0,\vec{x})).$$

The Green's functions are given by a formula analogous to (11.21)—instead of the $\hat{x}_H(t)$ operator, we now take the scalar field operator in the Heisenberg picture. If both states $|a\rangle$ and $|b\rangle$ are the vacuum state $|0\rangle$, then $E_a = E_b = 0$, and

$$\langle 0|T\left(\hat{\phi}(x_1)\ldots\hat{\phi}(x_n)\right)|0\rangle$$
$$= \mathcal{N}\int (d\phi'')(d\phi')\; \Psi_0^*[\phi'']\Psi_0[\phi'] \int_{\substack{\phi(T_i,\vec{x}) = \phi'(\vec{x}) \\ \phi(T_f,\vec{x}) = \phi''(\vec{x})}} [d\phi]\, \phi(x_1)\ldots\phi(x_n) e^{iS[\phi]}, \tag{11.28}$$

where $\Psi_0[\phi] = \langle\phi|0\rangle$ is the wave functional of the vacuum state. The time T_f is later and T_i earlier than any of the times x_k^0.

Unfortunately, in the most interesting cases the wave functional $\Psi_0[\phi]$ is not known. We circumvent this problem with the help of formula

$$\langle 0|T\left(\hat{\phi}(x_1)\ldots\hat{\phi}(x_n)\right)|0\rangle = \lim_{\substack{T_f\to\infty \\ T_i\to-\infty}} \frac{\langle\chi|e^{-iT_f\hat{H}}T\left(\hat{\phi}(x_1)\ldots\hat{\phi}(x_n)\right)e^{iT_i\hat{H}}|\eta\rangle}{\langle\chi|e^{-i(T_f-T_i)\hat{H}}|\eta\rangle}, \quad (11.29)$$

which appeared in Chap. 7, in the derivation of the Gell-Mann–Low formula precisely in order to get rid of the vacuum state $|0\rangle$. Next, we use the field theoretic version of formula (11.20) with

$$|a\rangle = e^{iT_i\hat{H}}|\eta\rangle, \qquad \langle b| = \langle\chi|e^{-iT_f\hat{H}}.$$

Because the exponentials with T_f and T_i on the r.h.s. cancel out, the numerator in (11.29) can be written as

$$\begin{aligned}&\langle\chi|e^{-iT_f\hat{H}}T\left(\hat{\phi}(x_1)\ldots\hat{\phi}(x_n)\right)e^{iT_i\hat{H}}|\eta\rangle \\ &\quad = \mathcal{N}\int(d\phi'')(d\phi')\,\chi^*[\phi'']\,\eta[\phi']\int_{\substack{\phi(T_i,\vec{x})=\phi'(\vec{x}) \\ \phi(T_f,\vec{x})=\phi''(\vec{x})}}[d\phi]\,\phi(x_1)\ldots\phi(x_n)e^{iS[\phi]},\end{aligned} \quad (11.30)$$

where $\chi[\phi''] = \langle\phi''|\chi\rangle$ and $\eta[\phi'] = \langle\phi'|\eta\rangle$.

For the denominator we have

$$\langle\chi|e^{-i(T_f-T_i)\hat{H}}|\eta\rangle = \mathcal{N}\int(d\phi'')(d\phi')\,\chi^*[\phi'']\,\eta[\phi']\int_{\substack{\phi(T_i,\vec{x})=\phi'(\vec{x}) \\ \phi(T_f,\vec{x})=\phi''(\vec{x})}}[d\phi]\,e^{iS[\phi]}. \quad (11.31)$$

The path integral representation of the generating functional for the Green's functions

$$Z[j] = \langle 0|T\exp\left(i\int d^4x\, j(x)\hat{\phi}(x)\right)|0\rangle$$

follows from the formulas (11.29)–(11.31):

$$Z[j] = \frac{\overline{Z}[j]}{\overline{Z}[0]}, \quad (11.32)$$

where

$$\overline{Z}[j] = \int(d\phi'')(d\phi')\,\chi^*[\phi'']\,\eta[\phi']\int_{\substack{\phi(-\infty,\vec{x})=\phi'(\vec{x}) \\ \phi(\infty,\vec{x})=\phi''(\vec{x})}}[d\phi]\,e^{iS[\phi]+i\int d^4x\, j(x)\hat{\phi}(x)}. \quad (11.33)$$

Here T_f and T_i have been replaced by ∞ and $-\infty$, respectively.

Let us show how one can recover formulas (7.50) and (7.31), on which the derivation of the perturbative expansion was based, starting from the path integral (11.33). In the first step we put

$$\eta[\phi] = \chi[\phi] = \exp\left(-\frac{1}{2}\int d^3x\, \phi(\vec{x})\, \sqrt{m_0^2 - \triangle}\, \phi(\vec{x})\right).$$

These wave functionals correspond to the choice $|\eta\rangle = |\chi\rangle = |0_I\rangle$ made in Sect. 7.1, see Exercise 6.6. Next, we use the following identity [9]

$$\int d^3x\, \left[\phi'(\vec{x})\sqrt{m_0^2 - \triangle}\, \phi'(\vec{x}) + \phi''(\vec{x})\sqrt{m_0^2 - \triangle}\, \phi''(\vec{x})\right]$$
$$= \lim_{\epsilon \to 0_+} [\epsilon \int d^4x\, e^{-\epsilon|x^0|}\phi(x^0, \vec{x})\sqrt{m_0^2 - \triangle}\, \phi(x^0, \vec{x})],$$

where $\phi(x^0, \vec{x})$ can be any function such that the integral on the r.h.s. exists and, moreover,

$$\lim_{x^0 \to \infty} \phi(x^0, \vec{x}) = \phi''(\vec{x}), \qquad \lim_{x^0 \to -\infty} \phi(x^0, \vec{x}) = \phi'(\vec{x}).$$

In order to check this identity, first we change the integration variable from x^0 to ϵx^0, next we split the integration range into subintervals $(-\infty, 0]$ and $[0, +\infty)$, then we take the limit $\epsilon \to 0_+$ separately in each subinterval, and note that $\int_0^\infty dx^0\, \exp(-x^0) = 1$.

In the next step we insert that identity on the r.h.s. of (11.33), and note that

$$\int (d\phi'')(d\phi') \int_{\substack{\phi(-\infty, \vec{x}) = \phi'(\vec{x}) \\ \phi(\infty, \vec{x}) = \phi''(\vec{x})}} [d\phi] \ldots = \int [d\phi] \ldots,$$

where $[d\phi] = \prod_{x \in M} d\phi(x)$. In the last path integral there are no restrictions on the ends of the paths. The resulting formula

$$\overline{Z}[j] = \lim_{\epsilon \to 0_+} \int [d\phi] \exp(iS[\phi]$$
$$+ i \int d^4x\, j(x)\phi(x) - \frac{1}{2}\epsilon \int d^4x\, e^{-\epsilon|x^0|}\phi\sqrt{m_0^2 - \triangle}\, \phi) \qquad (11.34)$$

contains the integration over all of the paths in the configuration space of the field, without any restriction on the ends of the paths.

Finally, we use the correspondence $\phi(x) \leftrightarrow -i\delta/\delta j(x)$ in order to write

$$\overline{Z}[j] = \exp\left[-iV\left(-i\frac{\delta}{\delta j(x)}\right)\right]\overline{Z}_0[j], \qquad (11.35)$$

where

$$\overline{Z}_0[j] = \lim_{\epsilon\to 0_+} \int [d\phi] \exp\left(i S_0[\phi] + i \int d^4x\, j\phi - \frac{1}{2}\epsilon \int d^4x\, e^{-\epsilon|x^0|}\phi \sqrt{m_0^2 - \triangle}\, \phi\right), \tag{11.36}$$

and

$$S_0[\phi] = \frac{1}{2}\int d^4x\, (\partial_\mu\phi\partial^\mu\phi - m_0^2\phi^2).$$

As we know from Chap. 6, expressions of the form $\delta^4/(\delta j(x))^4$ are ill-defined. The cure lies in introducing a regularization in the form of an integration with a test function g, see formula (7.33) in the case of $V = \lambda_0\phi^4/4!$. Henceforth we replace V in formula (11.35) by its regularized form V_g.

The functional $\overline{Z}_0[j]$ can be calculated explicitly. To this end, we write it in the form of the Gaussian integral,

$$\overline{Z}_0[j] = \lim_{\epsilon\to 0_+} \int [d\phi] \exp\left(-\frac{i}{2}\int d^4x d^4y\, \phi(x) O_\epsilon(x,y)\phi(y) + i\int d^4x\, j(x)\phi(x)\right),$$

where

$$O_\epsilon(x,y) = -\frac{\partial^2\delta(x-y)}{\partial x^\mu \partial y_\mu} + m_0^2\delta(x-y)$$
$$- \frac{1}{2} i\epsilon e^{-\epsilon|y^0|}\sqrt{m_0^2 - \triangle}\, \delta(x-y) - \frac{1}{2} i\epsilon e^{-\epsilon|x^0|}\sqrt{m_0^2 - \triangle}\, \delta(x-y),$$

and change the integration variable ϕ in the path integral to $\phi_1(x) = \phi(x) - \int d^4z\, O_\epsilon^{-1}(x,z)j(z)$, where $O_\epsilon^{-1}(x,z)$ is defined by the following equations:

$$\int d^4z\, O_\epsilon(x,z)O_\epsilon^{-1}(z,y) = \delta(x-y), \quad \int d^4z\, O_\epsilon^{-1}(x,z)O_\epsilon(z,y) = \delta(x-y). \tag{11.37}$$

Such a shift of the integration variable does not change the 'volume element', $[d\phi] = [d\phi_1]$, because $d\phi(x) = d\phi_1(x)$ for each fixed x, as follows from the fact that $\int d^4z\, O_\epsilon^{-1} j$ does not depend on ϕ_1. Therefore,

$$\overline{Z}_0[j] = \lim_{\epsilon\to 0_+} \int [d\phi_1] \exp\left(-\frac{i}{2}\int d^4x d^4y\, \phi_1(x) O_\epsilon(x,y)\phi_1(y)\right)$$
$$\exp\left(\frac{i}{2}\int d^4z d^4y\, j(z)O_\epsilon^{-1}(z,y)j(y)\right).$$

The path integral gives a non-vanishing constant $\mathcal{N}_0$, which does not depend on j. It cancels out in formula (11.32), because the same constant is also present in the denominator.

It remains to compute O_ϵ^{-1}. Because $O_\epsilon(x, y) = O_\epsilon(y, x)$, $O_\epsilon^{-1}(x, y)$ is also symmetric in x and y, and then it is sufficient to consider only one of Eqs. (11.37), for instance the first. Moreover, we may take the limit $\epsilon \to 0_+$ in two steps: in the first one we put $e^{-\epsilon|x^0|} = e^{-\epsilon|y^0|} = 1$ in $O_\epsilon(x, y)$, but we keep the ϵ's in front of the exponentials. Let us seek O_ϵ^{-1} in the Fourier form

$$O_\epsilon^{-1}(z, y) = (2\pi)^{-4} \int d^4k_1 d^4k_2 \, e^{ik_1 z + ik_2 y} \, \tilde{O}_\epsilon^{-1}(k_1, k_2),$$

and substitute the Fourier representation of $O_\epsilon(x, z)$ into the first equation (11.37),

$$O_\epsilon(x, z) = (2\pi)^{-4} \int d^4q \, e^{iq(x-z)} \big(-q^2 + m_0^2 - i\epsilon\sqrt{m_0^2 + \vec{q}^{\,2}} \,\big).$$

Simple calculations give

$$\tilde{O}_\epsilon^{-1}(k_1, k_2) = \frac{\delta(k_1 + k_2)}{k_1^2 - m_0^2 + i\epsilon\sqrt{m_0^2 + \vec{k}_1^{\,2}}}.$$

Thus, finally

$$\overline{Z}_0[j] = \mathcal{N}_0 \exp\left[-\frac{i}{2} \int d^4k_1 d^4k_2 \; \tilde{j}(k_1) \frac{\delta(k_1 + k_2)}{k_1^2 - m_0^2 + i0_+} \tilde{j}(k_2) \right], \tag{11.38}$$

where $\tilde{j}$ is the Fourier transform of j.

Comparing our present results for the scalar field with formula (7.54), obtained in Chap. 7, we see that $\overline{Z}_0[j] = Z_0[j]$ up to the constant $\mathcal{N}_0$. Furthermore, $\overline{Z}[j] = Z_I[j]$ if we take $V = \lambda_0\phi^4/4!$, compare formula (7.31). Thus, we have recovered the results for the generating functional obtained in Chap. 7 in the framework of the operator approach. This gives us a certain confidence in the path integral formulation, in spite of it lacking some mathematical rigor.

11.3 The Path Integral for Fermionic Fields

Field theoretical models of fundamental importance for physics, e.g., the standard model of particle physics, usually involve several kinds of fields, among them are fermionic ones. For this reason it is desirable to also have a path integral formulation for the quantum theory of fermionic fields, similar to the one presented above for the scalar field. This would provide a unified theoretical framework for investigating such models, complementary to the operator formulation.

Our main objective is a path integral formula for the Green's functions of the fermionic field, analogous to (11.28), or equivalently, for the pertinent generating functional. For concreteness we consider the Dirac field $\psi(x)$. The Green's functions are defined as the vacuum expectation values of time ordered products of the field operators in the Heisenberg picture. There is an innocent looking difference in the definition of the time ordered product (T-product) in bosonic and fermionic cases: in the latter any interchange of two factors results in the change of the sign of the T-product, i.e., the T-product is antisymmetric. For example,

$$\langle 0|T(\ldots\hat{\psi}^\alpha\hat{\psi}^\beta\ldots)|0\rangle = -\langle 0|T(\ldots\hat{\psi}^\beta\hat{\psi}^\alpha\ldots)|0\rangle.$$

The T-product of anticommuting operators $\hat{\psi}(t_i)$ is defined as follows

$$T\left(\hat{\psi}(t_1)\hat{\psi}(t_2)\ldots\hat{\psi}(t_n)\right)$$
$$= \sum_P \text{sign}(P)\,\Theta(t_{i_1}-t_{i_2})\Theta(t_{i_2}-t_{i_3})\ldots\Theta(t_{i_{n-1}}-t_{i_n})\hat{\psi}(t_{i_1})\hat{\psi}(t_{i_2})\ldots\hat{\psi}(t_{i_n}), \tag{11.39}$$

where we have omitted the bispinor indices and vectors $\vec{x}_i$. The sum is over the set of all permutations $(t_1, t_2, \ldots, t_n) \to (t_{i_1}, t_{i_2}, \ldots, t_{i_n})$, and sign$(P)$ is equal to +1 for even permutations, and -1 for odd permutations. The presence of the factor sign(P) is related to the fact that the components of the quantized Dirac field taken at spatially separated points anticommute. Without it we would get a contradiction. Let us take, for example, $t_1 > t_2$,

$$T\left(\hat{\psi}(t_1)\hat{\psi}(t_2)\right) = \hat{\psi}(t_1)\hat{\psi}(t_2) \neq 0.$$

On the other hand, if the T-product does not contain the sign factor, and $\hat{\psi}(t_1)$, $\hat{\psi}(t_2)$ anticommute, then

$$T\left(\hat{\psi}(t_1)\hat{\psi}(t_2)\right) = T\left(-\hat{\psi}(t_2)\hat{\psi}(t_1)\right) = -\hat{\psi}(t_1)\hat{\psi}(t_2),$$

in contradiction with the previous result for $T\left(\hat{\psi}(t_1)\hat{\psi}(t_2)\right)$.

We would like to have a formula similar to (11.28). Because the T-product present on the l.h.s. is antisymmetric, the classical fields in the product preceding the exponential on the r.h.s. have to anticommute with each other. Thus, we need a path integral over a set of anticommuting classical fields. Let us begin from integrals over a finite set of independent anticommuting elements $\theta_1, \ldots \theta_N$, where

$$\theta_i\theta_j + \theta_j\theta_i = 0.$$

Because in particular $\theta_i^2 = 0$, the set of expressions one can construct from these elements is rather small. There are $2^N - 1$ independent products, including the elements themselves, and the most general expression has the form

$$f(\theta_1, \theta_2, \ldots \theta_N) = c_0 + c_1\theta_1 + \cdots + c_N\theta_N + c_{12}\,\theta_1\theta_2 + \cdots + c_{12\ldots N}\,\theta_1\theta_2\ldots\theta_N, \tag{11.40}$$

where $c_0, c_i, c_{12}, \ldots$ are numbers. The set of all such expressions is called the Grassmann algebra, and $\theta_1, \ldots \theta_N$ are its generating elements. In the present case its dimension is finite, equal to 2^N. The integral we are seeking is a linear mapping which ascribes a number to each expression of the form (11.40) (by integral we mean here a definite one). Let us consider the integral of f over θ_1, traditionally denoted as $\int d\theta_1\, f$. There are only two kinds of terms we have to deal with: terms which contain θ_1 and terms which do not. As the value of the integral $\int d\theta_1\,\theta_1$ we may take an arbitrary number different from 0—it is just a normalization of the integral. Therefore, we assume that

$$\int d\theta_1\,\theta_1 = 1, \int d\theta_2\,\theta_2 = 1, \ldots, \int d\theta_N\,\theta_N = 1. \tag{11.41}$$

Apart from the linearity, we also assume that the integral is invariant under translations in the following sense: $\int d\theta_1\, f(\theta_1 + g, \theta_2, \ldots \theta_N) = \int d\theta_1\, f(\theta_1, \theta_2, \ldots, \theta_N)$, where g can be any expression which does not contain the element θ_1. This requirement corresponds to the identity $\int_{-\infty}^{\infty} dx\, f(x+a) = \int_{-\infty}^{\infty} dx\, f(x)$ for the ordinary definite integral over the whole real axis. The invariance under translations is achieved by assuming that

$$\int d\theta_k\, f(\ldots \not{\theta}_k \ldots) = 0 \tag{11.42}$$

for any expression f that does not contain θ_k. In particular, $\int d\theta_k = 0$ (in this case $f = 1$). Formulas (11.41) would lead to contradictions if not supplemented by another rule: the integration symbol $\int d\theta_k$ should be anticommuted with the generating elements until it is just in front of θ_k—only then may we apply (11.41). In order to see the contradiction, consider, for example, $\int d\theta_1\,\theta_1\theta_2 = \theta_2$. On the other hand, if we abandon the rule, $\int d\theta_1\,\theta_1\theta_2 = -\int d\theta_1\,\theta_2\theta_1 = -\theta_2$. With the rule adopted, we have $-\int d\theta_1\,\theta_2\theta_1 = \theta_2 \int d\theta_1\,\theta_1 = \theta_2$, as it should be.

Let us now take another Grassmann algebra, such that it can be regarded as a finite dimensional analogue of the Grassmann algebra that will appear when we come to the Dirac field. Now there are $4N$ independent generating elements denoted as follows $\psi^1, \ldots, \psi^N, \overline{\psi}_1, \ldots, \overline{\psi}_N, b^1, \ldots, b^N, \overline{b}_1, \ldots, \overline{b}_N$. It turns out that

$$\int \prod_{j=1}^{N} d\psi^j d\overline{\psi}_j \, \exp\left(\overline{\psi}_k A^k{}_l \psi^l + i\overline{\psi}_k b^k + i\overline{b}_k \psi^k\right) = \det \hat{A}\, \exp\left(\overline{b}_k (A^{-1})^k{}_l b^l\right). \tag{11.43}$$

Here, the N by N matrix $\hat{A} = [A^i{}_k]$ is nonsingular and its matrix elements $A^i{}_k$ are numbers. Derivation of formula (11.43) is left as Exercise 11.2(a).

Now let us turn to the Dirac field. The quantum theory of the free Dirac field has been constructed in Sect. 6.2. In the case of an interacting Dirac field we proceed analogously as in Sects. 7.1 and 7.2 for the real scalar field. Let us consider a model with a Lagrangian of the form

$$\mathcal{L} = \mathcal{L}_0(\overline{\psi}, \psi) - V(\overline{\psi}, \psi).$$

Here $\mathcal{L}_0$ is the free field part of the Lagrangian. It has the same form as the Lagrangian (6.64) of the free Dirac field. V is the interaction term. We do not need to specify its form. The generating functional for Green's functions is defined as follows

$$Z[\eta, \overline{\eta}] = \langle 0|T \exp\left(i \int d^4x \sum_{\alpha=1}^{4} (\overline{\eta}_\alpha \hat{\psi}^\alpha + \hat{\overline{\psi}}_\alpha \eta^\alpha) \right) |0\rangle, \tag{11.44}$$

where the external sources η and $\overline{\eta}$ are generating elements of a certain Grassmann algebra, $\hat{\psi}$ and $\hat{\overline{\psi}}$ are the Dirac field and its conjugate in the Heisenberg picture, and $|0\rangle$ is the vacuum state. The Green's functions are obtained by taking variational derivatives of Z with respect to η and $\overline{\eta}$ and putting $\eta = \overline{\eta} = 0$ afterwards. For example,

$$\langle 0|T(\hat{\psi}^\alpha(x)\, \hat{\overline{\psi}}_\beta(y))|0\rangle = \left. \frac{\delta^2 Z}{\delta\overline{\eta}_\alpha(x)\delta\eta^\beta(y)} \right|_{\eta=\overline{\eta}=0}.$$

The Gell-Mann–Low formula for the generating functional in the present case has the form

$$Z[\eta, \overline{\eta}] = \frac{\langle 0_I|T\left(\exp\left[i \int d^4x \sum_{\alpha=1}^4 (\overline{\eta}_\alpha \hat{\psi}_I^\alpha + \hat{\overline{\psi}}_{I\alpha}\eta^\alpha)\right] \exp(-i \int d^4x V_I(\overline{\psi}_I, \psi_I))\right)|0_I\rangle}{\langle 0_I|T \exp(-i \int d^4x V_I(\overline{\psi}_I, \psi_I))|0_I\rangle}. \tag{11.45}$$

This formula is used in order to express Z by the generating functional Z_0 for the Green's functions of the free Dirac field:

$$Z[\eta, \overline{\eta}] = \frac{Z_I[\eta, \overline{\eta}]}{Z_I[0, 0]}, \tag{11.46}$$

where

$$Z_I[\eta, \overline{\eta}] = \exp\left(-i \int d^4x \; V_I\left(i\frac{\delta}{\delta\eta}, -i\frac{\delta}{\delta\overline{\eta}} \right) \right) Z_0[\eta, \overline{\eta}], \tag{11.47}$$

and

$$Z_0[\eta, \overline{\eta}] = \langle 0_I | T(\exp[i \int d^4x \sum_{\alpha=1}^{4} (\overline{\eta}_\alpha \hat{\psi}_I^\alpha + \hat{\overline{\psi}}_{I\alpha} \eta^\alpha)]) | 0_I \rangle. \tag{11.48}$$

Thus, it suffices to provide the path integral representation for the generating functional Z_0.

The generating functional Z_0 can be calculated with the help of the free Dirac field version of Wick's formula. Such a formula can be obtained by repeating the calculations of Sect. 7.2 with the scalar field replaced by the free Dirac field, see Exercise 7.7. The result has the form

$$Z_0[\eta, \overline{\eta}] = \exp\big(-i \int d^4x d^4y\, \overline{\eta}_\alpha(x) S_F^\alpha{}_\beta(x-y) \eta^\beta(y)\big), \tag{11.49}$$

where

$$S_F^\alpha{}_\beta(x-y) = -i \langle 0_I | T(\psi_I^\alpha(x) \overline{\psi}_{I\beta}(y)) | 0_I \rangle = (\gamma^\mu \frac{\partial}{\partial x^\mu} - imI_4)^\alpha_\beta \Delta_F(x-y).$$

S_F is the inverse of the Dirac operator $i\gamma^\mu \partial_\mu - mI_4$, that is

$$\int d^4y\, (i\gamma^\mu \frac{\partial}{\partial x^\mu} - mI_4)^\gamma{}_\alpha \delta(x-y) S_F^\alpha{}_\beta(y-z) = \delta^\gamma_\beta \delta(x-z).$$

Therefore, the path integral representation for Z_0 is obtained from formula (11.43) by the following substitutions: $\hat{A}^{-1} \to -iS_F$, $b^k \to \eta^\beta(y)$, $\overline{b}_s \to \overline{\eta}_\alpha(x)$, $\psi^k \to \psi^\beta(y)$ and $\overline{\psi}_i \to \overline{\psi}_\alpha(x)$. The discrete indices i and k are replaced by multi-indices (α, x) and (β, y). Instead of $A^i{}_k$ we now have $(-\gamma^\mu \partial/\partial x^\mu - imI_4)^\gamma{}_\alpha \delta(x-y)$, and

$$Z_0[\eta, \overline{\eta}] = \mathcal{N}^{-1} \int [d\psi d\overline{\psi}] \exp\big[i \int d^4x\, (\mathcal{L}_0(\overline{\psi}, \psi) + \overline{\eta}_\alpha(x) \psi^\alpha(x) + \overline{\psi}_\alpha(x) \eta^\alpha(x))\big], \tag{11.50}$$

where

$$\mathcal{L}_0 = \overline{\psi}(i\gamma^\mu \frac{\partial}{\partial x^\mu} - mI_4)\psi.$$

The constant $\mathcal{N}$ is determined from the condition $Z_0[0, 0] = 1$:

$$\mathcal{N} = \int [d\psi d\overline{\psi}] \exp\big(i \int d^4x\, \mathcal{L}_0(\overline{\psi}, \psi)\big).$$

Finally, by inserting (11.50) for Z_0 in formula (11.47), we find the path integral representation for the model with interactions: the generating functional is given by formula (11.46), where

$$Z_I[\eta, \overline{\eta}] = \mathcal{N}^{-1} \int [d\psi d\overline{\psi}]$$

$$\exp\Big[i\int d^4x \left(\mathcal{L}_0(\overline{\psi}, \psi) - V(\overline{\psi}, \psi) + \overline{\eta}_\alpha(x)\psi^\alpha(x) + \overline{\psi}_\alpha(x)\eta^\alpha(x)\right)\Big]. \quad (11.51)$$

Note that the coefficient $\mathcal{N}^{-1}$ cancels out in formula (11.46).

Comparing the derivations of the path integral representation for the real scalar field, for the spinless particle, and for the Dirac field, we see that in the fermionic case it is indirect, in the sense that it has been obtained by rewriting the known formula (11.49) as the path integral, formula (11.50). There has been no reference to a Hilbert space, basis states like $|\phi\rangle$, or wave functionals. For a derivation analogous to the ones presented in Sects. 11.1 and 11.2 we would need a Grassmann analogue of the particle considered in Sect. 11.1 and its quantum mechanics. Such a Grassmann analogue should have trajectories in a space with anticommuting coordinates instead of x^i. It turns out that it can be constructed [9], and proceeding in full analogy with the bosonic case one can first obtain the path integral in the quantum mechanics of such a particle, and next its field theoretic generalization. Such a direct approach turns out to be rather complicated. Moreover, it is rather artificial because Grassmann analogues of ordinary particles have not been observed in Nature—one should not confuse such a Grassmann analogue with a real fermionic particle, e.g., an electron, which has an ordinary configuration space with commuting coordinates x^i.

Exercises

11.1 Compute the r.h.s. of formula (11.14) in the case of a one dimensional, non relativistic particle with the action $S[x(t)] = \int_{t'}^{t''} dt\; m\dot{x}^2(t)/2$. Compare the result with the formula

$$\langle x''|U(t'', t')|x'\rangle = \sqrt{\frac{m}{2\pi i\hbar(t''-t')}}\, \exp\Big[-i\frac{m(x''-x')^2}{2\hbar(t''-t')}\Big],$$

known from textbooks on quantum mechanics.

Hints: Consider the Fourier transform $\int_{-\infty}^{\infty} dx''\, e^{ikx''} R$, where R denotes the r.h.s. of formula (11.14). The Fourier transform of convolution of functions is equal to the product of the Fourier transforms of these functions.

11.2 (a) Prove formula (11.43).

Hints: 1. Using the translational invariance of the integral, replace ψ^k by $\psi^k + i(\hat{A}^{-1})^k{}_l b^l$ and $\overline{\psi}_k$ by $\overline{\psi}_k + i\overline{b}_j(\hat{A}^{-1})^j{}_k$ in order to simplify the exponent on the l.h.s. of formula (11.43).
2. Check that

$$\int \prod_{j=1}^{N} d\psi^j d\overline{\psi}_j \; \exp\left(\overline{\psi}_k A^k{}_l \psi^l\right) = \frac{1}{N!} \int \prod_{j=1}^{N} d\psi^j d\overline{\psi}_j \; \left(\overline{\psi}_k A^k{}_l \psi^l\right)^N = \det \hat{A}.$$

(b) The Grassmann elements θ'_k, θ_l, where $k, l = 1, 2, \ldots N$, are related by the formula $\theta'_k = A_{kl}\,\theta_l$. The matrix $\hat{A} = (A_{kl})$ is nonsingular, its matrix elements are numbers. The integrals over θ_k and θ'_l are defined by formulas (11.41) and (11.42). Check that the relation

$$d\theta'_1 \ldots d\theta'_N = (\det\hat{A})^{-1} d\theta_1 \ldots d\theta_N$$

is consistent with these definitions.

Hint: Start from $\int d\theta'_1 \ldots d\theta'_N \theta'_N \ldots \theta'_1 = 1$.

Chapter 12
The Perturbative Expansion for Non-Abelian Gauge Fields

Abstract The invariant volume element in the $SU(N)$ group (the Haar measure). The Faddeev–Popov–DeWitt determinant for a given gauge condition. The Faddeev–Popov ghost fields. The correct path integral representation of the Green's functions of local gauge-invariant operators. Feynman diagrams for the pure non-Abelian gauge field theory. The essential role of the gauge fixing term in the classical effective action. BRST invariance of the effective action and of the measure in the path integral. The Slavnov–Taylor identity for the generating functional of Green's functions.

We have considered in Chap. 4 the classical non-Abelian gauge fields. However, from a physical viewpoint, the quantum theory of these fields is much more important. As we know from the case of the renormalizable $:\phi_4^4:$ model, it is possible to develop, with some effort, a sensible perturbative expansion for the Green's functions. On the other hand, it is still practically impossible to construct an exact quantum version of the model. The same is true for the non-Abelian gauge fields, but here even the perturbative expansion is rather intricate. Its construction, completed around 1970, is regarded as one of the most outstanding achievements of theoretical physics in the second half of the 20th century. It clearly shows the sophisticated beauty of the non-Abelian gauge fields. In the present chapter, we construct the perturbative expansion and obtain the very important Slavnov–Taylor identities for the Green's functions of the quantized non-Abelian gauge fields. As the main tool we use the path integrals.

Because of the utmost importance of the quantized non-Abelian gauge fields for particle physics, an enormous effort has been put into non perturbative approaches to their theory. Many important results have been obtained in this direction, nevertheless it is clear that a lot of work and new ideas are still needed in order to get closer to the exact version of the quantum theory of these fields. Particularly hard is the most important problem, that of finding the particle spectrum. It is known as the problem of the confinement of gluons, and of quarks, when an interaction with quark fields is included. We do not touch these fascinating topics here.

H. Arodź and L. Hadasz, *Lectures on Classical and Quantum Theory of Fields*, Graduate Texts in Physics, DOI 10.1007/978-3-319-55619-2_12

12.1 The Faddeev–Popov–DeWitt Determinant

Trajectories of the classical non-Abelian gauge field of the $SU(N)$ type are represented by the matrix valued functions $\hat{A}_\mu(x)$ on Minkowski space-time M. For each $x \in M$ and $\mu = 0, 1, 2, 3,$ $\hat{A}_\mu(x)$ is an $N \times N$ Hermitian, traceless matrix. Note that here the hat just denotes the matrix, not a quantum operator in a Hilbert space—in the present chapter such operators will be denoted by the boldface $\hat{\mathbf{A}}_\mu$. The gauge fields related by the gauge transformation (4.23) are physically equivalent, that is they give identical values of all observables. It is quite natural to expect that in the path integral in the quantum theory of such fields just one gauge field from each class of the equivalent fields should appear, not all fields. To achieve this, we first introduce a gauge condition

$$F(\hat{A}_\mu) = 0, \tag{12.1}$$

so that in each class of physically equivalent fields there is exactly one gauge field that satisfies it. In other words, the condition

$$F(\hat{A}_\mu^\omega) = 0,$$

regarded as an equation for the $SU(N)$ matrix-valued gauge function $\omega(x)$, has exactly one solution for every fixed gauge field $\hat{A}_\mu$. The elements of the $SU(N)$ group in a vicinity of the unit matrix I_N can be parameterized by $N^2 - 1$ real parameters, let us denote them by t^a, $a = 1, ..., N^2 - 1$, which form a local coordinate system on the group. Therefore, the $SU(N)$ valued function $\omega(x)$ is equivalent to $N^2 - 1$ real valued functions $t^a(x)$. The gauge condition (12.1) should uniquely determine all these functions, hence it should be equivalent to $N^2 - 1$ independent equations for them. We shall write these equations as $F^a(\hat{A}_\mu^\omega)(x) = 0$.

We will use integration over the $SU(N)$ group regarded as a certain n-dimensional space, $n = N^2 - 1$. In the mathematical theory of Lie groups, such as the $SU(N)$ group, it is shown that one can introduce a volume element on the group, which in mathematics is called the Haar measure. We denote it as $dV(\omega)$, where $\omega \in SU(N)$. When the group elements are parameterized by t^a, such an infinitesimal volume element has the form

$$dV(\omega) = v(t^a)\, dt^1 \ldots dt^n, \tag{12.2}$$

where $v(t^a)$ is a certain positive function of the parameters. Furthermore, this volume element is invariant under the so called translations on the group, that is transformations of the form $\omega(t^a) \to \omega_0\omega(t^a)$ (the left translations), and $\omega(t^a) \to \omega(t^a)\omega_0$ (the right translations), where $\omega_0 \in SU(N)$. The coordinates of the group element $\omega_0\omega(t^a)$ are denoted as $\bar{t}^a$, hence $\omega_0\omega(t^a) = \omega(\bar{t}^a)$. Similarly, $\omega(t^a)\omega_0 = \omega(\underline{t}^a)$. The invariance of the volume element means that

$$dV(\omega_0\omega) = dV(\omega) = dV(\omega\omega_0),$$

or

$$v(\bar{t}^a)\, d\bar{t}^1 \ldots d\bar{t}^n = v(t^a)\, dt^1 \ldots dt^n = v(\underline{t}^a)\, d\underline{t}^1 \ldots d\underline{t}^n.$$

We shall not need the detailed form of the invariant volume element.

One way to eliminate the gauge equivalent fields from the path integral is to include in its integrand a functional Dirac delta of the form[1]

$$\delta[F(\hat{A})] = \prod_{a,x} \delta(F^a(\hat{A}_\mu(x)),$$

but then the result of the integration would in general depend on the choice of the gauge condition. This would not be satisfactory, because the choice of the gauge condition should not affect the expectation values of observables represented by gauge invariant operators. According to Faddeev and Popov, the functional Dirac delta should be inserted in the path integral indirectly, namely one should hide it in a numerical factor equal to 1, which certainly does not change the integral. Moreover, it obviously does not depend on the choice of the gauge condition. The factor 1 is constructed from the Dirac delta as follows

$$1 = \mathcal{M}[\hat{A}] \int [d\omega]\, \delta[F(\hat{A}^\omega)]. \tag{12.3}$$

Here $[d\omega] = \prod_x dV(\omega(x))$ is the measure (the infinitesimal volume element) in the space of the gauge functions $\omega(x)$. Thus, with each point $x \in M$ we associate the invariant volume element in the $SU(N)$ group. $\hat{A}^\omega$ denotes the gauge transformed field, i.e.,

$$\hat{A}^\omega_\mu(x) = \omega(x)\hat{A}_\mu(x)\omega^{-1}(x) + \frac{i}{g}\partial_\mu\omega(x)\, \omega^{-1}(x). \tag{12.4}$$

In the present chapter we use the rescaled gauge field introduced in Sect. 4.2 (below formula (4.33)) and denoted there as $\hat{B}_\mu$. $\mathcal{M}[\hat{A}]$ is a functional of the gauge field defined by formula (12.3). It is called the Faddeev–Popov–DeWitt determinant. Of course it depends on the choice of F, but it is gauge invariant, that is

$$\mathcal{M}[\hat{A}^{\omega_0}] = \mathcal{M}[\hat{A}]$$

for any gauge function $\omega_0(x)$. This follows from the invariance of the measure $dV(\omega)$, namely

$$1 = \mathcal{M}[\hat{A}^{\omega_0}] \int [d\omega]\, \delta[F((\hat{A}^{\omega_0})^\omega)] = \mathcal{M}[\hat{A}^{\omega_0}] \int [d(\omega_0\omega)]\, \delta[F(\hat{A}^{\omega_0\omega})]$$

$$\overset{(\omega'=\omega_0\omega)}{=} \mathcal{M}[\hat{A}^{\omega_0}] \int [d\omega']\, \delta[F(\hat{A}^{\omega'})] = \frac{\mathcal{M}[\hat{A}^{\omega_0}]}{\mathcal{M}[\hat{A}]}.$$

[1] We will often omit the space-time index μ of $\hat{A}_\mu$ in order to keep formulas transparent.

In the case of $\hat{A}_\mu$ obeying the gauge condition (12.1), the integral in (12.3) is determined by the form of the integrand in an arbitrarily small vicinity of the constant $\omega = I_N$. Let us parameterize $\omega(x)$ in such a vicinity as follows:

$$\omega(x) = I_N + ig\epsilon^a(x)\,\hat{T}_a + \mathcal{O}(\vec{\epsilon}^{\,2}),$$

where the matrices $\hat{T}_a$, with $a = 1, \ldots, N^2 - 1$, have been introduced in Sect. 4.2. Then, the volume element has the form $dV(\omega) = v(\epsilon^a)d^n\epsilon$, where we may replace $v(\epsilon^a)$ by $v(0)$, and normalize the $SU(N)$ volume element by putting $v(0) = 1$. Thus, as the measure $[d\omega]$ in (12.3) we take $[d\omega] = \prod_{x\in M} d^n\epsilon(x) \equiv [d\epsilon]$. This expression should be treated in the same spirit as the measures that appear in the path integrals.

Formula (12.3) also contains $F(\hat{A}^\omega)$. For ω in the vicinity of I_N

$$\begin{aligned}\hat{A}^\omega_\mu(x) &= \hat{A}_\mu(x) - \partial_\mu\hat{\epsilon}(x) + ig[\hat{\epsilon}(x), \hat{A}_\mu(x)] + \mathcal{O}(\vec{\epsilon}^{\,2}) \\ &= \hat{T}_b(A^b_\mu(x) - \partial_\mu\epsilon^b(x) - gf_{acb}\epsilon^a(x)A^c_\mu(x)) + \mathcal{O}(\vec{\epsilon}^{\,2}). \end{aligned} \tag{12.5}$$

The structure constants f_{acb} are antisymmetric in all indices, see Exercise 4.2.

Let us expand $F(\hat{A}^\omega)$ with respect to $\epsilon^a(x)$:

$$F^c(\hat{A}^\omega)(x) = F^c(\hat{A})(x) + \int d^4y d^4z\, \left.\frac{\delta F^c(\hat{A}^\omega)(x)}{\delta(\hat{A}^\omega_\mu)^b(y)}\right|_{\omega=I_N} \left.\frac{\delta(\hat{A}^\omega_\mu)^b(y)}{\delta\epsilon^a(z)}\right|_{\vec{\epsilon}=0} \epsilon^a(z) + \mathcal{O}(\vec{\epsilon}^{\,2}),$$

where $F^c(\hat{A})(x) = 0$ because $\hat{A}_\mu$ obeys condition (12.1). Using formula (12.5) we obtain

$$\left.\frac{\delta(\hat{A}^\omega_\mu)^b(y)}{\delta\epsilon^a(z)}\right|_{\vec{\epsilon}=0} = -\delta_{ab}\frac{\partial}{\partial z^\mu}\delta(y-z) - gf_{acb}A^c_\mu(y)\delta(y-z).$$

Therefore,

$$F^c(\hat{A}^\omega)(x) = \int d^4y\, M_{ca}(x,y)\epsilon^a(y) + \mathcal{O}(\vec{\epsilon}^{\,2}), \tag{12.6}$$

where

$$M_{ca}(x,y) = \left(\delta_{ab}\frac{\partial}{\partial y^\mu} - gf_{adb}A^d_\mu(y)\right)\frac{\delta F^c(\hat{A})(x)}{\delta A^b_\mu(y)}. \tag{12.7}$$

Formula (12.6) can be written in a concise form as

$$F(\hat{A}^\omega) = \hat{M}\epsilon + \mathcal{O}(\epsilon^2),$$

where the operator $\hat{M}$ has the matrix elements $M_{cx;ay} = M_{ca}(x,y)$. Our assumption about the uniqueness of the solution of the equation $F(\hat{A}^\omega) = 0$ implies that $\hat{M}$ is nonsingular (i.e., $\hat{M}^{-1}$ exists).

After these preparations we can compute $\mathcal{M}$. Definition (12.3) gives

$$1 = \mathcal{M}[\hat{A}] \int [d\epsilon]\, \delta[\hat{M}\epsilon + \mathcal{O}(\vec{\epsilon}^{\,2})] = \mathcal{M}[\hat{A}]\, (\det\hat{M})^{-1}.$$

Thus, for the gauge fields obeying gauge condition (12.1)

$$\mathcal{M}[\hat{A}] = \det\hat{M}.$$

It is of course not clear how to actually compute the determinant of $\hat{M}$. Luckily, one can evade this problem by using the infinite dimensional version of formula (11.43),

$$\det\hat{M} = \mathcal{N} \int [d\overline{c}dc]\, \exp\left(-i \int d^4x d^4y\, \overline{c}_a(x) M_{ab}(x, y) c^b(y)\right), \tag{12.8}$$

where $\overline{c}_a(x)$ and $c^b(y)$ are independent Grassmann fields, called antighost or ghost, respectively. The factor $-i$ in the exponent in (12.8) has been introduced for later convenience. The factor $\mathcal{N}$ is not important as it will not appear in the final formula for the generating functional for the Green's functions. The expression

$$S_{gh}[A, c, \overline{c}] = -\int d^4x d^4y\, \overline{c}_a(x) M_{ab}(x, y) c^b(y)$$

is often called the Faddeev–Popov–DeWitt action.

As an example, let us consider the Lorentz gauge condition

$$\partial_\mu A^{a\mu}(x) = 0. \tag{12.9}$$

In this case

$$\frac{\delta F^c(\hat{A})(x)}{\delta A^b_\mu(y)} = \delta_{bc} \frac{\partial \delta(x-y)}{\partial x_\mu},$$

$$M_{ab}(x, y) = -\delta_{ab} \frac{\partial^2 \delta(x-y)}{\partial x_\mu \partial x^\mu} + g f_{adb} A^d_\mu(y) \frac{\partial \delta(x-y)}{\partial x_\mu},$$

and

$$S_{gh} = \int d^4x \left(\overline{c}_a(x) \partial_\mu \partial^\mu c^a(x) - g f_{acb} \overline{c}_a(x) \partial^\mu (A^c_\mu(x) c^b(x))\right). \tag{12.10}$$

Introducing the covariant derivative of the ghost field,

$$(D_\mu c)^a(x) = \partial_\mu c^a(x) - g f_{acb} A^c_\mu(x) c^b(x),$$

we may write S_{gh} in the concise form

$$S_{gh} = \int d^4x \, \overline{c}_a(x) \, \partial^\mu[(D_\mu c)^a(x)]. \tag{12.11}$$

Note that the (anti-)ghost fields do not bear any spinor indices—they would yield spinless particles if regarded as relativistic quantum fields. A spin zero fermionic field violates the spin-statistics theorem, hence it cannot be regarded as a physical field. In our considerations it has appeared only as an auxiliary mathematical variable to be integrated over in formula (12.8).

The Lorentz gauge condition is used in applications of the non-Abelian gauge fields in particle physics. It should be noted that this condition is not perfect because among fields obeying it one can find gauge equivalent ones.[2] This is the so called Gribov problem with the gauge condition. It is also present for other choices of gauge condition. The gauge equivalent solutions of a gauge condition are called Gribov copies. The question of whether their presence has an influence on the physical predictions obtained within the perturbative approach to the quantized non-Abelian gauge fields, remains an open question. In the considerations below, in which we use the Lorentz condition, the Gribov copies are automatically included in the path integral because we sum over all the gauge fields that obey that condition.

12.2 The Generating Functional for Green's Functions

The Faddeev–Popov–DeWitt determinant is needed for the construction of the correct generating functional for Green's functions in the non-Abelian gauge theory with the classical action (4.33)

$$S_{YM}[A] = -\frac{1}{4} \int d^4x \; F^a_{\mu\nu} F^{a\mu\nu}.$$

Let us begin by writing an analogous to (11.28) path integral formula for the vacuum expectation value (correlation function) of the time ordered product of local gauge invariant operators $\mathbf{O}_1[\mathbf{A}](x_1), \ldots, \mathbf{O}_n[\mathbf{A}](x_n)$ (in the Heisenberg picture). Operator $\mathbf{O}[\mathbf{A}](x)$ is local if it is constructed from the non-Abelian gauge field operators $\mathbf{A}^a_\mu(x)$ and their derivatives at the point x.[3] Thus, we begin with

[2]Let us recall that we assume that $\omega(x) \to I_N$ when $|\vec{x}| \to \infty$. This condition excludes, for example, ω independent of x and different from I_N.

[3]For brevity, we write A instead of $\hat{A}$ if there is no risk of confusion.

$$\langle 0|T(\mathbf{O}_1[\mathbf{A}](x_1)\cdots \mathbf{O}_n[\mathbf{A}](x_n))|0\rangle$$
$$= \mathcal{N}\int (dA'')(dA')\, \Psi_0^*[A'']\Psi_0[A'] \int_{\substack{A(T_i)=A' \\ A(T_f)=A''}} [dA]\, e^{iS_{YM}[A]} \prod_{i=1}^{n} \mathcal{O}_i[A](x_i). \tag{12.12}$$

Here $\Psi_0[A]$ is the wave functional of the vacuum state of the gauge field. It is defined on the configuration space of the field, and A', A'' denote points in that infinite dimensional space. The single point A is represented by the set of functions $A^a_\mu(\vec{x})$, where $\vec{x} \in R^3$. The trajectory of the field may be denoted as $A(t)$—it is the set of functions $A^a_\mu(t, \vec{x})$. Note that it is the trajectory that is customarily adopted as the mathematical representation of the non-Abelian gauge field, and not $A = A^a_\mu(\vec{x})$—the field as such is a physical object. (dA) denotes the measure (the volume element) in the configuration space, $(dA) = \prod_{\vec{x}\in R^3} \prod_{a=1}^{N^2-1} \prod_{\mu=0}^{3} dA^a_\mu(\vec{x})$.

Formula (12.12) is not satisfactory because it contains the integral over all of the gauge fields, including the ones related by a gauge transformation. In order to improve it, we multiply the r.h.s. of (12.12) by 1 in the form (12.3), and change the order of the functional integrations by shifting the integral over $\omega(x)$ to the left. Next we change the integration variable from A to $B = A^\omega$. The action S_{YM}, the expressions $O_i[A](x_i)$, and $\mathcal{M}[A]$ are gauge invariant, hence we may simply replace A by B.

The measure $[dA]$ is also invariant, $[dA] = [dB]$. To see this, first notice that the gauge transformation does not change the space-time arguments or Lorentz indices of the field. Therefore, we need only to show the invariance of the N^2-1 dimensional volume element, that is the equality

$$\prod_{a=1}^{N^2-1} dB^a_\mu(x) = \prod_{a=1}^{N^2-1} dA^a_\mu(x).$$

Let us split the gauge transformation into the shift $\hat{A}_\mu \to \hat{A}_\mu + i\partial_\mu\omega\omega^{-1}$ and the 'rotation' $\hat{A}_\mu \to \hat{C}_\mu = \omega\hat{A}_\mu\omega^{-1}$. Neither of them changes the volume element. In the case of the shift, this follows from the fact that $\partial_\mu\omega\omega^{-1}$ does not depend on A^a_μ. The 'rotation' does not change the volume element because it leaves the lengths and angles unchanged. This can be seen from the invariance of the scalar product:

$$X_1^a X_2^a = 2\mathrm{tr}(\hat{X}_1\hat{X}_2) = 2\mathrm{tr}(\hat{Y}_1\hat{Y}_2) = Y_1^a Y_2^a,$$

where $\hat{X}_i = \omega\hat{Y}_i\omega^{-1}$.

After these steps formula (12.12) acquires the following form

$$\langle 0|T(\mathbf{O}_1[\mathbf{A}](x_1)\cdots\mathbf{O}_n[\mathbf{A}](x_n))|0\rangle = \mathcal{N}\int[d\omega]\int(dA'')(dA')\ \Psi_0^*[A'']\Psi_0[A']$$

$$\int_{\substack{B(T_i)=A'^{\omega^{-1}}\\ B(T_f)=A''^{\omega^{-1}}}}[dB]\ \mathcal{M}[\hat{B}]\ \delta[F(B)]\ e^{iS_{YM}[B]}\prod_{i=1}^{n}\mathcal{O}_i[B](x_i).$$

Now we change the integration variables $A' \to B' = A'^{\omega^{-1}}$ and $A'' \to B'' = A''^{\omega^{-1}}$. The measures (dA') and (dA'') are invariant for exactly the same reasons as $[dA]$. The wave functional Ψ_0 is assumed to be invariant up to multiplication by a phase factor[4] which can depend on ω. Such a phase factor cancels out in the product $\Psi_0^*\Psi_0$. Thus, the gauge function ω has been removed from all terms on the r.h.s. of the path integral formula. In consequence, the integral $\int[d\omega]$ has a constant integrand. This integral yields a constant (the total volume of the gauge group) which is canceled by an appropriate coefficient in the normalization factor $\mathcal{N}$. Thus, writing A instead of B everywhere, we finally have

$$\langle 0|T(\mathbf{O}_1[\mathbf{A}](x_1)\cdots\mathbf{O}_n[\mathbf{A}](x_n))|0\rangle = \mathcal{N}\int(dA'')(dA')\ \Psi_0^*[A'']\Psi_0[A']$$

$$\int_{\substack{A(T_i)=A'\\ A(T_f)=A''}}[dA]\ \mathcal{M}[\hat{A}]\ \delta[F(A)]\ e^{iS_{YM}[A]}\prod_{i=1}^{n}\mathcal{O}_i[A](x_i). \qquad (12.13)$$

The normalization factor $\mathcal{N}$ is determined from the condition $\langle 0|0\rangle = 1$, which corresponds to taking $n = 1$ and $\mathbf{O}_1[\mathbf{A}] = \mathbf{I}$. Formula (12.13) explicitly incorporates the gauge condition (12.1). It is clear from its derivation that the r.h.s. of it does not depend on the form of F, in spite of its appearance.

In order to construct the perturbative expansion we have to write the integrand in (12.13) in exponential form, from which we can read off the kinetic and interaction parts. For $\mathcal{M}$ we use formula (12.8) with the ghosts. The functional Dirac delta is dealt with by making use of the lack of dependence of the correlation function on the form of F. Let us replace the condition (12.1) by an auxiliary gauge condition of the form $F^a(\hat{A})(x) - \lambda^a(x) = 0$ with certain functions $\lambda^a(x)$. Because $\delta\lambda^a/\delta A^b_\mu = 0$, we see from formula (12.7) that $\hat{M}$, and in consequence $\mathcal{M}$, do not depend on these functions. On the r.h.s. of formula (12.13) they are present only in the factor $\delta[F(A) - \lambda]$. Next, we multiply both sides of formula (12.13) by $\exp(-i\int d^4x\ \lambda^a(x)\lambda^a(x)/2\alpha)$, and functionally integrate over λ^a. The real parameter α is often called the gauge parameter. On the r.h.s. we have the integral

$$\int[d\lambda]\ \delta[F(A) - \lambda]\ \exp\left(-\frac{i}{2\alpha}\int d^4x\ \lambda^a(x)\lambda^a(x)\right) = \exp(iS_{gf}[A]),$$

[4]This is an assumption because we are not able to compute Ψ_0, nor to prove that there exists exactly one vacuum state.

where

$$S_{gf}[A] = -\frac{1}{2\alpha}\int d^4x\ F^a(\hat{A})(x)F^a(\hat{A})(x).$$

On the l.h.s. we obtain a constant factor. We divide by it and include it in the factor $\mathcal{N}$. Thus, we have obtained, from (12.13), the following formula

$$\langle 0|T(\mathbf{O}_1[\mathbf{A}](x_1)\cdots\mathbf{O}_n[\mathbf{A}](x_n))|0\rangle =$$

$$\mathcal{N}\int(dA'')(dA')\ \Psi_0^*[A'']\Psi_0[A']\int_{\substack{A(T_i)=A' \\ A(T_f)=A''}}[dA][d\overline{c}dc]\ e^{iS[A,c,\overline{c}]}\prod_{i=1}^{n}\mathcal{O}_i[A](x_i), \tag{12.14}$$

where

$$S[A,c,\overline{c}] = S_{YM}[A] + S_{gh}[c,\overline{c}] + S_{gf}[A]. \tag{12.15}$$

$S_{gf}[A]$ is called the gauge fixing term, and $S[A,c,\overline{c}]$ the classical effective action.
Now it should be clear that we may take as the generating functional

$$Z[j,\xi,\overline{\xi}] = \frac{\overline{Z}[j,\xi,\overline{\xi}]}{\overline{Z}[0,0,0]}, \tag{12.16}$$

where

$$\overline{Z}[j,\xi,\overline{\xi}] = \int(dA'')(dA')\ \Psi_0^*[A'']\Psi_0[A']$$

$$\int_{\substack{A(T_i)=A' \\ A(T_f)=A''}}[dA][d\overline{c}dc]\ e^{iS[A,c,\overline{c}]+i\int d^4x\,(j_\mu^a(x)A^{a\mu}(x)+\overline{c}_a(x)\xi^a(x)+\overline{\xi}_a(x)c^a(x))}. \tag{12.17}$$

Suitable combinations of the derivatives $-i\delta/\delta j_\mu^a(x)$ acting on Z will give, after putting $j_\mu^a = 0$, $\xi^a = 0$ and $\overline{\xi}_a = 0$, formulas for the vacuum expectation values of the time ordered products of the components of the gauge field. Formula (12.17) also contains Grassmann type external sources ξ^a and $\overline{\xi}_a$ for the ghost fields. They anticommute with the ghost fields, and with themselves. The derivatives $-i\delta/\delta\overline{\xi}_a$ and $i\delta/\delta\xi^a$ will give Green's functions in which the ghost fields are also present. Such more general Green's functions are in principle not needed, because the ghost fields are not physical fields, but auxiliary variables introduced in order to write the Faddeev–Popov–DeWitt determinant in an exponential form. Nevertheless, corresponding to them internal vertices and internal lines will appear in the perturbative expansion anyway, and they have to be taken into account when discussing, e.g., the renormalizability of the model. Therefore, it is useful to consider graphs in which the ghosts appear as external lines.

Note that formula (12.17) can also be applied in the case of the free electromagnetic field: one should put $f_{abc} = 0$ and restrict the values of the Latin indices to just 1. Then A_μ^1 can be identified with the electromagnetic field. The ghost fields are needed only if $\det\hat{M}$ depends on the gauge field, because in the opposite case it is a constant that does not matter. Formula (12.7) with $f_{abc} = 0$ shows that the dependence on the Abelian gauge field is possible only if $F(A)$ is not linear in A_μ, for example, $F(A)(x) = (\partial_\mu - A_\mu(x))A^\mu(x)$ (Exercise 12.1). In electrodynamics such gauge conditions are not used in practice, because then even the quantum theory of the free field would become quite complicated. The most popular gauge conditions: Lorentz ($\partial_\mu A^\mu = 0$), Coulomb ($\partial_i A^i = 0$) and temporal ($A^0 = 0$), are all linear in A^μ.

The considerations presented above that have forced us to introduce the ghost fields, can be regarded as a spectacular example of the usefulness of the path integrals in field theory. There had been some earlier suggestions about the presence of ghost fields in the quantum theory of gauge fields, but only with the use of the path integrals came a clear recognition of this fact.

12.3 Feynman Diagrams

The derivation of Feynman diagrams for the non-Abelian gauge fields is based on the formulas (12.16) and (12.17) for the generating functional. The ghost part is taken in the form (12.11)—we adopt the Lorentz gauge condition (12.9). We divide the action S into the free and the interaction parts, and formally expand Z in powers of the interaction. The calculations are very similar to those presented in detail in Sect. 11.2 (below formula (11.33)) in the case of the scalar field. Therefore, we will skip details of calculations and present only the main points.

The functional $\Psi_0[A]$ is replaced by the wave functional of the vacuum state of the free non-Abelian gauge field[5]

$$\Phi_0[A] = \mathcal{N}_0 \exp\left(-\frac{1}{2}\int d^3x\, A^{ai}\sqrt{-\Delta}\, A^{ai}\right). \tag{12.18}$$

The same trick as in Sect. (11.2) gives

$$\begin{aligned}\overline{Z}[j,\xi,\overline{\xi}] = \lim_{\epsilon\to 0_+} &\int [dA][d\overline{c}dc] \exp\Big[iS[A,c,\overline{c}] \\ &- \frac{\epsilon}{2}\int d^4x\, e^{-\epsilon|x^0|} A^{ai}\sqrt{-\Delta}\, A^{ai} + i\int d^4x\, (j^a_\mu A^{a\mu} + \overline{c}_a\xi^a + \overline{\xi}_a c^a)\Big].\end{aligned} \tag{12.19}$$

[5] By the free non-Abelian gauge field we mean the field A^a_μ with the action that does not contain the self-interactions present in the full Yang–Mills action. Such self-interactions are switched off by equating the structure constants with zero, $f_{abc} = 0$. The resulting model contains the collection of $N^2 - 1$ independent free gauge fields of the Abelian type, and it is not invariant under the full $SU(N)$ gauge group.

By definition, the interaction part S_{int} of the action S contains all the terms that are proportional to g or g^2. Thus,

$$S_{int}[A, c, \overline{c}] = \int d^4x \, \big(g f_{abc} \partial_\mu A^a_\nu A^{b\mu} A^{c\nu} - \frac{g^2}{4} f_{abc} f_{ade} \, A^b_\mu A^c_\nu A^{d\mu} A^{e\nu} - g f_{acb} \, \overline{c}_a \partial_\mu (c^b A^{c\mu})\big). \quad (12.20)$$

In this part of the action we replace the fields with the appropriate functional derivatives with respect to the external currents:

$$A^a_\mu(x) \to -i \frac{\delta}{\delta j^{a\mu}(x)}, \quad c^a(x) \to -i \frac{\delta}{\delta \overline{\xi}_a(x)}, \quad \overline{c}_a(x) \to i \frac{\delta}{\delta \xi^a(x)}.$$

Then we may write

$$\overline{Z}[j, \xi, \overline{\xi}] = e^{i S_{int}[-i \frac{\delta}{\delta j}, -i \frac{\delta}{\delta \overline{\xi}}, i \frac{\delta}{\delta \xi}]} \Big(Z_0[j] \, Z_0[\xi, \overline{\xi}] \Big), \quad (12.21)$$

where

$$Z_0[j] = \lim_{\epsilon \to 0_+} \int [dA] \, \exp(i \int d^4x \; j^a_\mu A^{a\mu}) \, \exp\Big(\frac{i}{2} \int d^4x \, [\partial_\mu A^a_\nu \, \partial^\nu A^{a\mu} - \partial_\mu A^a_\nu \, \partial^\mu A^{a\nu} - \frac{1}{\alpha} \partial_\mu A^a_\mu \, \partial^\nu A^{a\nu} + i\epsilon e^{-\epsilon|x^0|} A^{ai} \sqrt{-\Delta} \, A^{ai}]\Big),$$

and

$$Z_0[\xi, \overline{\xi}] = \int [d\overline{c} dc] \, \exp\Big(i \int d^4x \, (\overline{c}_b \partial^\mu \partial_\mu c^b + \overline{c}\xi + \overline{\xi} c) \Big).$$

The Gaussian path integrals on the r.h.s.'s of these formulas can be calculated in the same way as shown in Chap. 11. The formula for $Z_0[j]$ can be rewritten in the form

$$Z_0[j] = \lim_{\epsilon \to 0_+} \int [dA] \, \exp\Big(\frac{i}{2} \int d^4x d^4y \; A^a_\mu(x) \mathcal{O}^{\mu\nu}_{\epsilon ab}(x, y) A^b_\nu(y) + i \int d^4x \; j^a_\mu(x) A^{a\mu}(x)\Big),$$

where

$$\mathcal{O}^{\mu\nu}_{\epsilon ab}(x, y) = \delta_{ab} \Big[\eta^{\mu\nu} \frac{\partial^2 \delta(x-y)}{\partial x_\lambda \partial x^\lambda} - (1 - \frac{1}{\alpha}) \frac{\partial^2 \delta(x-y)}{\partial x_\mu \partial x^\nu} - i\epsilon (\eta^{\mu\nu} - \delta^{\mu 0} \delta^{\nu 0}) e^{-\epsilon|x^0|} \sqrt{-\Delta} \, \delta(x-y) \, \Big].$$

The substitution $A = A' - \mathcal{O}_\epsilon^{-1} j$ (all indices suppressed) transforms the integral into a pure Gaussian integral which yields a constant independent of j. It turns out that in the limit $\epsilon \to 0_+$

$$Z_0[j] = \mathcal{N} \exp\left[\frac{i}{2}\int d^4x d^4y\; j_\mu^a(x)\,(\mathcal{O}^{-1})_{ab}^{\mu\nu}(x-y)\; j_\nu^b(y)\right], \tag{12.22}$$

where

$$(\mathcal{O}^{-1})_{ab}^{\mu\nu}(x-y) = -\frac{\delta_{ab}}{(2\pi)^4}\int d^4k\, \frac{e^{-ik(x-y)}}{k^2+i0_+}\left[\eta^{\mu\nu} - (1-\alpha)\frac{k^\mu k^\nu}{k^2+i0_+}\right], \tag{12.23}$$

and $\mathcal{N}$ is a constant. It follows from formula (12.23), that the free propagator of the gauge field has the following form

$$D_{ab}^{\mu\nu}(k) = \delta_{ab}\frac{i}{k^2+i0_+}\,[-\eta^{\mu\nu} + (1-\alpha)\frac{k^\mu k^\nu}{k^2+i0_+}]. \tag{12.24}$$

In Feynman diagrams it is represented by a wavy line, see Fig. 12.1. The exponent in formula (12.22) is symmetric with respect to the interchange $(x, a, \mu) \leftrightarrow (y, b, \nu)$ (in particular because $D_{ab}^{\mu\nu}(k) = D_{ab}^{\mu\nu}(-k)$), therefore we do not have to put an arrow on such lines.

Note that the presence of the gauge fixing term is crucial for the existence of $\mathcal{O}^{-1}$. The absence of this term in the action S would correspond to the limit $\alpha \to \infty$, but then formula (12.23) becomes meaningless. The choice $\alpha = 0$ is called the Landau gauge. It makes sense once we decide to work only within the perturbative approach—on the level of the action we may take α arbitrarily close to 0, but not equal to. In the Landau gauge the propagator is transverse, that is

$$k_\mu D_{ab}^{\mu\nu}(k) = 0.$$

The choice $\alpha = 1$ is called the Feynman gauge.

In the case of the functional $Z_0[\xi, \bar{\xi}]$ similar calculations give

$$Z_0[\xi, \bar{\xi}] = \mathcal{N}_1 \exp\left[i\int d^4x d^4y\; \bar{\xi}_a(x)\,(\mathcal{O}_1^{-1})_b^a(x-y)\; \xi^b(y)\right], \tag{12.25}$$

where

$$(\mathcal{O}_1^{-1})_b^a(x-y) = -\frac{\delta_{ab}}{(2\pi)^4}\int d^4p\, \frac{e^{-ip(x-y)}}{p^2+i0_+}. \tag{12.26}$$

As the free propagator of the ghost fields we take

$$\Delta_{ab}(p) = \delta_{ab}\frac{i}{p^2+i0_+}. \tag{12.27}$$

Fig. 12.1 The ghost and the gauge field propagators

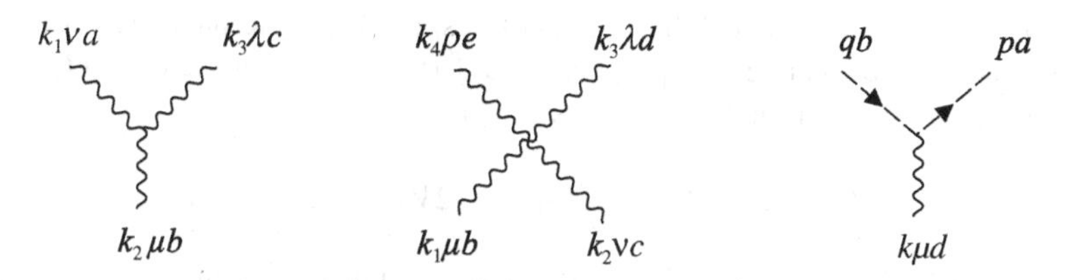

Fig. 12.2 The internal vertices of the $SU(N)$ non-Abelian gauge theory

It is represented graphically by the dashed line with an arrow, see Fig. 12.1. The arrow points to that end at which there was the external source ξ. Thus, such an arrow does not show the flow of four-momentum as it was the case in Fig. 8.9.

The S_{int} part of the action gives the internal vertices of the Feynman diagrams, see Fig. 12.2. The first vertex in that figure corresponds to the first term on the r.h.s. of formula (12.20). All three legs of this vertex bear indices of the same kind, therefore when connecting such a vertex with the rest of the diagram we can do it in 6 ways (if it were a scalar field instead of $\hat{A}_\mu$ this would give the combinatorial factor 3!). Summing all 6 possibilities we obtain the full, symmetric 3-leg vertex with a contribution of the form

$$\frac{ig}{(2\pi)^2} f_{abc}\delta(k_1+k_2+k_3)\,[(k_1-k_3)^\mu\eta^{\lambda\nu}+(k_3-k_2)^\nu\eta^{\mu\lambda}+(k_2-k_1)^\lambda\eta^{\mu\nu}]. \quad (12.28)$$

The linear dependence on the four-momenta k_i reflects the presence of the derivative $\partial_\mu A^a_\nu$ in the pertinent term in (12.20).

In the case of the 4-leg vertex, there are 24 ways to connect it with the rest of the diagram. Summing them all we obtain the full, symmetric 4-leg vertex

$$-i\frac{g^2}{(2\pi)^4}\delta(k_1+k_2+k_3+k_4)\,[f_{abc}f_{ade}(\eta^{\mu\lambda}\eta^{\nu\rho}-\eta^{\mu\rho}\eta^{\lambda\nu})$$
$$+\,f_{ace}f_{adb}(\eta^{\mu\rho}\eta^{\nu\lambda}-\eta^{\mu\nu}\eta^{\lambda\rho})+f_{abe}f_{adc}(\eta^{\mu\lambda}\eta^{\nu\rho}-\eta^{\mu\nu}\eta^{\lambda\rho})]. \quad (12.29)$$

The third vertex in Fig. 12.2 corresponds to the ghost term in S_{int}. The analytical expression associated with it has the form

$$\frac{ig}{(2\pi)^2} f_{adb}\;\delta(k+p-q)\;p^\mu. \quad (12.30)$$

The linear dependence on p^μ reflects the presence of the derivative ∂_μ in the ghost term in (12.20).

All the coupling constants in the action (12.20) are dimensionless. This fact suggests that the perturbative expansion in powers of S_{int} is renormalizable. Such expectation is corroborated by a calculation of the superficial degree of divergence of 1PI graphs. Let V_3, V_4 and V_{gh} denote the numbers of internal vertices shown in Fig. 12.2 (starting from the left), n and I the number of, respectively, external and internal lines corresponding to the gauge field propagator $D^{\mu\nu}_{ab}$ (the wavy lines), and n_{gh} (I_{gh})—the number of external (internal) ghost lines. Then,

$$3V_3 + 4V_4 + V_{gh} = 2I + n, \quad 2V_{gh} = 2I_{gh} + n_{gh}.$$

The number of independent loops and the superficial degree of divergence are given by the formulas

$$L = I + I_{gh} - V_3 - V_4 - V_{gh} + 1, \quad \omega = 4L + V_3 + V_{gh} - 2I - 2I_{gh}$$

(each vertex with three legs introduces one power of a four-momentum, see formulas (12.28) and (12.30)). It follows from these formulas that

$$\omega = 4 - n - n_{gh}. \tag{12.31}$$

Thus, ω depends only on the number of external legs, similarly as in the case of the renormalizable $\lambda_0\phi_4^4$ model.

Note that according to formula (12.31), the diagrams that have $n = 0$ and $n_{gh} = 4$ are logarithmically divergent. The corresponding counterterm would have the general form $(\overline{c}c)^2$. Because there is no term of this kind in the action (12.20), the presence of this counterterm in the effective action would pose a problem—it would signal that the deep analysis carried out in the Sects. 12.1 and 12.2 was not precise enough. Luckily, this is not the case. Two of the external lines in all diagrams with $n = 0, n_{gh} = 4$ have arrows pointing outward from the diagrams. Therefore, the two internal vertices from which these two external lines start are proportional to the fixed external four-momenta, c.f. formula (12.30) and the last vertex in Fig. 12.2. It follows that the superficial degree of divergence is in fact smaller by 2, i.e., it is equal to -1, and the controversial counterterm is not needed.

12.4 BRST Invariance and the Slavnov–Taylor Identities

The classical effective action $S[A, c, \overline{c}]$, formula (12.15), is not gauge invariant by its construction—it was precisely our goal in Sect. 12.1 to eliminate the freedom of performing the gauge transformations. However, in 1975 C. Becchi, A. Rouet, R. Stora, and independently I. V. Tyutin, discovered that this action is invariant with respect to rather special transformations, which are usually written in the following form:

$$A'^a_\mu(x) = A^a_\mu(x) + \delta A^a_\mu(x), \;\; c'^a(x) = c^a(x) + \delta c^a(x), \;\; \bar{c}'_a(x) = \bar{c}_a(x) + \delta \bar{c}_a(x), \tag{12.32}$$

where

$$\delta A^a_\mu(x) = \alpha\theta\,(D_\mu c)^a(x), \quad \delta c^a(x) = \frac{1}{2}\alpha g \theta f_{abd} c^b c^d, \quad \delta \bar{c}_a(x) = \theta\, \partial_\mu A^{a\mu}.$$

Here θ is a Grassmann element. By assumption, it anticommutes with the ghost fields. Because $\theta^2 = 0$, the above form of the transformations is the exact one, in spite of the notation which might suggest that, e.g., δA^a_μ is an infinitesimal contribution (which it in fact is not). It turns out that these transformations leave invariant the Lagrangian that corresponds to the action S,

$$\mathcal{L} = -\frac{1}{4} F^a_{\mu\nu} F^{a\mu\nu} - \frac{1}{2\alpha} \partial_\mu A^{a\mu} \partial_\nu A^{a\nu} + \bar{c}_a(x)\, \partial_\mu[(D^\mu c)^a(x)]. \tag{12.33}$$

Actually $\mathcal{L}$ is not the simplest BRST invariant object. There exist other invariants: $\mathcal{L}_{YM} = -F^a_{\mu\nu} F^{a\mu\nu}/4$, as well as

$$I^a_1(x) = \alpha g f_{abd} c^b(x) c^d(x)/2, \quad I^{a\mu}_2 = \alpha (D^\mu c)^a(x)$$

(Exercise 12.3). Their presence facilitates checking the invariance of $\mathcal{L}$.

The measure $[dA][d\bar{c}dc]$ is also invariant with respect to the BRST transformations. In order to demonstrate this, it is sufficient to consider the products

$$\prod_{a=1}^{N^2-1} dA^a_\mu(x) \prod_{b=1}^{N^2-1} d\bar{c}_b(x) \prod_{d=1}^{N^2-1} dc^d(x)$$

with an arbitrary fixed $x \in M$, $\mu = 0, 1, 2, 3$. Because θ is a constant,

$$\prod_{b=1}^{N^2-1} d\bar{c}'_b(x) = \prod_{b=1}^{N^2-1} d\bar{c}_b(x).$$

Next,

$$\prod_{d=1}^{N^2-1} dc'^d(x) = (\det \hat{J})^{-1} \prod_{d=1}^{N^2-1} dc^d(x),$$

where the matrix elements of the $N^2 - 1$ by $N^2 - 1$ matrix $\hat{J}$ have the form $J_{ab} = \delta_{ab} + \alpha g \theta f_{adb} c^d(x)$. In the derivation of this formula the antisymmetry of f_{adb} was used. Because $\theta^2 = 0$, $\det\hat{J} = 1 + \alpha g \theta f_{ada} c^d(x) = 1$. Furthermore,

$$\prod_{a=1}^{N^2-1} dA'^a_\mu(x) \prod_{d=1}^{N^2-1} dc'^d(x) = \det\hat{J}\ (\det\hat{J})^{-1} \prod_{a=1}^{N^2-1} dA^a_\mu(x) \prod_{d=1}^{N^2-1} dc^d(x)$$
$$= \prod_{a=1}^{N^2-1} dA^a_\mu(x) \prod_{d=1}^{N^2-1} dc^d(x).$$

The BRST invariance of the Lagrangian, and of the measure, implies certain identities for the Green's functions, called the Slavnow–Taylor identities. It is convenient to first obtain the Slavnov–Taylor identity for a certain generating functional. The Green's functions will be considered next. Let us introduce an extended generating functional $\mathcal{Z}$,

$$\mathcal{Z}[j, \xi, \overline{\xi}, H, \overline{K}] = \frac{\overline{\mathcal{Z}}[j, \xi, \overline{\xi}, H, \overline{K}]}{\overline{\mathcal{Z}}[0]}, \tag{12.34}$$

where

$$\overline{\mathcal{Z}}[j, \xi, \overline{\xi}, H, \overline{K}] = \lim_{\epsilon\to 0_+} \int [dA][d\overline{c}dc] \exp\Big[iS[A, c, \overline{c}]$$
$$- \frac{\epsilon}{2}\int d^4x\, e^{-\epsilon|x^0|} A^{ai}\sqrt{-\Delta}A^{ai}\Big] \exp\Big[i\int d^4x\ \big(j^a_\mu(x)A^{a\mu}(x) + \overline{c}_a(x)\xi^a(x)$$
$$+\overline{\xi}_a(x)c^a(x) + H_a(x)I^a_1(x) + \overline{K}^\mu_a(x)I^a_{2\mu}\big)\Big]. \tag{12.35}$$

Here $H_a(x)$ and $\overline{K}^\mu_a(x)$ are new external sources. $\overline{K}^\mu_a$ is of Grassmann type like ξ^a and $\overline{\xi}_a$. The expression

$$N[A, c, \overline{c}] =$$
$$\exp\Big[i\int d^4x \left(j^a_\mu(x)A^{a\mu}(x) + \overline{c}_a(x)\xi^a(x) + \overline{\xi}_a(x)c^a(x) - \frac{\epsilon}{2}\, e^{-\epsilon|x^0|}A^{ai}\sqrt{-\Delta}A^{ai}\right)\Big],$$

which is a part of formula (12.35), is not invariant under the transformations (12.32):

$$N[A', c', \overline{c}'] = N[A, c, \overline{c}] + i\theta \lim_{\epsilon\to 0_+} \int d^4x\Big[j^{a\mu}I^a_{2\mu}$$
$$+ \xi^a\partial_\mu A^{a\mu} - \overline{\xi}_a I^a_1 + \frac{i}{2}\epsilon e^{-\epsilon|x^0|}(I^{ai}_2\sqrt{-\Delta}A^{ai} + A^{ai}\sqrt{-\Delta}I^{ai}_2)\Big]\, N[A, c, \overline{c}].$$

The last term on the r.h.s. vanishes in the limit $\epsilon \to 0$. Note that $I^a_1(x)$, $I^a_{2\mu}(x)$, and $A^{a\mu}(x)$ can be replaced by the variational derivatives $-i\delta/\delta H_a(x)$, $-i\delta/\delta\overline{K}^\mu_a(x)$, and $-i\delta/\delta j^a_\mu(x)$, respectively.

The Slavnov–Taylor identity for $\mathcal{Z}$ follows from the fact that in the path integral giving this functional, formula (12.35), we may perform a nonsingular change of the integration variables, in particular the change given by formulas (12.32). Because of the invariance of the integration measure, and of the whole integrand except

$N[A, c, \overline{c}]$, we have the following identity

$$\overline{\mathcal{Z}}[j, \xi, \overline{\xi}, H, \overline{K}] = \overline{\mathcal{Z}}[j, \xi, \overline{\xi}, H, \overline{K}] + \theta \int d^4x \left[j^{a\mu}(x) \frac{\delta \overline{\mathcal{Z}}}{\delta \overline{K}^{\mu}_{a}(x)} + \xi^a \partial_\mu \frac{\delta \overline{\mathcal{Z}}}{\delta j^a_\mu(x)} - \overline{\xi}_a \frac{\delta \overline{\mathcal{Z}}}{\delta H_a(x)} \right].$$

From here, dividing by $\overline{\mathcal{Z}}[0]$, we obtain the Slavnov–Taylor identity for the generating functional $\mathcal{Z}$:

$$\int d^4x \left(j^{a\mu}(x) \frac{\delta \mathcal{Z}}{\delta \overline{K}^{\mu}_{a}(x)} + \xi^a \partial_\mu \frac{\delta \mathcal{Z}}{\delta j^a_\mu(x)} - \overline{\xi}_a \frac{\delta \mathcal{Z}}{\delta H_a(x)} \right) = 0. \tag{12.36}$$

The identities for Green's functions are generated from (12.36) by taking variational derivatives with respect to j^a_μ, ξ^a and $\overline{\xi}_a$, and putting zero for all of the external sources, including H_a and $\overline{K}^{\mu}_{a}$. Note that such identities involve Green's functions, which are the vacuum expectation values of the time ordered products of not only the fields $\mathbf{A}^a_\mu(x)$, $\mathbf{c}^a(c)$, $\overline{\mathbf{c}}_a(x)$, but also the composite fields $\mathbf{I}^a_1(x)$ and $\mathbf{I}^a_{2\mu}(x)$.

The Slavnov–Taylor identities encode, on the level of Green's functions, the fact that the Lagrangian (12.33) has a very specific form. This form of the Lagrangian is the consequence of the gauge invariance of the original Yang–Mills Lagrangian $\mathcal{L}_{YM}$. Therefore, violation of these identities would imply violation of gauge invariance.[6]

It is also clear from the remarks above that the renormalized Green's functions should obey the Slavnov–Taylor identities. In order to achieve this, the various counterterms which are introduced in the process of removing the UV divergences have to be interrelated in the appropriate manner. With such restrictions on the counterterms, the renormalization of non-Abelian gauge theories is quite nontrivial. The proof of the renormalizability of these theories, provided by G. 't Hooft and M. Veltman around 1970, requires in particular a rather special regularization, called the dimensional regularization.

The quantum non-Abelian gauge field is asymptotically free— the Gell-Mann–Low β function turns out to be negative, at least for small values of the coupling constant g. Therefore the perturbative results are trustworthy only at very large four-momenta. Unfortunately, nuclear phenomena and the structure of hadrons belong to the realm of (relatively) low four-momenta physics, where the perturbative results are not reliable.

[6]One should distinguish between gauge invariance and gauge independence. This last term, often used in literature, refers to the lack of dependence on the concrete choice of a gauge condition (12.1).

Exercises

12.1 Find the form of the Faddeev–Popov–DeWitt action in the case of the free Abelian gauge field with the 't Hooft–Veltman non-linear gauge condition $(\partial_\mu - A_\mu)A^\mu = 0$.
Hint: $\delta F(x)/\delta A_\mu(y) = \partial^\mu \delta(x-y) - 2A^\mu(x)\delta(x-y)$

12.2 The gauge condition $n^\mu A_\mu^a(x) = 0$, where $n = (n^\mu)$ is a constant non-vanishing four-vector, encompasses the Coulomb, the temporal ($A_0^a = 0$) and other popular gauge conditions. Show that with this gauge condition one can obtain formula (12.14) in which the ghost fields are absent.
Hint: Notice that $M_{ca}(x, y)$ contains the expression $n^\mu A_\mu^d(y)$, which is equal to $\lambda^d(y)$ when we consider the auxiliary gauge condition $F^a(A)(x) = \lambda^a(x)$ used in the derivation of formula (12.14). Therefore, the factor $\mathcal{M}[\hat{A}]$ in formula (12.13), in which we now have $\delta[F(A) - \lambda]$ instead of $\delta[F(A)]$, can be replaced by $\mathcal{M}[\hat{A}]|_{nA^a=\lambda^a}$. This factor does not depend on A_μ^a, hence it can be omitted (in fact it is canceled by a factor in $\mathcal{N}$).

12.3 Check that $\mathcal{L}_{YM}$, I_1^a, and $I_{2\mu}^a$ are invariant with respect to the BRST transformations.
Hint: In the case of $\mathcal{L}_{YM}$ first prove that $\delta\hat{F}_{\mu\nu} = i\alpha g\theta\,[\hat{F}_{\mu\nu}, \hat{c}]$, where $\hat{c} = \hat{T}_a c^a$. Next, write $F_{\mu\nu}^a F^{a\mu\nu}/4$ as $\mathrm{tr}(\hat{F}_{\mu\nu}\hat{F}^{\mu\nu})/2$ and check that $\mathrm{tr}(\delta\hat{F}_{\mu\nu}\hat{F}^{\mu\nu}) = 0$.

Chapter 13
The Simplest Supersymmetric Models

Abstract The generating elements and their (anti-)commutation relations in the $N = 1$ superalgebra. Multiplets of quantum states generated by elements of the superalgebra. An example of a supersymmetric Lagrangian with free fields. The notions of superspace, superfield, and chiral superfield. The Wess–Zumino model and the Feynman diagrams for it. Examples of the mutual cancellation of ultraviolet divergences. The supersymmetric gauge theory. The $N = 2$ extended supersymmetry. A glossary of formulas used in the analysis of supersymmetric models.

The BRST invariance of the classical effective action for the non-Abelian gauge fields is an example of a symmetry with the parameters of the transformation belonging to the Grassmann algebra. Such symmetries, called supersymmetries, have become increasingly popular in field theory in their own right. Below we present three examples of supersymmetric models: a free field model, the so called Wess–Zumino model and (with less details) supersymmetric models with gauge fields.

13.1 The Superalgebra

A superalgebra includes, besides the bosonic generators of the Poincaré group $\mathcal{P}$, at least one spinor generator $\hat{Q}$ with two components. We will discuss in this chapter only four-dimensional Minkowski space-time so that the simplest possibility is to take $\hat{Q}$ to be the right-handed Weyl spinor $\hat{Q}_\alpha$, $\alpha = 1, 2$, of the Grassmann type (see Chap. 5). This will lead us to the so called $N = 1$ supersymmetry. Our goal will be thus: to determine the allowed form of the superalgebra containing—besides the generators of $\mathcal{P}$—the generator $\hat{Q}_\alpha$ and its conjugate $\hat{\bar{Q}}^{\dot{\alpha}}$ (notice that—to conform with most of the literature on supersymmetry—we have changed the notation for the conjugate spinor from a 'star' to a 'bar').

H. Arodź and L. Hadasz, *Lectures on Classical and Quantum Theory of Fields*, Graduate Texts in Physics, DOI 10.1007/978-3-319-55619-2_13

Consider first the commutator $[\hat{P}^\mu, \hat{Q}_\alpha]$. It is a spinor quantity, so let us assume that[1]:

$$[\hat{P}^\mu, \hat{Q}_\alpha] = c\sigma^\mu_{\alpha\dot{\beta}}\hat{\bar{Q}}^{\dot{\beta}} \tag{13.1}$$

with some complex constant c. Consequently, upon conjugation of both sides

$$[\hat{P}^\mu, \hat{\bar{Q}}^{\dot{\beta}}] = -c^*\tilde{\sigma}^{\mu\dot{\beta}\gamma}\hat{Q}_\gamma. \tag{13.2}$$

Using (13.1), (13.2), the Jacobi identity

$$[\hat{P}^\mu, [\hat{P}^\nu, \hat{Q}_\alpha]] + [\hat{P}^\nu, [\hat{Q}_\alpha, \hat{P}^\mu]] + [\hat{Q}_\alpha, [\hat{P}^\mu, \hat{P}^\nu]] = 0 \tag{13.3}$$

and the relation $[\hat{P}^\mu, \hat{P}^\nu] = 0$, we get

$$|c|^2\,(\sigma^\mu\tilde{\sigma}^\nu + \sigma^\nu\tilde{\sigma}^\mu) = 0,$$

so that $c = 0$, and we have

$$[\hat{P}^\mu, \hat{Q}_\alpha] = \left[\hat{P}^\mu, \hat{\bar{Q}}^{\dot{\beta}}\right] = 0. \tag{13.4}$$

In the spinor representation (see (5.19)) the Dirac matrices read

$$\gamma^\mu = \begin{pmatrix} 0 & \sigma^\mu \\ \tilde{\sigma}^\mu & 0 \end{pmatrix}, \qquad [\gamma^\mu, \gamma^\nu] = \begin{pmatrix} \sigma^{\mu\nu} & 0 \\ 0 & \tilde{\sigma}^{\mu\nu} \end{pmatrix}.$$

It then follows from (5.17) that under a Lorentz transformation with an antisymmetric, infinitesimal $\omega_{\mu\nu}$, the generator Q_α transforms as

$$\hat{Q}'_\alpha = (1 + \tfrac{1}{2}\omega_{\mu\nu}\sigma^{\mu\nu})_\alpha{}^\beta\,\hat{Q}_\beta = \hat{Q}_\alpha + \frac{i}{2}\omega_{\mu\nu}\left[\hat{M}^{\mu\nu}, \hat{Q}_\alpha\right],$$

so that

$$\left[\hat{M}^{\mu\nu}, \hat{Q}_\alpha\right] = -i(\sigma^{\mu\nu})_\alpha{}^\beta\,\hat{Q}_\beta. \tag{13.5}$$

A similar derivation gives

$$\left[\hat{M}^{\mu\nu}, \hat{\bar{Q}}^{\dot{\alpha}}\right] = -i(\tilde{\sigma}^{\mu\nu})^{\dot{\alpha}}{}_{\dot{\beta}}\,\hat{\bar{Q}}^{\dot{\beta}}. \tag{13.6}$$

Consider now the anticommutator $\{\hat{Q}_\alpha, \hat{Q}^\beta\}$. It is clearly a bosonic object and the transformation properties of $\hat{Q}_\alpha$ under Poincaré transformations constrain it to be proportional to $(\sigma_{\mu\nu})_\alpha{}^\beta\,\hat{M}^{\mu\nu}$. In view of (13.4)

[1] See Sect. 13.7 for the notation and conventions.

$$\left[\hat{P}^{\mu}, \{\hat{Q}_{\alpha}, \hat{Q}^{\beta}\}\right] = 0$$

while $[\hat{P}^{\mu}, \hat{M}^{\nu\rho}] \neq 0$ (see Sect. 10.1), so that the proportionality constant must vanish and we have

$$\{\hat{Q}_{\alpha}, \hat{Q}^{\beta}\} = \{\hat{\bar{Q}}_{\dot{\alpha}}, \hat{\bar{Q}}^{\dot{\beta}}\} = 0. \tag{13.7}$$

Finally, $\{\hat{Q}_{\alpha}, \hat{\bar{Q}}_{\dot{\beta}}\} \propto \sigma^{\mu}_{\alpha\dot{\beta}} \hat{P}_{\mu}$. The proportionality constant can be adjusted at will by appropriately rescaling the generators, and we take

$$\{\hat{Q}_{\alpha}, \hat{\bar{Q}}_{\dot{\beta}}\} = 2\sigma^{\mu}_{\alpha\dot{\beta}} \hat{P}_{\mu}. \tag{13.8}$$

This relation has very interesting consequences. Since

$$\sigma^{\mu}\tilde{\sigma}^{\nu} = \eta^{\mu\nu} I_2 + 2\sigma^{\mu\nu}$$

and $\sigma^{\mu\nu}$ are traceless, we have

$$\text{tr}(\sigma^{\mu}\tilde{\sigma}^{\nu}) = 2\eta^{\mu\nu}.$$

From (13.8) we thus get

$$(\tilde{\sigma}^{\nu})^{\dot{\beta}\alpha}\{\hat{Q}_{\alpha}, \hat{\bar{Q}}_{\dot{\beta}}\} = 2\,\text{tr}(\tilde{\sigma}^{\nu}\sigma^{\mu})\hat{P}_{\mu} = 4\hat{P}^{\nu}.$$

Taking $\nu = 0$ we have for any state $|\psi\rangle \neq 0$

$$\begin{aligned}
\langle\psi|\hat{P}_0|\psi\rangle &= \frac{1}{4}\langle\psi|\hat{Q}_1\hat{\bar{Q}}_{\dot{1}} + \hat{Q}_2\hat{\bar{Q}}_{\dot{2}} + \hat{\bar{Q}}_{\dot{1}}\hat{Q}_1 + \hat{\bar{Q}}_{\dot{2}}\hat{Q}_2|\psi\rangle \\
&= \frac{1}{4}\langle\psi|\hat{Q}_{\alpha}(\hat{Q}_{\alpha})^{\dagger} + (\hat{Q}_{\alpha})^{\dagger}\hat{Q}_{\alpha}|\psi\rangle \geq 0.
\end{aligned} \tag{13.9}$$

Here we have taken into account the fact that the Grassmann conjugation of the Weyl spinors can finally be reduced to Hermitian conjugation. Thus in any supersymmetric theory

$$\langle 0|\hat{P}_0|0\rangle = 0 \;\Leftrightarrow\; \hat{Q}_{\alpha}|0\rangle = 0 \tag{13.10}$$

and all states have non-negative energy.

13.2 Supersymmetry Multiplets

Let us have a look at the consequences of the superalgebra restricted to the subspace of single particle states. The spatial components of the operator $\hat{M}_{\mu\nu}$, generators of the rotations in the three dimensional space, are often denoted as

$$\hat{M}_{23} = -\hat{M}_{32} = \hat{J}^1, \qquad \hat{M}_{31} = -\hat{M}_{13} = \hat{J}^2, \qquad \hat{M}_{12} = -\hat{M}_{21} = \hat{J}^3,$$

or equivalently $\hat{J}^i = \frac{1}{2}\epsilon^{ijk}\hat{M}_{jk}$. It follows from the commutation relation satisfied by the operators $\hat{P}_\mu$ and $\hat{M}_{\mu\nu}$, see (10.12), (10.17), and (10.19), that $\hat{P}^\mu$, $\hat{J}^3$ and $\hat{\vec{J}}^{\,2}$ commute with each other if restricted to the subspace of states with vanishing momentum $\hat{P}^i$. As a basis in this subspace we take the states $|\vec{0}, s, s_3\rangle$ such that

$$\begin{aligned} \hat{P}^\mu|\vec{0}, s, s_3\rangle &= m\delta^\mu_0|\vec{0}, s, s_3\rangle, \\ \hat{\vec{J}}^{\,2}|\vec{0}, s, s_3\rangle &= s(s+1)|\vec{0}, s, s_3\rangle, \\ \hat{J}^3|\vec{0}, s, s_3\rangle &= s_3|\vec{0}, s, s_3\rangle, \end{aligned} \tag{13.11}$$

where $m > 0$ is the mass of the particle, which is assumed to be positive. Now define rescaled generators,

$$\hat{a}_\alpha = \frac{1}{\sqrt{2m}}\hat{Q}_\alpha, \qquad \hat{a}^\dagger_\alpha = \frac{1}{\sqrt{2m}}\hat{\bar{Q}}_{\dot{\alpha}}. \tag{13.12}$$

In the particle's rest frame, their algebra (with the form which follows from (13.7) and (13.8)) is isomorphic to the algebra of two fermionic creation and annihilation operators,

$$\{\hat{a}_\alpha, \hat{a}^\dagger_\beta\} = \delta^\beta_\alpha, \qquad \{\hat{a}_\alpha, \hat{a}_\beta\} = \{\hat{a}^\dagger_\alpha, \hat{a}^\dagger_\beta\} = 0. \tag{13.13}$$

We can construct their representation on the space spanned by the vectors $|\vec{0}, s, s_3\rangle$ as follows. Suppose that $|\vec{0}, s', s'_3\rangle$ is an eigenstate of $\hat{P}^\mu$, $\hat{\vec{J}}^{\,2}$ and $\hat{J}^3$. Then either $\hat{a}_1|\vec{0}, s', s'_3\rangle = 0$ or, thanks to (13.4), (13.5), $|\vec{0}, s, s_3\rangle = \hat{a}_1|\vec{0}, s', s'_3\rangle$ is also an eigenstate of these operators (although corresponding to different eigenvalues of the latter two). Moreover, from the relation $\hat{a}_1^2 = \frac{1}{2}\{\hat{a}_1, \hat{a}_1\} = 0$ it follows that

$$\hat{a}_1|\vec{0}, s, s_3\rangle = 0$$

An analogous argument for $\hat{a}_2$ shows that we can always choose $|\vec{0}, s, s_3\rangle$ to be annihilated by $\hat{a}_\alpha$, $\alpha = 1, 2$. We shall denote the state satysfying (13.11) and annihilated by a_α, $\alpha = 1, 2$, by $|\vec{0}, s, s_3\rangle_0$.

From each of the states $|\vec{0}, s, s_3\rangle_0$ we then construct three more states with the same mass,

$$\hat{a}^\dagger_\alpha|\vec{0}, s, s_3\rangle_0, \qquad \hat{a}^\dagger_2\hat{a}^\dagger_1|\vec{0}, s, s_3\rangle_0.$$

Equation (13.6) implies that

$$[\hat{J}^i, \hat{a}^\dagger_\alpha] = \frac{1}{2}(\sigma^i\hat{a}^\dagger)_\alpha$$

from which it follows that

$$\hat{J}^3\hat{a}_1^\dagger|\vec{0}, s, s_3\rangle = (s_3 + \tfrac{1}{2})\hat{a}_1^\dagger|\vec{0}, s, s_3\rangle, \qquad \hat{J}^3\hat{a}_2^\dagger|\vec{0}, s, s_3\rangle = (s_3 - \tfrac{1}{2})\hat{a}_2^\dagger|\vec{0}, s, s_3\rangle, \tag{13.14}$$

and

$$\hat{J}^3\hat{a}_2^\dagger\hat{a}_1^\dagger|\vec{0}, s, s_3\rangle = s_3\hat{a}_2^\dagger\hat{a}_1^\dagger|\vec{0}, s, s_3\rangle. \tag{13.15}$$

Notice also that

$$[\hat{J}^1 - i\hat{J}^2, \hat{a}_1^\dagger] = \hat{a}_2^\dagger, \qquad [\hat{J}^1 + i\hat{J}^2, \hat{a}_2^\dagger] = \hat{a}_1^\dagger, \qquad [\hat{J}^1 + i\hat{J}^2, \hat{a}_1^\dagger] = [\hat{J}^1 - i\hat{J}^2, \hat{a}_2^\dagger] = 0. \tag{13.16}$$

From the relations above one may show that in general, starting with a $2s + 1$ component multiplet with spin s, and acting on it with $\hat{a}_\alpha^\dagger$ we generate a spin $(s + \frac{1}{2})$ multiplet, a spin $(s - \frac{1}{2})$ multiplet and one more spin s multiplet. Thus a general massive representation has $4(2s + 1)$ basis states, half of which are bosonic and half fermionic. In particular, starting from the scalar $|\vec{0}, 0, 0\rangle_0$ we get an $s = \frac{1}{2}$ doublet

$$\hat{a}_1^\dagger|\vec{0}, 0, 0\rangle_0 = |\vec{0}, \tfrac{1}{2}, \tfrac{1}{2}\rangle, \qquad \hat{a}_2^\dagger|\vec{0}, 0, 0\rangle_0 = |\vec{0}, \tfrac{1}{2}, -\tfrac{1}{2}\rangle,$$

and a second scalar

$$|\vec{0}, 0, 0\rangle' = \hat{a}_2^\dagger\hat{a}_1^\dagger|\vec{0}, 0, 0\rangle_0.$$

In the $s = \frac{1}{2}$ case we get

$$\begin{aligned}
\hat{a}_1^\dagger|\vec{0}, \tfrac{1}{2}, \tfrac{1}{2}\rangle_0 &= |\vec{0}, 1, 1\rangle, & \hat{a}_2^\dagger|\vec{0}, \tfrac{1}{2}, \tfrac{1}{2}\rangle_0 &= \frac{1}{\sqrt{2}}\left(|\vec{0}, 1, 0\rangle + |\vec{0}, 0, 0\rangle\right), \\
\hat{a}_2^\dagger|\vec{0}, \tfrac{1}{2}, -\tfrac{1}{2}\rangle_0 &= |\vec{0}, 1, -1\rangle, & \hat{a}_1^\dagger|\vec{0}, \tfrac{1}{2}, -\tfrac{1}{2}\rangle_0 &= \frac{1}{\sqrt{2}}\left(|\vec{0}, 1, 0\rangle - |\vec{0}, 0, 0\rangle\right), \\
\hat{a}_2^\dagger\hat{a}_1^\dagger|\vec{0}, \tfrac{1}{2}, \tfrac{1}{2}\rangle_0 &= |\vec{0}, \tfrac{1}{2}, \tfrac{1}{2}\rangle', & \hat{a}_2^\dagger\hat{a}_1^\dagger|\vec{0}, \tfrac{1}{2}, -\tfrac{1}{2}\rangle_0 &= |\vec{0}, \tfrac{1}{2}, -\tfrac{1}{2}\rangle'.
\end{aligned}$$

13.3 Representation of Supersymmetry in a Space of Fields

An important feature of the superalgebra is that it can be realized in a field theory and its generators may be represented in terms of integrals of conserved, local currents,

$$\hat{Q}_\alpha = \int d^3x\, \hat{j}_\alpha^0(x), \qquad \partial_\mu \hat{j}_\alpha^\mu(x) = 0.$$

The currents may in turn be expressed as local products of fields (for an example see Exercise 13.2).

Let ξ^α and $\bar{\xi}_{\dot\alpha}$ denote Grassmann (anticommuting) parameters, satisfying

$$\{\xi^\alpha, \xi^\beta\} = \{\xi^\alpha, \bar{\xi}_{\dot\beta}\} = \{\bar{\xi}_{\dot\alpha}, \bar{\xi}_{\dot\beta}\} = 0,$$

which are supposed to anticommute with the supersymmetry generators, and to commute with the generators of the Poincaré group,

$$\begin{aligned}
\{\xi^\alpha, \hat{Q}_\beta\} = \{\xi^\alpha, \hat{\bar{Q}}^{\dot\beta}\} = [\xi^\alpha, \hat{P}^\mu] = [\xi^\alpha, \hat{M}^{\mu\nu}] = 0,\\
\{\bar{\xi}_{\dot\alpha}, Q_\beta\} = \{\bar{\xi}_{\dot\alpha}, \hat{\bar{Q}}^{\dot\beta}\} = [\bar{\xi}_{\dot\alpha}, \hat{P}^\mu] = [\bar{\xi}_{\dot\alpha}, \hat{M}^{\mu\nu}] = 0.
\end{aligned} \tag{13.17}$$

Using them, and another set of constant Grassmann parameters η^α and $\overline{\eta}_{\dot\alpha}$, we can rewrite the superalgebra using only commutators,

$$\begin{aligned}
[\hat{P}^\mu, \xi\hat{Q}] &= [\hat{P}^\mu, \bar{\xi}\hat{\bar{Q}}] = 0,\\
\left[\hat{M}^{\mu\nu}, \xi\hat{Q}\right] &= -i\,\xi\sigma^{\mu\nu}\hat{Q}, \quad [\hat{M}^{\mu\nu}, \bar{\xi}\hat{\bar{Q}}] = -i\,\bar{\xi}\bar{\sigma}^{\mu\nu}\hat{\bar{Q}},\\
[\xi\hat{Q}, \eta\hat{Q}] = [\bar{\xi}\hat{\bar{Q}}, \bar{\eta}\hat{\bar{Q}}] &= 0, \quad [\xi\hat{Q}, \bar{\eta}\hat{\bar{Q}}] = 2(\xi\sigma^\mu\bar{\eta})\hat{P}_\mu,
\end{aligned} \tag{13.18}$$

where, in the adopted conventions, $\xi\hat{Q} = \xi^\alpha\hat{Q}_\alpha$, $\bar{\xi}\hat{\bar{Q}} = \bar{\xi}_{\dot\alpha}\hat{\bar{Q}}^{\dot\alpha}$, e.t.c.

As was already discussed in Chap. 10, any quantum field in the Heisenberg picture $\hat{u}(x)$ transforms under a symmetry transformation, represented by a unitary operator U, as

$$\hat{u}'(x) = U^\dagger\hat{u}(x)U.$$

For a supersymmetry transformation parameterized by ξ and $\bar{\xi}$

$$U = U(\xi, \bar{\xi}) = e^{i(\xi\hat{Q}+\hat{\bar{Q}}\bar{\xi})},$$

and, up to the terms linear in ξ and $\bar{\xi}$,

$$\delta_\xi\hat{u}(x) \equiv \hat{u}'(x) - \hat{u}(x) = -i[\xi\hat{Q} + \bar{\xi}\hat{\bar{Q}}, \hat{u}(x)]. \tag{13.19}$$

The form of $\delta_\xi\hat{u}(x)$ must be consistent with the algebra (13.18). In particular, for two subsequent SUSY transformations, (13.19) and the Jacobi identity for commutators

$$[A, [B, C]] + [B, [C, A]] + [C, [A, B]] = 0$$

give

$$[\delta_\eta, \delta_\xi]\hat{u}(x) \equiv (\delta_\eta\delta_\xi - \delta_\xi\delta_\eta)\hat{u}(x) = -\left[[\eta\hat{Q} + \bar{\eta}\hat{\bar{Q}}, \xi\hat{Q} + \bar{\xi}\hat{\bar{Q}}], \hat{u}(x)\right].$$

Using (13.18), we thus get

$$[\delta_\eta, \delta_\xi]\hat{u}(x) = 2(\xi\sigma^\mu\bar{\eta} - \eta\sigma^\mu\bar{\xi})[\hat{P}_\mu, \hat{u}(x)]. \tag{13.20}$$

Moreover, if $\hat{u}(x)$ is scalar under translations, then (see Chap. 10, (10.24))

$$[\hat{P}_\mu, \hat{u}(x)] = -i\partial_\mu\hat{u}(x), \tag{13.21}$$

and we arrive at the condition

$$[\delta_\eta, \delta_\xi]\hat{u}(x) = 2i(\eta\sigma^\mu\bar{\xi} - \xi\sigma^\mu\bar{\eta})\partial_\mu\hat{u}(x). \tag{13.22}$$

This is in fact a sufficient condition for the consistency of (13.19) with (13.18).

Equation (13.22) was derived for the quantum field $\hat{u}(x)$, but its right hand side also makes perfect sense when $u(x)$ is a classical field. Our goal now will be to find the simplest possible set of classical fields $u_i(x)$, and to define their supersymmetric variations $\delta_\xi u_i(x)$ so that (13.22) is satisfied. Moreover, we shall require the classical action functional for the fields $u_i(x)$ to be invariant when we replace $u_i(x)$ with $u_i(x) + \delta_\xi u_i(x)$.

As we have learned in the previous Section, the simplest supersymmetric multiplet (obtained by starting from the state $|\vec{0}, 0, 0\rangle$) contains two states with total spin $s = 0$, and a doublet of states with the total spin $s = \frac{1}{2}$ and its two possible s_3 components. Thus, we may try to construct its field theoretic realization in a model containing a classical Weyl spinor field ψ and a complex scalar field φ. Let us start by postulating—guided by the dimensional analysis—a transformation law for the scalar field $\varphi(x)$. The form of the action functionals for the scalar and Weyl fields show that (in the system of units $\hbar = c = 1$) their dimensions read

$$[\varphi(x)] = \text{cm}^{-1}, \qquad [\psi(x)] = [\bar{\psi}(x)] = \text{cm}^{-\frac{3}{2}}.$$

Now, $[\hat{P}_\mu] = \text{cm}^{-1}$ so that equation (13.8) gives

$$[\hat{Q}] = [\hat{\bar{Q}}] = \text{cm}^{-\frac{1}{2}}$$

and (since $\xi\hat{Q}$ and $\bar{\xi}\hat{\bar{Q}}$ have to be dimensionless for $U(\xi)$ to make sense)

$$[\xi] = [\bar{\xi}] = \text{cm}^{\frac{1}{2}}.$$

Because an infinitesimal transformation is linear in the transformation parameters and $[\delta_\xi\varphi] = [\varphi]$, it must therefore be of the form

$$\delta_\xi\varphi = a\xi\psi + b\bar{\xi}\bar{\psi} \tag{13.23}$$

where a and b are complex constants to be determined from (13.22), and $\bar{\psi} = (\psi)^*$.

Since $\delta_\xi \psi$ also has the dimension $\text{cm}^{-\frac{3}{2}}$ and we assume that φ and ψ are the only fields in the constructed model, the only choice for $\delta_\xi \psi$ is

$$\delta_\xi \psi_\alpha = c\sigma^\mu_{\alpha\dot{\alpha}} \bar{\xi}^{\dot{\alpha}} \partial_\mu \varphi \tag{13.24}$$

with a constant c, and consequently

$$\delta_\xi \bar{\psi}^{\dot{\beta}} = \epsilon^{\dot{\beta}\dot{\alpha}} \left(\delta_\xi \psi_\alpha\right)^* = -c^* (\tilde{\sigma}^\mu)^{\dot{\beta}\beta} \xi_\beta \partial_\mu \varphi^* \tag{13.25}$$

where the formula

$$\epsilon^{\beta\alpha} \epsilon^{\dot{\beta}\dot{\alpha}} \sigma^\mu_{\alpha\dot{\alpha}} = (\tilde{\sigma}^\mu)^{\dot{\beta}\beta}$$

was used. Equations (13.23), (13.24), and (13.25) give

$$\delta_\eta \delta_\xi \varphi = ac(\xi\sigma^\mu \bar{\eta})\partial_\mu \varphi + bc^* (\eta\sigma^\mu \bar{\xi})\partial_\mu \varphi^*,$$

and the consistency condition

$$[\delta_\eta, \delta_\xi]\varphi(x) = 2i(\eta\sigma^\mu \bar{\xi} - \xi\sigma^\mu \bar{\eta})\partial_\mu \varphi(x)$$

holds if

$$ac = -2i, \qquad b = 0. \tag{13.26}$$

Furthermore (for θ being a constant spinor, introduced here to avoid explicitly writing down the indices)

$$\delta_\eta \delta_\xi \, \theta\psi = ca(\theta\sigma^\mu \bar{\xi})\big(\eta\partial_\mu \psi\big),$$

and, using (13.26) with an appropriate Fierz identity (see Exercise 13.5), we get

$$\delta_\eta \delta_\xi (\theta\psi) = 2i(\eta\sigma^\mu \bar{\xi})(\theta\partial_\mu \psi) - i(\eta\sigma_\nu \bar{\xi})(\theta\sigma^\nu \tilde{\sigma}^\mu \partial_\mu \psi). \tag{13.27}$$

The consistency condition

$$[\delta_\eta, \delta_\xi]u(x) = 2i(\eta\sigma^\mu \bar{\xi} - \xi\sigma^\mu \bar{\eta})\partial_\mu u(x)$$

is satisfied for $u(x) = \psi$ (and, by complex conjugation, for $u(x) = \bar{\psi}$) if and only if the second term on the r.h.s. of (13.27) vanishes, or, equivalently, ψ obeys the equation of the motion of a free, massless Weyl field,

$$\tilde{\sigma}^\mu \partial_\mu \psi = 0. \tag{13.28}$$

The consistency of field variations with the SUSY algebra is just a necessary condition for the SUSY invariance of a given field theoretic model. We also need to check whether the pertinent action functional is invariant. In view of (13.28), we shall discuss the theory of a non–interacting Weyl spinor and a massless, free, complex scalar with a Lagrangian of the form

$$\mathcal{L}^{(1)} = \partial_\mu \varphi^* \partial^\mu \varphi + \frac{i}{2}\left(\bar{\psi}\bar{\sigma}^\mu \partial_\mu \psi - \partial_\mu \bar{\psi}\bar{\sigma}^\mu \psi\right). \tag{13.29}$$

Using (13.23), (13.24), (13.25), and taking into account (13.26) we get

$$\begin{aligned}\delta_\xi \mathcal{L}^{(1)} &= (a + ic^*)\xi \partial^\mu \psi\, \partial_\mu \varphi^* + (a^* - ic)\bar{\xi}\partial_\mu \bar{\psi}\partial^\mu \varphi \\ &+ \frac{i}{2}\partial_\mu \left[c\left(2\bar{\psi}\bar{\sigma}^{\nu\mu}\bar{\xi}\partial_\nu \varphi + \bar{\psi}\bar{\xi}\partial^\mu \varphi\right) + c^*\left(2\xi\sigma^{\nu\mu}\psi\, \partial_\nu \varphi^* - \xi\psi\, \partial^\mu \varphi^*\right)\right]. \end{aligned} \tag{13.30}$$

Thus, the Lagrangian itself is invariant if, and only if, $a = c = 0$, but then the supersymmetry transformations are trivial. Fortunately, the presence of a total derivative in $\delta_\xi \mathcal{L}^{(1)}$ does not spoil the invariance of the action functional. Therefore we take $a = -ic^*$. Equation (13.26) then gives $|c|^2 = 2$, and choosing the phase factor conveniently we finally get

$$a = -\sqrt{2}, \qquad c = i\sqrt{2}.$$

The action

$$S = \int d^4x\, \mathcal{L}^{(1)}$$

with the Lagrangian given by (13.29) is thus invariant under SUSY transformations of the form

$$\delta_\xi \varphi(x) = -\sqrt{2}\,\xi\psi(x), \qquad \delta_\xi \psi(x) = i\sqrt{2}\,\sigma^\mu \bar{\xi}\,\partial_\mu \varphi(x), \tag{13.31}$$

and (13.31) ‘close on shell’, that is, the consistency conditions (13.22) are satisfied provided the spinor field obeys the equation (13.28).

Let us count the number of (functional) degrees of freedom of the fields involved. If we do not take into account the equations of motion (i.e. “off shell”), we have two (real) bosonic and four fermionic degrees of freedom (remember that φ and ψ_α are complex). If we now impose the equations of motion, then—since the e.o.m. for φ, the massless Klein-Gordon equation, is of second order and its solutions are determined by two arbitrary functions, say $\varphi(t,\vec{x})\big|_{t=0}$ and $\partial_t \varphi(t,\vec{x})\big|_{t=0}$, while to determine a solution of the first order Dirac equation one only needs to specify $\psi_\alpha(t,\vec{x})\big|_{t=0}$—we have two bosonic and two fermionic d.o.f. To match the degrees of freedom in the off shell case we have to introduce another complex scalar field, whose equation of motion is trivial,

$$F(x) = 0,$$

in order not to spoil the counting of the degrees of freedom on shell. The modified Lagrangian thus reads

$$\mathcal{L} = \partial_\mu \varphi^* \partial^\mu \varphi + \frac{i}{2}\left(\bar{\psi}\bar{\sigma}^\mu \partial_\mu \psi - \partial_\mu \bar{\psi}\bar{\sigma}^\mu \psi\right) + F^* F. \tag{13.32}$$

Consequently, $[F] = \text{cm}^{-2}$, and we can modify the transformation law of the spinor field to be

$$\delta_\xi \psi(x) = i\sqrt{2}\sigma^\mu \bar{\xi} \partial_\mu \varphi(x) + d\,\xi F(x) \tag{13.33}$$

with some constant d, and postulate (matching the dimensions of fields)

$$\delta_\xi F(x) = g\bar{\xi}\tilde{\sigma}^\mu \partial_\mu \psi(x), \tag{13.34}$$

where g is yet another constant. Thus, we have

$$\begin{aligned}\delta_\eta \delta_\xi(\theta\psi) &= -2i(\theta\sigma^\mu\bar{\xi})(\eta\partial_\mu\psi) + dg(\bar{\eta}\tilde{\sigma}^\mu\partial_\mu\psi)(\theta\xi) \\ &= 2i(\eta\sigma^\mu\bar{\xi})(\theta\partial_\mu\psi) - i(\eta\sigma_\nu\bar{\xi})(\theta\sigma^\nu\tilde{\sigma}^\mu\partial_\mu\psi) + \frac{1}{2}gd(\xi\sigma_\nu\bar{\eta})(\theta\sigma^\nu\tilde{\sigma}^\mu\partial_\mu\psi),\end{aligned}$$

where we used the Fierz identity (13.96) and (13.88). Therefore,

$$\begin{aligned}&[\delta_\eta, \delta_\xi](\theta\psi) \\ &\quad = 2i\left(\eta\sigma^\mu\bar{\xi} - \xi\sigma^\mu\bar{\eta}\right)(\theta\partial_\mu\psi) + \frac{1}{2}(gd + 2i)\left(\xi\sigma_\nu\bar{\eta} - \eta\sigma_\nu\bar{\xi}\right)(\theta\sigma^\nu\tilde{\sigma}^\mu\partial_\mu\psi).\end{aligned}$$

Consequently, if

$$gd = -2i, \tag{13.35}$$

then we get

$$[\delta_\eta, \delta_\xi]\psi_\alpha = 2i\left(\eta\sigma^\mu\bar{\xi} - \xi\sigma^\mu\bar{\eta}\right)\partial_\mu\psi_\alpha$$

without using the equations of motion. Similarly,

$$\delta_\eta\delta_\xi F = gc\,\bar{\xi}\tilde{\sigma}^\mu\sigma^\nu\bar{\eta}\,\partial_\mu\partial_\nu\varphi + 2i\,\eta\sigma^\mu\bar{\xi}\,\partial_\mu F = gc\,\bar{\xi}\bar{\eta}\,\partial_\mu\partial^\mu\varphi + 2i\,\eta\sigma^\mu\bar{\xi}\,\partial_\mu F,$$

where (13.35) was employed, so that, without using the equations of motion, we get the closure of the supersymmetry algebra on the auxiliary field F:

$$[\delta_\eta, \delta_\xi]F = 2i(\eta\sigma^\mu\bar{\xi} - \xi\sigma^\mu\bar{\eta})\partial_\mu F.$$

Finally,

$$\begin{aligned}\delta_\xi\mathcal{L} &= \delta_\xi\mathcal{L}^{(1)} + g\,\bar{\xi}\tilde{\sigma}^\mu\partial_\mu\psi\,F^* - g^*\xi\sigma^\mu\partial_\mu\bar{\psi}\,F \\ &\quad + \frac{id}{2}\left[\xi\sigma^\mu\partial_\mu\bar{\psi}\,F - \xi\sigma^\mu\bar{\psi}\,\partial_\mu F\right] + \frac{id^*}{2}\left[\bar{\xi}\tilde{\sigma}^\mu\partial_\mu\psi\,F^* - \bar{\xi}\tilde{\sigma}^\mu\psi\,\partial_\mu F^*\right] \\ &= \delta_\xi\mathcal{L}^{(1)} + (g^* - id)\xi\sigma^\mu\bar{\psi}\,\partial_\mu F - (g + id^*)\bar{\xi}\tilde{\sigma}^\mu\psi\,\partial_\mu F^* \\ &\quad + \partial_\mu\left[\left(g + \tfrac{id^*}{2}\right)\bar{\xi}\tilde{\sigma}^\mu\psi\,F^* - \left(g^* - \tfrac{id}{2}\right)\xi\sigma^\mu\bar{\psi}\,F\right].\end{aligned}$$

The action

$$\int d^4x\, \mathcal{L}$$

is thus invariant provided

$$g = -id^*. \tag{13.36}$$

The final form of the SUSY variations, keeping the action with the Lagrangian (13.32) invariant and satisfying the classical counterpart of the consistency condition (13.22), reads

$$\begin{aligned}
\delta_\xi \varphi(x) &= -\sqrt{2}\,\xi\psi(x), \\
\delta_\xi \psi(x) &= i\sqrt{2}\,\sigma^\mu\bar{\xi}\partial_\mu\varphi(x) - \sqrt{2}\,\xi F(x), \\
\delta_\xi F(x) &= i\sqrt{2}\,\bar{\xi}\tilde{\sigma}^\mu\partial_\mu\psi(x).
\end{aligned} \tag{13.37}$$

13.4 The Superspace

The field $\hat{\varphi}(x)$ can be viewed as an operator $\hat{\varphi}(0)$ translated from 0 to an arbitrary space-time point x,

$$\hat{\varphi}(x) = \mathrm{e}^{ix^\mu\hat{P}_\mu}\hat{\varphi}(0)\mathrm{e}^{-ix^\mu\hat{P}_\mu}. \tag{13.38}$$

The presence of the generators, $\hat{Q}$ and $\hat{\bar{Q}}$, allows us to define a more general object: the superfield

$$\hat{S}(x,\theta,\bar{\theta}) = \mathrm{e}^{i(x^\mu\hat{P}_\mu+\theta^\alpha\hat{Q}_\alpha+\hat{\bar{Q}}_{\dot{\alpha}}\bar{\theta}^{\dot{\alpha}})}\hat{\varphi}(0)\mathrm{e}^{-i(x^\mu\hat{P}_\mu+\theta^\alpha\hat{Q}_\alpha+\hat{\bar{Q}}_{\dot{\alpha}}\bar{\theta}^{\dot{\alpha}})} \tag{13.39}$$

(do not confuse it with the action functional), where θ^α and $\bar{\theta}^{\dot{\alpha}}$ are Grassmann variables. The set of variables $(x, \theta, \bar{\theta})$ defines a structure called the superspace.

Definition (13.39) allows us to find the form of a translation (parameterized by the commuting variable a and the Grassmann numbers ξ and $\bar{\xi}$) on the superspace. We have

$$\hat{S}(x',\theta',\bar{\theta}') = \mathrm{e}^{i(a^\mu\hat{P}_\mu+\xi^\alpha\hat{Q}_\alpha+\hat{\bar{Q}}_{\dot{\alpha}}\bar{\xi}^{\dot{\alpha}})}\,\hat{S}(x,\theta,\bar{\theta})\,\mathrm{e}^{-i(a^\mu\hat{P}_\mu+\xi^\alpha\hat{Q}_\alpha+\hat{\bar{Q}}_{\dot{\alpha}}\bar{\xi}^{\dot{\alpha}})}. \tag{13.40}$$

The Baker–Campbell–Hausdorff formula

$$e^A e^B = e^{A+B+\frac{1}{2}[A,B]}$$

with A and B such that $[A, [A, B]] = [B, [A, B]] = 0$, then gives

$$\exp\left(i(a^\mu \hat{P}_\mu + \xi^\alpha \hat{Q}_\alpha + \hat{\bar{Q}}_{\dot\alpha}\bar{\xi}^{\dot\alpha})\right) \exp\left(i(x^\mu \hat{P}_\mu + \theta^\alpha \hat{Q}_\alpha + \hat{\bar{Q}}_{\dot\alpha}\bar{\theta}^{\dot\alpha})\right) =$$
$$\exp\left(i\left[(x^\mu + a^\mu + i\xi^\alpha \sigma^\mu_{\alpha\dot\alpha}\bar{\theta}^{\dot\alpha} - i\theta^\alpha \sigma^\mu_{\alpha\dot\alpha}\bar{\xi}^{\dot\alpha})\hat{P}_\mu + (\theta^\alpha + \xi^\alpha)\hat{Q}_\alpha + \hat{\bar{Q}}_{\dot\alpha}(\bar{\theta}^{\dot\alpha} + \bar{\xi}^{\dot\alpha})\right]\right)$$

so that

$$\begin{aligned} x' &= x + a + i\xi\sigma\bar{\theta} - i\theta\sigma\bar{\xi}, \\ \theta' &= \theta + \xi, \\ \bar{\theta}' &= \bar{\theta} + \bar{\xi}. \end{aligned} \tag{13.41}$$

By definition, the superfield which transforms as a scalar with respect to the translation (13.41) satisfies

$$\hat{S}'(x', \theta', \bar{\theta}') = \hat{S}(x, \theta, \bar{\theta}). \tag{13.42}$$

From (13.40)

$$\hat{S}'(x, \theta, \bar{\theta}) = e^{-i(a^\mu \hat{P}_\mu + \xi^\alpha \hat{Q}_\alpha + \hat{\bar{Q}}_{\dot\alpha}\bar{\xi}^{\dot\alpha})}\, \hat{S}'(x', \theta', \bar{\theta}')\, e^{i(a^\mu \hat{P}_\mu + \xi^\alpha \hat{Q}_\alpha + \hat{\bar{Q}}_{\dot\alpha}\bar{\xi}^{\dot\alpha})},$$

and, using (13.42), we have, up to the terms linear in $a, \xi, \bar{\xi}$

$$\begin{aligned} \hat{S}'(x, \theta, \bar{\theta}) &= e^{-i(a^\mu \hat{P}_\mu + \xi^\alpha \hat{Q}_\alpha + \hat{\bar{Q}}_{\dot\alpha}\bar{\xi}^{\dot\alpha})}\, \hat{S}(x, \theta, \bar{\theta})\, e^{i(a^\mu \hat{P}_\mu + \xi^\alpha \hat{Q}_\alpha + \hat{\bar{Q}}_{\dot\alpha}\bar{\xi}^{\dot\alpha})} \\ &= \hat{S}(x, \theta, \bar{\theta}) - ia^\mu \left[\hat{P}_\mu, \hat{S}(x, \theta, \bar{\theta})\right] \\ &\quad -i\left[\xi^\alpha \hat{Q}_\alpha, \hat{S}(x, \theta, \bar{\theta})\right] - i\left[\hat{\bar{Q}}_{\dot\alpha}\bar{\xi}^{\dot\alpha}, \hat{S}(x, \theta, \bar{\theta})\right]. \end{aligned} \tag{13.43}$$

From (13.41) and (13.42) we also have

$$\begin{aligned} \hat{S}'(x, \theta, \bar{\theta}) &= \hat{S}(x - a - i\xi\sigma\bar{\theta} + i\theta\sigma\bar{\xi}, \theta - \xi, \bar{\theta} - \bar{\xi}) \\ &= \hat{S}(x, \theta, \bar{\theta}) - \left(a^\mu + i\xi\sigma^\mu\bar{\theta} - i\theta\sigma^\mu\bar{\xi}\right)\frac{\partial}{\partial x^\mu}\hat{S}(x, \theta, \bar{\theta}) - \xi^\alpha \frac{\partial}{\partial \theta^\alpha}\hat{S}(x, \theta, \bar{\theta}) \\ &\quad -\bar{\xi}^{\dot\alpha}\frac{\partial}{\partial \bar{\theta}^{\dot\alpha}}\hat{S}(x, \theta, \bar{\theta}). \end{aligned}$$

Comparing this result with (13.43) we get

$$i\left[\hat{P}^\mu, \hat{S}(x, \theta, \bar{\theta})\right] = \frac{\partial}{\partial x^\mu}\hat{S}(x, \theta, \bar{\theta}), \tag{13.44}$$

$$i\left[\xi^\alpha \hat{Q}_\alpha, \hat{S}(x, \theta, \bar{\theta})\right] = \xi^\alpha\left(\frac{\partial}{\partial \theta^\alpha} + i\sigma^\mu_{\alpha\dot\alpha}\bar{\theta}^{\dot\alpha}\frac{\partial}{\partial x^\mu}\right)\hat{S}(x, \theta, \bar{\theta}) \equiv \xi^\alpha Q_\alpha \hat{S}(x, \theta, \bar{\theta}),$$

$$i\left[\hat{\bar{Q}}_{\dot\alpha}\bar{\xi}^{\dot\alpha}, \hat{S}(x, \theta, \bar{\theta})\right] = \bar{\xi}^{\dot\alpha}\left(\frac{\partial}{\partial \bar{\theta}^{\dot\alpha}} + i\theta^\alpha \sigma^\mu_{\alpha\dot\alpha}\frac{\partial}{\partial x^\mu}\right)\hat{S}(x, \theta, \bar{\theta}) \equiv \bar{Q}_{\dot\alpha}\bar{\xi}^{\dot\alpha}\hat{S}(x, \theta, \bar{\theta}),$$

where

$$Q_\alpha = \frac{\partial}{\partial \theta^\alpha} + i\sigma^\mu_{\alpha\dot{\alpha}}\bar{\theta}^{\dot{\alpha}}\frac{\partial}{\partial x^\mu}, \qquad \bar{Q}_{\dot{\alpha}} = -\frac{\partial}{\partial \bar{\theta}^{\dot{\alpha}}} - i\theta^\alpha\sigma^\mu_{\alpha\dot{\alpha}}\frac{\partial}{\partial x^\mu} \tag{13.45}$$

are differential operators generating supersymmetric transformations on the space of scalar superfields.

We can expand the superfield $\hat{S}(x, \theta, \bar{\theta})$ in a power series in θ and $\bar{\theta}$. Since the square of a Grassmann variable is zero, this expansion terminates after a few terms

$$\begin{aligned}\hat{S}(x, \theta, \bar{\theta}) = {}& \hat{\varphi}(x) + \theta\hat{\psi}(x) + \bar{\theta}\hat{\bar{\psi}}(x) + \theta\theta\hat{F}(x) + \bar{\theta}\bar{\theta}\hat{G}(x) \\ & + (\theta\sigma^\mu\bar{\theta})\hat{v}_\mu(x) + \bar{\theta}\bar{\theta}\,\theta\hat{\lambda}(x) + \theta\theta\,\bar{\theta}\hat{\bar{\lambda}}(x) + \theta\theta\,\bar{\theta}\bar{\theta}\,\hat{D}(x).\end{aligned} \tag{13.46}$$

The SUSY variations of the component fields $\hat{\varphi}, \hat{\psi}, \ldots, \hat{D}$ can now be computed by comparing the formula

$$\begin{aligned}\delta_\xi S(x, \theta, \bar{\theta}) = {}& \delta_\xi\hat{\varphi}(x) + \theta\delta_\xi\hat{\psi}(x) + \bar{\theta}\delta_\xi\hat{\bar{\psi}}(x) + \theta\theta\delta_\xi\hat{F}(x) + \bar{\theta}\bar{\theta}\delta_\xi\hat{G}(x) \\ & + (\theta\sigma^\mu\bar{\theta})\delta_\xi\hat{v}_\mu(x) + \bar{\theta}\bar{\theta}\,\theta\delta_\xi\hat{\lambda}(x) + \theta\theta\,\bar{\theta}\delta_\xi\hat{\bar{\lambda}}(x) + \theta\theta\,\bar{\theta}\bar{\theta}\,\delta_\xi\hat{D}(x)\end{aligned}$$

with (see (13.43) and (13.44))

$$\begin{aligned}\delta_\xi\hat{S}(x, \theta, \bar{\theta}) &= -i\left[\xi^\alpha\hat{Q}_\alpha, \hat{S}(x, \theta, \bar{\theta})\right] - i\left[\hat{\bar{Q}}_{\dot{\alpha}}\bar{\xi}^{\dot{\alpha}}, \hat{S}(x, \theta, \bar{\theta})\right] \\ &= -(\xi^\alpha Q_\alpha + \bar{Q}_{\dot{\alpha}}\bar{\xi}^{\dot{\alpha}})\hat{S}(x, \theta, \bar{\theta}),\end{aligned} \tag{13.47}$$

where Q_α and $\bar{Q}_{\dot{\alpha}}$ in the last line are given by (13.45). In particular,

$$\delta_\xi\hat{D}(x) = \frac{i}{2}\frac{\partial}{\partial x^\mu}\left(\xi\sigma^\mu\hat{\bar{\lambda}}(x) - \hat{\lambda}(x)\sigma^\mu\bar{\xi}\right) \tag{13.48}$$

—the supersymmetric variation of the $\theta\theta\,\bar{\theta}\bar{\theta}$ term of the superfield (which is customarily named the D-term) is a total derivative.

The general superfield $\hat{S}$ contains four scalar, four Weyl, and one vector field. One can construct superfields with a smaller number of components, transforming into each other under the SUSY transformations. To this end, we may use a supersymmetric covariant derivative D_β, a first order (in θ^α) differential operator such that its action on a superfield does not change its transformation properties, i.e.

$$\begin{aligned}\delta_\xi D_\beta\hat{S}(x, \theta, \bar{\theta}) &\equiv -(\xi^\alpha Q_\alpha + \bar{Q}_{\dot{\alpha}}\bar{\xi}^{\dot{\alpha}})D_\beta\hat{S}(x, \theta, \bar{\theta}) \\ &= -D_\beta(\xi^\alpha Q_\alpha + \bar{Q}_{\dot{\alpha}}\bar{\xi}^{\dot{\alpha}})\hat{S}(x, \theta, \bar{\theta}) \equiv D_\beta\delta_\xi\hat{S}(x, \theta, \bar{\theta}),\end{aligned}$$

which is equivalent to requiring

$$\{Q_\alpha, D_\beta\} = \{\bar{Q}_{\dot{\alpha}}, D_\beta\} = 0.$$

An operator satisfying these conditions is of the form

$$D_\beta = \frac{\partial}{\partial\theta^\beta} - i\sigma^\mu_{\beta\dot{\beta}}\bar{\theta}^{\dot{\beta}}\frac{\partial}{\partial x^\mu}. \tag{13.49}$$

Similarly, a first order differential operator in $\bar{\theta}^{\dot{\alpha}}$, anticommuting with Q_α and $\bar{Q}_{\dot{\alpha}}$ has the form

$$\bar{D}_{\dot{\beta}} = \frac{\partial}{\partial\bar{\theta}^{\dot{\beta}}} - i\theta^\beta\sigma^\mu_{\beta\dot{\beta}}\frac{\partial}{\partial x^\mu}. \tag{13.50}$$

The chiral superfield (an object important enough to deserve a separate 'name' $\hat{\Phi}$) is a superfield satisfying

$$\bar{D}_{\dot{\beta}}\hat{\Phi}(x,\theta,\bar{\theta}) = 0.$$

Let

$$y^\mu = x^\mu - i\theta^\alpha\sigma^\mu_{\alpha\dot{\alpha}}\bar{\theta}^{\dot{\alpha}}.$$

We have

$$\bar{D}_{\dot{\beta}}y^\mu = -i\theta^\beta\sigma^\nu_{\beta\dot{\beta}}\frac{\partial}{\partial x^\nu}x^\mu - i\frac{\partial}{\partial\bar{\theta}^{\dot{\beta}}}\theta^\alpha\sigma^\mu_{\alpha\dot{\alpha}}\bar{\theta}^{\dot{\alpha}} = -i\theta^\beta\sigma^\nu_{\beta\dot{\beta}}\delta^\mu_\nu + i\theta^\alpha\sigma^\mu_{\alpha\dot{\alpha}}\delta^{\dot{\alpha}}_{\dot{\beta}} = 0$$

and

$$\bar{D}_{\dot{\beta}}\theta^\alpha = 0, \qquad \bar{D}_{\dot{\beta}}\bar{\theta}^{\dot{\alpha}} = \delta^{\dot{\alpha}}_{\dot{\beta}}.$$

Changing variables and defining

$$\hat{\Phi}(x,\theta,\bar{\theta}) = \hat{\hat{\Phi}}(y,\theta,\bar{\theta}),$$

we get

$$\begin{aligned}\bar{D}_{\dot{\beta}}\hat{\Phi}(x,\theta,\bar{\theta}) &= \bar{D}_{\dot{\beta}}\hat{\hat{\Phi}}(y,\theta,\bar{\theta}) \\ &= \left(\bar{D}_{\dot{\beta}}y^\mu\frac{\partial}{\partial y^\mu} + \bar{D}_{\dot{\beta}}\theta^\alpha\frac{\partial}{\partial\theta^\alpha} + \bar{D}_{\dot{\beta}}\bar{\theta}^{\dot{\alpha}}\frac{\partial}{\partial\bar{\theta}^{\dot{\alpha}}}\right)\hat{\hat{\Phi}}(y,\theta,\bar{\theta}) \\ &= \frac{\partial}{\partial\bar{\theta}^{\dot{\beta}}}\hat{\hat{\Phi}}(y,\theta,\bar{\theta}).\end{aligned}$$

Consequently, the chiral superfield is an arbitrary function of θ and y,

$$\hat{\Phi}(x,\theta,\bar{\theta}) = \hat{\hat{\Phi}}(y,\theta).$$

Writing it as a power series in θ we get

$$\hat{\hat{\Phi}}(y,\theta) = \hat{\varphi}(y) + \sqrt{2}\,\theta^\alpha \hat{\psi}_\alpha(y) + \theta^\alpha\theta_\alpha\,\hat{F}(y), \tag{13.51}$$

and further

$$\begin{aligned}
\hat{\varphi}(x - i\theta\sigma^\mu\bar{\theta}) &= \hat{\varphi}(x) - i(\theta\sigma^\mu\bar{\theta})\partial_\mu\hat{\varphi}(x) - \frac{1}{2}(\theta\sigma^\mu\bar{\theta})(\theta\sigma^\nu\bar{\theta})\partial_\mu\partial_\nu\hat{\varphi}(x)\\
&= \hat{\varphi}(x) - i(\theta\sigma^\mu\bar{\theta})\partial_\mu\hat{\varphi}(x) - \frac{1}{4}\,\theta\theta\,\bar{\theta}\bar{\theta}\partial_\mu\partial^\mu\hat{\varphi}(x),\\
\sqrt{2}\theta^\alpha\hat{\psi}_\alpha(x - i\theta\sigma^\mu\bar{\theta}) &= \sqrt{2}\theta^\alpha\hat{\psi}_\alpha(x) - i\sqrt{2}\theta^\alpha(\theta\sigma^\mu\bar{\theta})\partial_\mu\hat{\psi}_\alpha(x)\\
&= \sqrt{2}\theta^\alpha\hat{\psi}_\alpha(x) + \frac{i}{\sqrt{2}}\theta\theta\,\partial_\mu\hat{\psi}(x)\tilde{\sigma}^\mu\bar{\theta},\\
\theta\theta\hat{F}(x - i\theta\sigma^\mu\bar{\theta}) &= \theta\theta\hat{F}(x).
\end{aligned} \tag{13.52}$$

Using (13.47) with $\hat{\Phi}$ substituted for $\hat{S}$, we see that the SUSY transformations of the fields $\hat{\varphi}$, $\hat{\psi}$ and $\hat{F}$, obtained in this section have forms coinciding with (13.37).

13.5 The Wess–Zumino Model

Let us define the classical, chiral superfield

$$\Phi(x,\theta,\bar{\theta}) = \varphi(y) + \sqrt{2}\,\theta^\alpha\psi_\alpha(y) + \theta^\alpha\theta_\alpha\,F(y), \tag{13.53}$$

where φ and F are classical, complex scalar fields and ψ is a classical (Grassmann type) Weyl field. As in the previous section $y^\mu = x^\mu - i\theta\sigma^\mu\theta$ is a formal argument, and the fields $\varphi(x)$, $\psi(x)$ and $F(x)$ may be obtained as in (13.52).

Since the covariant derivative satisfies the Leibniz rule, the product of chiral superfields is again a chiral superfield,

$$\bar{D}_{\dot{\alpha}}\Phi_i = 0,\ i = 1,2, \quad \Rightarrow \quad \bar{D}_{\dot{\alpha}}(\Phi_1\Phi_2) = 0. \tag{13.54}$$

The SUSY variation of the $\theta\theta$ coefficient (the F-term) in the expansion of a chiral field is a total derivative (see (13.52), (13.51) and (13.37)). We can thus construct a SUSY invariant expression by integrating over the space-time expressions of the form

$$a_1\Phi\big|_{\theta\theta} + a_2\Phi\Phi\big|_{\theta\theta} + a_3\Phi\Phi\Phi\big|_{\theta\theta} + \ldots$$

with $|_{\theta\theta}$ denoting the $\theta\theta$ component and with constant a_i. As we shall see in a moment, this gives the mass and the interaction terms in the action functional; to obtain the kinetic term more work is needed.

It follows from definitions (13.49), (13.50), that the covariant derivatives satisfy an algebra of the form

$$\{D_\alpha, D_\beta\} = \{\bar{D}_{\dot\alpha}, \bar{D}_{\dot\beta}\} = 0,$$
$$\{D_\alpha, \bar{D}_{\dot\beta}\} = -2i\sigma^\mu_{\alpha\dot\beta}\frac{\partial}{\partial x^\mu}, \tag{13.55}$$

and consequently $\bar{D}_{\dot\alpha}(\bar{D}\bar{D}) = 0$. We may now use an antichiral field, i.e.,

$$\bar{\Phi}(x,\theta,\bar\theta) = \varphi^*(x + i\theta\sigma\bar\theta) + \sqrt{2}\,\bar\theta\bar\psi(x + i\theta\sigma\bar\theta) + \bar\theta\bar\theta\, F^*(x + i\theta\sigma\bar\theta), \tag{13.56}$$

satisfying $D_\alpha\bar\Phi = 0$, to construct a chiral field

$$\Phi_K = \frac{1}{4}\Phi(\bar{D}\bar{D})\bar\Phi \tag{13.57}$$

which will finally yield the kinetic term in a SUSY invariant action functional.

It is customary to replace the $|_{\theta\theta}$ operation with an integration over the Grassmann variables θ^α. Let us define

$$\int d^2\theta = \int d\theta^1 d\theta^2, \quad \int d^2\bar\theta = \int d\bar\theta^{\dot 2} d\bar\theta^{\dot 1},$$

hence

$$\int d^2\theta\,\theta\theta = \int d^2\bar\theta\,\bar\theta\bar\theta = 2. \tag{13.58}$$

From (13.52) we see that

$$\int d^4x\,\Phi(x,\theta,\bar\theta)\big|_{\theta\theta} = \frac{1}{2}\int d^4x d^2\theta\,\Phi(x,\theta,\bar\theta).$$

Let us notice that the difference between $\bar{D}_{\dot\alpha}$ and $\partial/\partial\bar\theta^{\dot\alpha}$ is an ordinary space-time derivative multiplied by a coefficient which does not depend on $\bar\theta$. Since the rules of Grassmann integration make it equivalent to the differentiation,

$$\int d\eta\, f(\eta) = \frac{\partial}{\partial\eta} f(\eta),$$

we have

$$\int d^4x d^2\theta\,\Phi(\bar{D}\bar{D})\bar\Phi = \int d^4x d^2\theta\,\bar{D}\bar{D}(\Phi\bar\Phi) = \int d^4x d^2\theta\,\frac{\partial}{\partial\bar\theta^{\dot\alpha}}\frac{\partial}{\partial\bar\theta_{\dot\alpha}}\Phi\bar\Phi$$
$$= \frac{1}{2}\int d^4x d^2\theta d^2\bar\theta\,\Phi\bar\Phi.$$

The integral appearing on the l.h.s. of this identity is SUSY invariant, thanks to the chirality of $\Phi(\bar{D}\bar{D})\bar{\Phi}$. On the other hand, even if $\Phi\bar{\Phi}$ is no longer a chiral superfield, the SUSY invariance of the integral on the r.h.s. can be inferred from (13.48).

Before we finally construct an action functional containing only the fields which build up the chiral field Φ and its conjugate $\bar{\Phi}$, it is useful to perform a dimensional analysis. We have already seen that θ and $\bar{\theta}$ carry the dimensions $\mathrm{cm}^{\frac{1}{2}}$, φ and ψ have dimensions cm^{-1} and $\mathrm{cm}^{-\frac{3}{2}}$, respectively, so that the whole superfield Φ has the dimension cm^{-1}. From (13.58) it then follows that the measures $d^2\theta$ and $d^2\bar{\theta}$ have the dimensions cm^{-1} each, the dimension of the measure $d^4xd^2\theta$ is thus cm^3 and the dimension of $d^4xd^2\theta d^2\bar{\theta}$ is equal to cm^2.

In order to construct a (perturbatively) renormalizable theory, we have to build a dimensionless action which contains no coupling constants with positive length dimensions. The only possibility (the so-called Wess–Zumino action) with the super-kinetic term quadratic in the fields is

$$
\begin{aligned}
S_{WZ} &= \frac{1}{2}\int d^4xd^2\theta\left(\frac{1}{2}\Phi\bar{D}^2\bar{\Phi} + \frac{1}{2}m\Phi^2 + \frac{1}{3}g\Phi^3\right) + \text{c.c.} \\
&= \frac{1}{8}\int d^4xd^2\theta d^2\bar{\theta}\ \Phi\bar{\Phi} + \frac{1}{2}\int d^4xd^2\theta\left(\frac{1}{2}m\Phi^2 + \frac{1}{3}g\Phi^3\right) + \text{c.c.} \qquad (13.59)
\end{aligned}
$$

with the coupling constants: m having dimension cm^{-1} (or, equivalently, mass dimension $+1$) and dimensionless g. In terms of the component, scalar fields $\varphi(x)$, $F(x)$ and $\psi_\alpha(x)$ we have

$$
S_{WZ} = \int d^4x\,\mathcal{L}_{WZ},
$$

where

$$
\begin{aligned}
\mathcal{L}_{WZ} = \partial_\mu\varphi^*\partial^\mu\varphi + i\bar{\psi}\tilde{\sigma}^\mu\partial_\mu\psi + F^*F + m(F\varphi + F^*\varphi^*) - \tfrac{m}{2}(\psi\psi + \bar{\psi}\bar{\psi}) \\
+ g\left(F\varphi\varphi + F^*\varphi^*\varphi^* - \varphi\psi\psi - \varphi^*\bar{\psi}\bar{\psi}\right) + (4-\text{div}). \qquad (13.60)
\end{aligned}
$$

It is slightly more convenient to rewrite (13.60) in terms of the Majorana spinor

$$
\Psi_M = \begin{pmatrix}\psi_\alpha \\ \bar{\psi}^{\dot{\alpha}}\end{pmatrix}.
$$

Since in the spinor representation

$$
\gamma_5 = \begin{pmatrix} I_2 & 0 \\ 0 & -I_2\end{pmatrix},
$$

we have

$$
\psi\psi = \frac{1}{2}\bar{\Psi}_M(1+\gamma_5)\Psi_M, \qquad \bar{\psi}\bar{\psi} = \frac{1}{2}\bar{\Psi}_M(1-\gamma_5)\Psi_M,
$$

and (13.60) takes the form

$$\mathcal{L}_{WZ} = \partial_\mu \varphi^* \partial^\mu \varphi + m(F\varphi + F^*\varphi^*) + \frac{1}{2}\bar{\Psi}_M (i\partial\!\!\!/ - m)\Psi_M + F^*F$$
$$+ g\left(F\varphi\varphi + F^*\varphi^*\varphi^*\right) - \frac{g}{2}\varphi\bar{\Psi}_M(1+\gamma_5)\Psi_M - \frac{g}{2}\varphi^*\bar{\Psi}_M(1-\gamma_5)\Psi_M. \qquad (13.61)$$

The equations of motion for the F and F^* fields are purely algebraic

$$\begin{aligned} F &= -m\varphi^* - g(\varphi^*)^2, \\ F^* &= -m\varphi - g\varphi^2. \end{aligned} \qquad (13.62)$$

Inserting (13.62) back into (13.61) we arrive at the Wess–Zumino model Lagrangian expressed entirely in terms of the fields $\varphi(x)$, $\Psi_M(x)$ and their conjugates,

$$\begin{aligned} \mathcal{L}_{WZ} &= \partial_\mu \varphi^* \partial^\mu \varphi - m^2 \varphi^* \varphi + \frac{1}{2}\bar{\Psi}_M (i\partial\!\!\!/ - m)\Psi_M \\ &- \frac{g}{2}\varphi\bar{\Psi}_M(1+\gamma_5)\Psi_M - \frac{g}{2}\varphi^*\bar{\Psi}_M(1-\gamma_5)\Psi_M \\ &- mg\left(\varphi^*\varphi^2 + (\varphi^*)^2\varphi\right) - g^2(\varphi^*)^2\varphi^2. \end{aligned} \qquad (13.63)$$

The price one pays for this simplification is—as we already know—that the obtained action is SUSY invariant only on shell, i.e. only when the equations of motion for the spinor field are taken into account.

Let us now sketch the Feynman rules of the theory which is obtained upon quantization of the Wess–Zumino model. The first line of (13.63) defines the free action, and consequently in the interaction picture the operators $\hat{\varphi}(x)$ and $\hat{\Psi}_M(x)$, together with their Hermitian conjugates, satisfy the Klein–Gordon and Dirac equations

$$\begin{aligned} \partial_\mu \partial^\mu \hat{\varphi}(x) + m^2 \hat{\varphi}(x) &= 0, \\ (i\partial\!\!\!/ - m)\hat{\Psi}_M(x) &= 0. \end{aligned}$$

One can then show, analogously as in Chap. 6 for the real scalar field, that the complex scalar field operator $\hat{\varphi}(x)$ can be represented as

$$\hat{\varphi}(x) = \int \frac{d^3k}{\sqrt{2(2\pi)^3\omega(k)}} \left(\mathrm{e}^{-ikx}\hat{a}(\vec{k}) + \mathrm{e}^{ikx}\hat{b}^\dagger(\vec{k})\right),$$

where $k^0 = \omega(\vec{k}) = \sqrt{\vec{k}^2 + m^2}$ and

$$[\hat{a}(k), \hat{a}^\dagger(\vec{q})] = [\hat{b}(k), \hat{b}^\dagger(\vec{q})] = \delta(\vec{k} - \vec{q}), \qquad [\hat{a}(k), \hat{b}(\vec{q})] = [\hat{a}(k), \hat{b}^\dagger(\vec{q})] = 0.$$

This gives the φ field propagator

$$\langle 0_I | T\left(\hat{\varphi}(x)\hat{\varphi}^\dagger(y)\right) |0_I\rangle = \Delta_F(x-y) = \int \frac{d^4k}{(2\pi)^4} \mathrm{e}^{-ik(x-y)} \frac{i}{k^2 - m^2 + i0_+}. \tag{13.64}$$

The operator $\hat{\Psi}_M(x)$ can be constructed with the help of the formulae (5.68) and (5.69) by 'promoting' the generating elements of the Grassmann algebra $c_\epsilon^\lambda(\vec{k})$ to be operators, acting in the appropriate Hilbert space and satisfying the usual anticommutation relations. Calculating the propagator for the Majorana field we get

$$\langle 0_I | T\left(\hat{\Psi}_M(x)\hat{\bar{\Psi}}_M(y)\right) |0_I\rangle = S_F(x-y) = \int \frac{d^4k}{(2\pi)^4} \mathrm{e}^{-ik(x-y)} \frac{i(\not{k} + m)}{k^2 - m^2 + i0_+}. \tag{13.65}$$

Graphically, we shall denote the scalar field propagator with a wavy line, directed from the $\varphi^\dagger$ to the φ field and the Majorana field propagator by a solid line. In the latter case the arrow just points in the direction in which momentum flows (Fig. 13.1).

The second and the third line of (13.63) allow us to read off the interaction vertices. Up to the factor $(2\pi)^4$ and the four-momentum conservation Dirac delta, they are specified in Fig. 13.2.

Let us end this section by presenting two examples of a phenomenon which makes supersymmetric models especially interesting—the cancellation of some divergences appearing in the Green's functions due to the opposite signs of contributions from

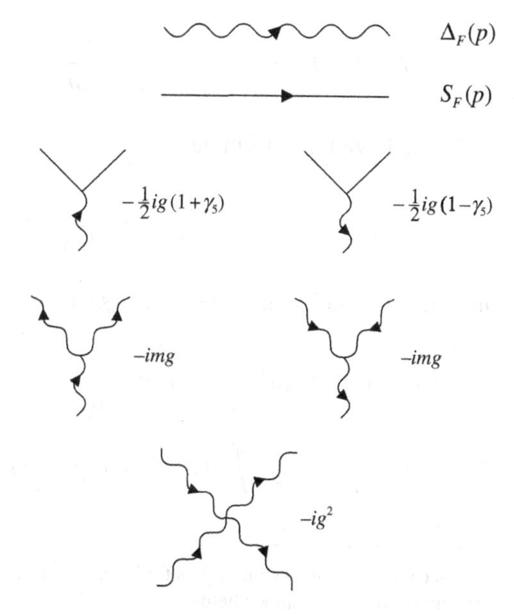

Fig. 13.1 Propagators in the Wess–Zumino model

Fig. 13.2 Vertices in the Wess–Zumino model

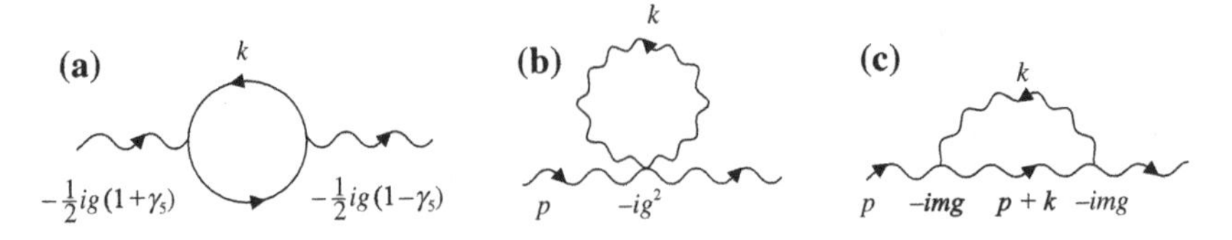

Fig. 13.3 The one-loop graphs contributing to the scalar field self-energy

the bosonic and fermionic fields. Our first example is the one-loop correction to the scalar field self-energy, given by the contributions from the graphs in the Fig. 13.3. The superficial degree of divergence of the first two graphs is equal (in the four-dimensional space-time) to two—if we try to calculate the integrals by restricting the modulus of the integration momentum by a cut-off Λ, they diverge like Λ^2. In the dimensional regularization the contribution from the first graph is[2]

$$
\begin{aligned}
I_a(p) &= 2(-1)\frac{g^2}{4}\int d^Dk\,\frac{\mathrm{Tr}(1-\gamma_5)(\not k+m)(1+\gamma_5)(\not k+\not p+m)}{(k^2-m^2+i0_+)\left((k+p)^2-m^2+i0_+\right)}\\
&= -4g^2\int d^Dk\,\frac{k(k+p)}{(k^2-m^2+i0_+)\left((k+p)^2-m^2+i0_+\right)},
\end{aligned}
$$

while the second graph yields

$$
I_b = 4g^2\int d^Dk\,\frac{1}{k^2-m^2+i0_+}.
$$

Consequently,

$$
I_a(p)+I_b = 4g^2\int d^Dk\,\frac{p(k+p)-m^2}{(k^2-m^2+i0_+)\left((k+p)^2-m^2+i0_+\right)}.
$$

Using Feynman's formula

$$
\frac{1}{ab} = \int_0^1 \frac{dx}{[ax+b(1-x)]^2}
$$

in order to combine the denominators we get

$$
\begin{aligned}
I_a(p)+I_b &= 4g^2\int_0^1 dx\int d^Dk\,\frac{p(k+p)-m^2}{\left((k+xp)^2+x(1-x)p^2-m^2\right)^2}\\
&= 4g^2\int_0^1 dx\,\left((1-x)p^2-m^2\right)\int\frac{d^Dq}{\left[q^2+x(1-x)p^2-m^2\right]^2},
\end{aligned}
$$

[2] Notice the additional symmetry factor 2 which appears for the Majorana spinors and would not be present for the Dirac spinor fields.

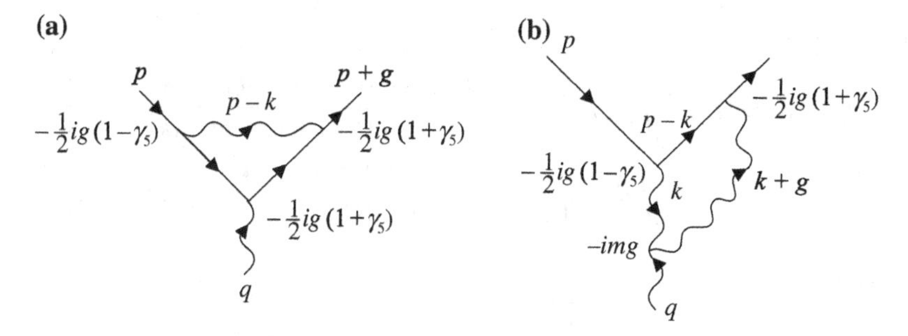

Fig. 13.4 The one-loop graphs contributing to the fermion-scalar interaction vertex

where in the last line we have changed the integration variable $k \to q = k - xp$ and neglected the term which is odd in q. The resulting integral diverges for large $|q| = \Lambda$ only as $\log \Lambda$—the quadratic divergences of Fig. 13.3a, b have canceled each other.

The contribution to the scalar field self-energy depicted by the graph (c) in Fig. 13.3 is already only logarithmically divergent and reads

$$
\begin{aligned}
I_c(p) &= 4g^2m^2 \int d^Dk \, \frac{1}{(k^2 - m^2 + i0_+)\left((k+p)^2 - m^2 + i0_+\right)} \\
&= 4g^2m^2 \int_0^1 dx \, (1-x)p^2 \int \frac{d^Dq}{\left[q^2 + x(1-x)p^2 - m^2\right]^2}.
\end{aligned}
$$

The quantity

$$
I_a(p) + I_b + I_c(p) = 4g^2p^2 \int_0^1 dx(1-x) \int \frac{d^Dq}{\left[q^2 + x(1-x)p^2 - m^2\right]^2}
$$

calculated at zero external momentum, $p = 0$, gives the one-loop correction to the scalar field mass. From this formula follows the remarkable result, that in the Wess–Zumino model this correction actually vanishes.

A similar mechanism in the minimal supersymmetric extension of the Standard Model (MSSM) allows one to solve the so called hierarchy problem: it ‘protects’ the mass of the Higgs particle from receiving large, physically unacceptable perturbative corrections.

Our second example is the one-loop correction to the $\varphi\bar{\Psi}_M\Psi_M$ interaction vertex. Two of the four contributions are depicted in Fig. 13.4 (we encourage the reader to draw and analyze the remaining two contributions).

We have

$$
\begin{aligned}
I_a(p,q) &= g^3 \int d^Dk \, \frac{(1+\gamma_5)(\not{k} + \not{q} + m)(1+\gamma_5)(\not{k} + m)(1-\gamma_5)}{\left((k-p)^2 - m^2\right)\left((k+q)^2 - m^2\right)\left(k^2 - m^2\right)} \\
&= 4mg^3 \int d^Dk \, \frac{(1+\gamma_5)\not{k}}{\left((k-p)^2 - m^2\right)\left((k+q)^2 - m^2\right)\left(k^2 - m^2\right)},
\end{aligned}
$$

and

$$
\begin{aligned}
I_b(p,q) &= 2mg^3 \int d^Dk \, \frac{(1+\gamma_5)(\not{p}-\not{k}+m)(1-\gamma_5)}{\left((k-p)^2-m^2\right)\left((k+q)^2-m^2\right)\left(k^2-m^2\right)} \\
&= 4mg^3 \int d^Dk \, \frac{(1+\gamma_5)(\not{p}-\not{k})}{\left((k-p)^2-m^2\right)\left((k+q)^2-m^2\right)\left(k^2-m^2\right)}.
\end{aligned}
$$

The sum

$$
I_a(p,q)+I_b(p,q) = 4mg^3 \int d^Dk \, \frac{(1+\gamma_5)\not{p}}{\left((k-p)^2-m^2\right)\left((k+q)^2-m^2\right)\left(k^2-m^2\right)}
$$

is finite for $D=4$.

13.6 More Advanced Topics

Supersymmetric models with gauge fields

The Wess–Zumino model, even if providing illustration of important features of supersymmetric field models, does not contain gauge fields. Such a field appears as a component of the real superfield $V(x,\theta,\bar{\theta})$,

$$
V(x,\theta,\bar{\theta}) = \left(V(x,\theta,\bar{\theta})\right)^*, \tag{13.66}
$$

which contains, among others, the term

$$
\theta\sigma^\mu\bar{\theta}\, v_\mu(x)
$$

with $v_\mu(x)$ being a real four-vector. Under the transformation

$$
V \to V + \Phi + \Phi^*, \tag{13.67}
$$

where Φ is a chiral superfield, this component transforms as $v_\mu \to v_\mu + \partial_\mu(2Im\phi)$. Theory which contains the real superfield and is invariant under transformation (13.67) is therefore a supersymmetric version of an Abelian gauge theory.

A specific choice of the chiral superfield in (13.67) (the Wess–Zumino gauge) allows to reduce the real superfield to the form

$$
V_{\mathrm{WZ}} = \theta\sigma^\mu\bar{\theta}\, v_\mu(x) + i\theta\theta\bar{\theta}\bar{\lambda}(x) - i\bar{\theta}\bar{\theta}\theta\lambda(x) + \frac{1}{2}\theta\theta\bar{\theta}\bar{\theta}D(x).
$$

Physical components of "vector supermultiplet" are thus: the real scalar D (which turns out to be an auxiliary field and can be eliminated from the action by solving algebraic equations of motion), the Weyl fermion λ and the real vector field v_μ.

To construct a kinetic term for the vector field v_μ we need to act on V with some covariant derivatives. A suitable choice of the superfield which yields such a kinetic term turns out to be

$$W_\alpha = -\frac{1}{4}\bar{D}\bar{D}D_\alpha V.$$

Using the anticommutation relations satisfied by covariant derivatives and definition of (anti)chiral superfield it is immediate to check that W_α is invariant under the generalized gauge transformation (13.67). To calculate its components one can therefore work in any gauge, in particular in the Wess–Zumino one. Moreover, since W_α is a chiral field, the SUSY invariant Lagrangian can be constructed as a $d^2\theta$ integral of some of its power. The reader is encouraged to check that

$$\int d^2\theta\, W^\alpha W_\alpha = -\frac{1}{2} f_{\mu\nu} f^{\mu\nu} + 2i\lambda\sigma^\mu\partial_\mu\bar{\lambda} + \frac{i}{4}\epsilon^{\mu\nu\rho\sigma} f_{\mu\nu} f_{\rho\sigma},$$

where $f_{\mu\nu} = \partial_\mu v_\nu - \partial_\nu v_\mu$. The last, imaginary term has the form of four-dimensional divergence,

$$\frac{1}{2}\epsilon^{\mu\nu\rho\sigma} f_{\mu\nu} f_{\rho\sigma} = \partial_\mu \left(\epsilon^{\mu\nu\rho\sigma} v_\nu f_{\rho\sigma}\right),$$

but it contains also derivatives of the gauge field. Therefore, in general it should not be omitted.

In order to construct a supersymmetric version of non-Abelian gauge theory we replace a single real superfield V with a matrix superfield

$$V(x, \theta, \bar{\theta}) = V_a(x, \theta, \bar{\theta}) T^a$$

where T^a are generators of the desired gauge group in the adjoint representation. The non-Abelian analog of the W_α superfield reads

$$W_\alpha = -\frac{1}{4}\bar{D}\bar{D}\left(\mathrm{e}^{-2gV} D_\alpha\, \mathrm{e}^{2gV}\right)$$

where g is the gauge coupling constant. Under non-Abelian gauge transformation

$$\mathrm{e}^{2gV} \to \mathrm{e}^{i\Lambda^\dagger}\mathrm{e}^{2gV}\mathrm{e}^{-i\Lambda}$$

with a chiral superfield $\Lambda = \Lambda_a T^a$, W_α transforms covariantly, that is

$$W_\alpha \to \mathrm{e}^{i\Lambda}\, W_\alpha\, \mathrm{e}^{-i\Lambda}.$$

If we now combine the coupling constant g with a real parameter Θ (the so called "theta angle") into the complex coupling constant

$$\tau = \frac{\Theta}{2\pi} + \frac{4\pi i}{g^2},$$

then the Lagrangian for the non-Abelian gauge field can be written as

$$\mathcal{L}_{\text{gauge}} = \frac{1}{32\pi} Im \left(\tau \int d^2\theta \, \text{tr} \, W^\alpha W_\alpha \right). \tag{13.68}$$

Since W_α is a chiral field, this Lagrangian is manifestly invariant under supersymmetry transformations. The r.h.s. of (13.68), after expanding in components and evaluating the Grassmannian integral, reproduces the Yang-Mills Lagragian supplemented with the Lagrangian of the spinor λ_α covariantly coupled to the gauge field v_μ, as well as the term

$$\frac{\Theta g^2}{32\pi} \epsilon^{\mu\nu\rho\sigma} \, \text{tr} \, F_{\mu\nu} F_{\rho\sigma}.$$

As in the Abelian case this last term can be written as a four-divergence.

The non-linear sigma model

When we do not require theory to be renormalizable (e.g. we are discussing an effective theory which emerges from a quantum renormalizable one as a certain approximation to it) the potential containing the chiral superfield need not to be at most cubic polynomial. The most general, real, supersymmetric Lagrangian for chiral superfields Φ^a and their conjugate anti-chiral fields $\bar{\Phi}^{\bar{a}}$ which contains at most first order space-time derivatives has then the form

$$\mathcal{L}_\sigma = \int d^2\theta d^2\bar{\theta} \, K(\Phi^a, \bar{\Phi}^{\bar{a}}) + \int d^2\theta \, W(\Phi^a) + \int d^2\bar{\theta} \, \overline{W}(\bar{\Phi}^{\bar{a}}). \tag{13.69}$$

Here $K(\Phi^a, \bar{\Phi}^{\bar{a}})$ with $a, \bar{a} = 1, \ldots, n$ must be a real superfield which is the case provided that the function $K(z^a, \bar{z}^{\bar{a}})$ with arguments being complex numbers satisfies

$$\bar{K}(z^a, \bar{z}^{\bar{a}}) = K(\bar{z}^{\bar{a}}, z^a).$$

After expanding $K(\Phi^a, \bar{\Phi}^{\bar{a}})$ in powers of θ and $\bar{\theta}$ and dropping terms being four-dimensional divergences we obtain among others the term

$$\sum_{a,\bar{a}} \left. \frac{\partial^2 K}{\partial \phi^a \partial \bar{\phi}^{\bar{a}}} \right|_{\theta,\bar{\theta}=0} \partial_\mu \phi^a \partial^\mu \bar{\phi}^{\bar{a}}. \tag{13.70}$$

Notice that if we view $\left(\phi^a, \bar{\phi}^{\bar{a}}\right)$ as complex coordinates on some "internal space", then the line element on such a space would have the form

$$ds^2 = \sum_{a,\bar{a}} G_{a\bar{a}} d\phi^a d\bar{\phi}^{\bar{a}}.$$

Allowing for a dependence of $(\phi^a, \bar{\phi}^{\bar{a}})$ on space-time variable x^μ we get

$$ds^2 = \sum_{a,\bar{a}} G_{a\bar{a}} \partial_\mu \phi^a \partial_\nu \bar{\phi}^{\bar{a}} dx^\mu dx^\nu.$$

Comparing the last expression with (13.70) we see that it is legitimate to consider $G_{a\bar{a}} = \left.\frac{\partial^2 K}{\partial\phi^a \partial\bar{\phi}^{\bar{a}}}\right|_{\theta,\bar{\theta}=0}$ as a metric on this space. This is a special kind of a metric, since the metric tensor $G_{a\bar{a}}$ is obtained by differentiating a single function K. It is called Kähler metric, and the function $K(\phi^a, \bar{\phi}^{\bar{a}})$ is addressed to as a Kähler potential. The role of $G_{a\bar{a}}$ as a metric is even more transparent when we explicitly write down all the terms of the Lagrangian (13.69) which contain derivatives. They can be presented in the form:

$$\mathcal{L}_\sigma^{(1)} = G_{a\bar{a}} \left(\partial_\mu \phi^a \partial^\mu \bar{\phi}^{\bar{a}} + \tfrac{i}{2} D_\mu \psi^a \sigma^\mu \bar{\psi}^{\bar{a}} - \tfrac{i}{2} \psi^a \sigma^\mu D_\mu \bar{\psi}^{\bar{a}} \right),$$

where

$$D_\mu \psi^a = \partial_\mu \psi^a + \Gamma^a_{bc} \partial_\mu \phi^b \psi^c$$

is a covariant derivative with Γ^a_{bc} being the Christoffel symbol of the metric $G_{a\bar{a}}$.

Extended supersymmetry

New, interesting theories appear when we add to the algebra discussed in Sect. 13.1 a second pair of fermionic generators $\hat{Q}_\alpha$, $\hat{\bar{Q}}^{\dot{\beta}}$, thus constructing an $N = 2$ extended supersymmetry. Commutation relations of the second pair of supersymmetry generators with the operators of momentum and angular momentum are of the same form as the commutators of the first pair, while for the anticommutators of $\hat{Q}^I_\alpha$, $\hat{\bar{Q}}^{J\dot{\beta}}$, $I, J = 1, 2$ we postulate

$$\begin{aligned}
\{\hat{Q}^I_\alpha, \hat{\bar{Q}}^J_{\dot{\beta}}\} &= 2\delta^{IJ} \sigma^\mu_{\alpha\dot{\beta}} \hat{P}_\mu, \\
\{\hat{Q}^I_\alpha, \hat{Q}^J_\beta\} &= Z\epsilon^{IJ} \epsilon_{\alpha\beta}, \\
\{\hat{\bar{Q}}^I_{\dot{\alpha}}, \hat{\bar{Q}}^J_{\dot{\beta}}\} &= Z^* \epsilon^{IJ} \epsilon_{\dot{\alpha}\dot{\beta}}
\end{aligned} \tag{13.71}$$

where $\epsilon^{IJ} = -\epsilon^{JI}$ and Z, called the central charge, is an operator which commutes with all elements of the $N = 2$ algebra.

To obtain representation of this algebra we must allow all the fields composing the chiral and the real multiplet to transform into each other, creating in this way an

$N = 2$ "vector multiplet" which contains, besides the auxiliary fields, a vector field, two spinor fields and a complex scalar field.[3]

Given a gauge grup, the gauge invariant, renormalizable Lagrangian for such a multiplet can be constructed in a unique way and it reads:

$$\mathcal{L}_{\text{YM}}^{N=2} = \frac{1}{32\pi} Im \left(\tau \int d^2\theta \, \text{Tr} \, W^\alpha W_\alpha \right) + \int d^2\theta d^2\bar{\theta} \, \text{Tr} \, \mathbf{\Phi}^\dagger e^{2gV} \mathbf{\Phi}, \qquad (13.72)$$

where $\mathbf{\Phi} = \Phi^a T^a$. The possibility of chiral fields transforming in the representation of the gauge group different that the adjoint one is excluded by the fact that the chiral and the gauge fields transform into each other under the transformations generated by $\hat{Q}^i_\alpha$.

There is no way to distinguish in an invariant way between the generators $\hat{Q}^1_\alpha$ and $\hat{Q}^2_\alpha$. It is reflected by the fact that the $N = 2$ SUSY algebra is invariant under "rotation"

$$\hat{Q}^I_\alpha \to U^I{}_J \, \hat{Q}^J_\alpha$$

where U is a 2×2 unitary, unimodular matrix. This $\text{SU}(2)_\text{R}$ symmetry must be therefore respected by the Lagrangian (13.72). This requirement fixes the relative coefficient between the two terms appearing in (13.72) and does not allow for any terms of the form

$$\int d^2\theta \, W(\Phi).$$

In literature one can find an avalanche of nontrivial results concerning effective, low energy actions for supersymmetric models with a variety of gauge groups and fields, as well as relations of supersymmetric models with matrix models, topological field theories, and geometry of both Riemann surfaces and higher dimensional complex manifolds. It is a very interesting branch of mathematical physics.

13.7 Notation and Conventions

The notation traditionally used in the modern analysis of supersymmetric models can be somewhat cumbersome. In this section we have gathered the main definitions with the hope that such a glossary will be helpful.

For the antisymmetric symbol with two indices we choose

$$\epsilon_{12} = \epsilon_{\dot{1}\dot{2}} = 1, \qquad \epsilon^{12} = \epsilon^{\dot{1}\dot{2}} = -1 \qquad (13.73)$$

[3]There is a second possibility: one can obtain a representation of $N = 2$ algebra on fields composing two chiral multiplets, creating in this way the so called hypermultiplet, but we shall not discuss it here.

which gives

$$\epsilon^{\alpha\beta}\epsilon_{\beta\gamma} = \delta^{\alpha}_{\gamma}. \tag{13.74}$$

The ϵ symbol is used to raise and lower the spinor indices:

$$\psi^{\alpha} = \epsilon^{\alpha\beta}\psi_{\beta}, \qquad \bar{\chi}^{\dot{\alpha}} = \epsilon^{\dot{\alpha}\dot{\beta}}\bar{\chi}_{\dot{\beta}}, \tag{13.75}$$

and consequently

$$\psi_{\alpha} = \epsilon_{\alpha\beta}\psi^{\beta}, \qquad \bar{\chi}_{\dot{\alpha}} = \epsilon_{\dot{\alpha}\dot{\beta}}\bar{\chi}^{\dot{\beta}}. \tag{13.76}$$

Let θ^{α} be a constant spinor (of the Grassmann type). By definition, we have

$$\frac{\partial}{\partial\theta^{\beta}}\theta^{\alpha} = \delta^{\alpha}_{\beta} \tag{13.77}$$

which gives

$$\frac{\partial}{\partial\theta^{\beta}}\theta_{\alpha} = \epsilon_{\alpha\gamma}\frac{\partial}{\partial\theta^{\beta}}\theta^{\gamma} = \epsilon_{\alpha\beta}. \tag{13.78}$$

Similarly

$$\frac{\partial}{\partial\theta_{\beta}}\theta^{\alpha} = \epsilon^{\alpha\beta}, \qquad \epsilon_{\alpha\beta}\frac{\partial}{\partial\theta_{\beta}} = -\frac{\partial}{\partial\theta^{\alpha}}. \tag{13.79}$$

Analogous formulae hold for the conjugated spinors (with dotted indices).

We have chosen the 'NW–SE' (north west–south east) convention for the product of the spinors,

$$\psi\chi \equiv \psi^{\alpha}\chi_{\alpha} = -\chi_{\alpha}\psi^{\alpha} = \chi^{\alpha}\psi_{\alpha} = \chi\psi, \tag{13.80}$$

and the 'SW–NE' convention for the product of the conjugated spinors

$$\bar{\psi}\bar{\chi} \equiv \bar{\psi}_{\dot{\alpha}}\bar{\chi}^{\dot{\alpha}} = -\bar{\chi}^{\dot{\alpha}}\bar{\psi}_{\dot{\alpha}} = \bar{\chi}_{\dot{\alpha}}\bar{\psi}^{\dot{\alpha}} = \bar{\chi}\bar{\psi}. \tag{13.81}$$

Conjugation is defined as follows

$$(\psi_{\alpha})^* = \bar{\psi}_{\dot{\alpha}}, \qquad (\chi^{\alpha})^* = \bar{\chi}^{\dot{\alpha}}. \tag{13.82}$$

It reverses the order in products,

$$(\psi\chi)^* = (\psi^{\alpha}\chi_{\alpha})^* = \bar{\chi}_{\dot{\alpha}}\bar{\psi}^{\dot{\alpha}} = \bar{\chi}\bar{\psi}, \tag{13.83}$$

and changes a complex number into its complex conjugate.

The three Pauli matrices $\sigma^i = -\sigma_i$ and the 2×2 identity matrix I_2 can be assembled into two "matrix four-vectors"

$$(\sigma^{\mu}) = (I_2, \sigma^i), \qquad (\tilde{\sigma}^{\mu}) = (I_2, -\sigma^i) = (\sigma_{\mu}). \tag{13.84}$$

Thus, $\sigma^0 = I_2$. Let us also define

$$\sigma^{\mu\nu} = \frac{1}{4}\left(\sigma^\mu \tilde{\sigma}^\nu - \sigma^\nu \tilde{\sigma}^\mu\right), \qquad \tilde{\sigma}^{\mu\nu} = \frac{1}{4}\left(\tilde{\sigma}^\mu \sigma^\nu - \tilde{\sigma}^\nu \sigma^\mu\right). \tag{13.85}$$

These matrices have the following structure of indices:

$$(\sigma^\mu)_{\alpha\dot{\beta}}, \qquad (\tilde{\sigma}^\mu)^{\dot{\alpha}\beta}, \qquad (\sigma^{\mu\nu})_\alpha{}^\beta, \qquad (\tilde{\sigma}^{\mu\nu})^{\dot{\alpha}}{}_{\dot{\beta}}. \tag{13.86}$$

$\psi\sigma^\mu\bar{\chi} \equiv \psi^\alpha \sigma^\mu_{\alpha\dot{\beta}} \bar{\chi}^{\dot{\beta}}$ is a vector under the Poincaré transformations, $\psi\sigma^{\mu\nu}\chi$ is an (anti-symmetric) tensor, etc.

With (13.73) and the definitions (13.84), it is also immediate to check the identity

$$\tilde{\sigma}^{\mu\dot{\alpha}\alpha} = \epsilon^{\alpha\beta}\epsilon^{\dot{\alpha}\dot{\beta}}\sigma^\mu_{\beta\dot{\beta}}, \tag{13.87}$$

which also gives

$$\begin{aligned} \psi\sigma^\mu\bar{\chi} \equiv \psi^\alpha\sigma^\mu_{\alpha\dot{\alpha}}\bar{\chi}^{\dot{\alpha}} &= \epsilon^{\alpha\beta}\psi_\beta\sigma^\mu_{\alpha\dot{\alpha}}\epsilon^{\dot{\alpha}\dot{\beta}}\bar{\chi}_{\dot{\beta}} \\ &= -\bar{\chi}_{\dot{\beta}}\epsilon^{\beta\alpha}\epsilon^{\dot{\beta}\dot{\alpha}}\sigma^\mu_{\alpha\dot{\alpha}}\psi_\beta = -\bar{\chi}_{\dot{\beta}}\tilde{\sigma}^{\mu\dot{\beta}\beta}\psi_\beta = -\bar{\chi}\tilde{\sigma}^\mu\psi. \end{aligned} \tag{13.88}$$

The form of the Pauli matrices implies that under conjugation

$$\left(\sigma^\mu_{\alpha\dot{\beta}}\right)^* = \sigma^\mu_{\beta\dot{\alpha}} \tag{13.89}$$

so that

$$(\psi\sigma^\mu\bar{\chi})^* = \left(\psi^\alpha\sigma^\mu_{\alpha\dot{\beta}}\bar{\chi}^{\dot{\beta}}\right)^* = \chi^\beta\sigma^\mu_{\beta\dot{\alpha}}\bar{\psi}^{\dot{\alpha}} = \chi\sigma^\mu\bar{\psi}. \tag{13.90}$$

Similarly,

$$\begin{aligned} (\psi\sigma^\mu\tilde{\sigma}^\nu\chi)^* &= \bar{\chi}_{\dot{\gamma}}\tilde{\sigma}^{\nu\dot{\gamma}\beta}\sigma^\mu_{\beta\dot{\alpha}}\bar{\psi}^{\dot{\alpha}} = \bar{\chi}\tilde{\sigma}^\nu\sigma^\mu\bar{\psi}, \\ (\psi\sigma^{\mu\nu}\chi)^* &= \bar{\chi}\tilde{\sigma}^{\nu\mu}\bar{\psi} = -\bar{\chi}\tilde{\sigma}^{\mu\nu}\bar{\psi}, \\ \left(\bar{\chi}\tilde{\sigma}^{\mu\nu}\bar{\psi}\right)^* &= \bar{\chi}\sigma^{\nu\mu}\bar{\psi} = -\bar{\chi}\sigma^{\mu\nu}\bar{\psi}. \end{aligned} \tag{13.91}$$

Finally, let $\Psi_D = \begin{pmatrix}\psi_\alpha \\ \bar{\chi}^{\dot{\alpha}}\end{pmatrix}$ be an arbitrary Dirac spinor. In the spinor representation its charge conjugate is of the form $\Psi_D^c = \begin{pmatrix}\chi_\alpha \\ \bar{\psi}^{\dot{\alpha}}\end{pmatrix}$. The Majorana condition $\Psi_M = \Psi_M^c$ thus gives $\psi = \chi$, so that in the spinor representation the Majorana spinor has the form

$$\Psi_M = \begin{pmatrix}\psi_\alpha \\ \bar{\psi}^{\dot{\alpha}}\end{pmatrix}. \tag{13.92}$$

Exercises

13.1 In Sect. 13.2 we have discussed the representation of the superalgebra on the one-particle, massive states. We want to repeat this analysis for massless states. They can be chosen to satisfy

$$\hat{P}^\mu|p,\lambda\rangle = p^\mu|p,\lambda\rangle, \qquad \hat{W}^\mu|p,\lambda\rangle = \lambda\, p^\mu|p,\lambda\rangle$$

where $(p^\mu) = (E,0,0,E)$, $\hat{W}^\mu$ is the Pauli–Lubanski four-vector,

$$\hat{W}^\mu = \frac{1}{2}\epsilon^{\mu\nu\rho\lambda}\hat{P}_\nu\hat{M}_{\rho\lambda}$$

and $\lambda = \frac{1}{E}(\vec{J}\cdot\vec{P})$ is the helicity. Show that one can always choose the state $|p,\lambda\rangle$ so that $\hat{Q}_\alpha|p,\lambda\rangle = 0$, $\alpha = 1,2$, $\hat{\bar{Q}}_{\dot{1}}|p,\lambda\rangle = 0$, and that the only other state in the supersymmetric multiplet is

$$\frac{1}{\sqrt{4E}}\hat{\bar{Q}}_{\dot{2}}|p,\lambda\rangle.$$

What is the helicity of this state?

13.2 Find the form of a conserved current which exists thanks to an invariance of the action functional defined by the Lagrangian (13.32) under the transformations (13.37).

13.3 Check the validity of the relations

- $\theta^\alpha\theta^\beta = -\frac{1}{2}\epsilon^{\alpha\beta}\theta^\gamma\theta_\gamma \equiv -\frac{1}{2}\epsilon^{\alpha\beta}\,\theta\theta$,
- $\bar{\theta}^{\dot{\alpha}}\bar{\theta}^{\dot{\beta}} = \frac{1}{2}\epsilon^{\dot{\alpha}\dot{\beta}}\bar{\theta}_{\dot{\gamma}}\bar{\theta}^{\dot{\gamma}} \equiv \frac{1}{2}\epsilon^{\dot{\alpha}\dot{\beta}}\,\bar{\theta}\bar{\theta}$,
- $\epsilon^{\alpha\beta}\epsilon^{\dot{\alpha}\dot{\beta}}\sigma^\nu_{\beta\dot{\beta}} = (\tilde{\sigma}^\nu)^{\dot{\alpha}\alpha}$.

Using them prove the identities

$$(\theta\sigma^\mu\bar{\theta})(\theta\sigma^\nu\bar{\theta}) = \tfrac{1}{2}\,\theta\theta\,\bar{\theta}\bar{\theta}\,\eta^{\mu\nu}, \qquad \theta\sigma^\mu\bar{\theta}\,\theta\partial_\mu\psi(x) = -\tfrac{1}{2}\theta\theta\,\partial_\mu\psi(x)\sigma^\mu\bar{\theta}.$$

13.4 Taking into account that σ^μ form a basis of a (complex) vector space of 2×2 matrices, show the basic Fierz identity:

$$\delta^\beta_\alpha\delta^{\dot{\gamma}}_{\dot{\delta}} = \frac{1}{2}\sigma^\nu_{\alpha\dot{\delta}}\tilde{\sigma}^{\dot{\gamma}\beta}_\nu. \tag{13.93}$$

13.5 Contracting both sides of (13.93) with $\xi^\alpha\psi_\beta\bar{\chi}_{\dot{\gamma}}\bar{\eta}^{\dot{\delta}}$ show that

$$(\xi\psi)(\bar{\chi}\bar{\eta}) = -\frac{1}{2}(\xi\sigma^\nu\bar{\eta})(\bar{\chi}\tilde{\sigma}_\nu\psi) = \frac{1}{2}(\xi\sigma^\nu\bar{\eta})(\psi\sigma_\nu\bar{\chi}). \tag{13.94}$$

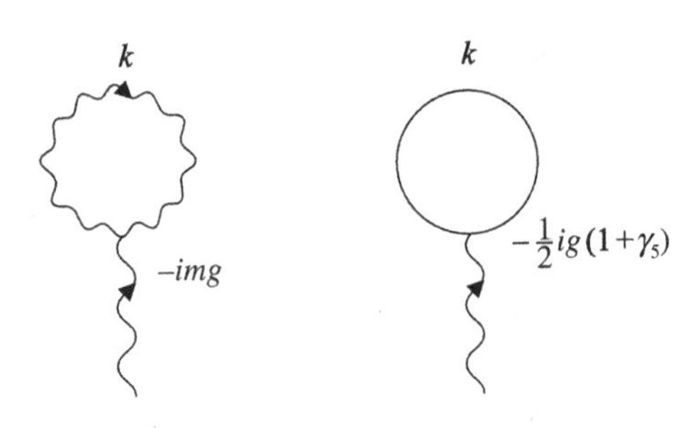

Fig. 13.5 The tadpole diagrams in the Wess–Zumino model

13.6 Contracting both sides of (13.93) with $\sigma^{\mu}_{\rho\dot{\gamma}}\eta^{\alpha}\psi_{\beta}\theta^{\rho}\bar{\xi}^{\dot{\delta}}$ show that

$$
\begin{aligned}
(\eta\psi)(\theta\sigma^{\mu}\bar{\xi}) &= -\frac{1}{2}(\eta\sigma_{\nu}\bar{\xi})(\theta\sigma^{\mu}\tilde{\sigma}^{\nu}\psi) = -\frac{1}{2}(\eta\sigma^{\mu}\bar{\xi})(\theta\psi) - (\eta\sigma_{\nu}\bar{\xi})(\theta\sigma^{\mu\nu}\psi) \\
&= -(\eta\sigma^{\mu}\bar{\xi})(\theta\psi) + \frac{1}{2}(\eta\sigma_{\nu}\bar{\xi})(\theta\sigma^{\nu}\tilde{\sigma}^{\mu}\psi). \qquad (13.95)
\end{aligned}
$$

13.7 Similarly as in Exercise 13.6, demonstrate that

$$
\begin{aligned}
(\theta\xi)(\bar{\eta}\tilde{\sigma}^{\mu}\psi) &= -\frac{1}{2}(\bar{\eta}\tilde{\sigma}_{\nu}\xi)(\theta\sigma^{\nu}\tilde{\sigma}^{\mu}\psi) = -\frac{1}{2}(\bar{\eta}\tilde{\sigma}^{\mu}\xi)(\theta\psi) - (\bar{\eta}\tilde{\sigma}_{\nu}\xi)(\theta\sigma^{\nu\mu}\psi) \\
&= -(\bar{\eta}\tilde{\sigma}^{\mu}\xi)(\theta\psi) + \frac{1}{2}(\bar{\eta}\tilde{\sigma}_{\nu}\xi)(\theta\sigma^{\mu}\tilde{\sigma}^{\nu}\psi). \qquad (13.96)
\end{aligned}
$$

13.8 Show that in the Wess–Zumino model the sum of the contributions from the 'tadpole' diagrams plotted in Fig. 13.5 vanishes.

Chapter 14
Anomalies

Abstract The splitting of the massless $(1+1)$-dimensional Dirac field into right- and left-handed components. The quantization of the right- and left-handed fields. Construction of the Hamiltonian and of the $U(1)$ current operator. The non-conservation of the $U(1)$ current in the presence of an external Abelian gauge field. Derivation of the $U(1)$ anomaly equation. Cancelation of anomalies. Non-invariance of the fermionic path integral measure under the axial $U(1)$ transformations. Derivation of the $U(1)$ anomaly equation in the path integral formulation of the quantum theory of the massless Dirac Dirac field in four-dimensional Euclidean space. The index of the Dirac operator.

The term 'anomaly' in quantum field theory refers to a case where a conservation law is lost on the way between classical and quantum versions of the theory. This phenomenon was discovered in 1969 (S. Adler, W. A. Bardeen, J.S. Bell, R. Jackiw), and it came as a surprise. Now it is rather well understood. Heuristically, the presence of an anomaly is a direct consequence of the fact that in order to obtain quantum observables it does not suffice just to replace classical fields by their quantum counterparts in the pertinent formulas. A careful approach to defining quantum observables involves a regularization, appropriate subtractions and removal of the regularization. Moreover, all this should be done in a physically relevant Hilbert space. Such a procedure can give surprising results. The phenomenon of anomalies is a very important example of that.

14.1 A Simple Example of an Anomaly

The model that we analyze below is distinguished by its mathematical simplicity.[1] It is related to a model first considered by J. Schwinger (see, e.g., Sect. 11.3 in [8]), but is significantly simpler, because we include only external gauge fields which by assumption have a fixed form.

---

[1]It is certainly simple when compared with other models, nonetheless we consider a system that has an infinite number of degrees of freedom. Simplicity is a relative notion.

H. Arodź and L. Hadasz, *Lectures on Classical and Quantum Theory of Fields*, Graduate Texts in Physics, DOI 10.1007/978-3-319-55619-2_14

We consider a massless Dirac field $\psi(t,x)$ in a one-dimensional space, $x \in R^1$. The time variable has the usual range, $t \in R^1$, hence the variables (t,x) can be regarded as coordinates on the plane R^2. This plane is pseudoeuclidean because the metric tensor has the Minkowski form: $\eta_{12} = \eta_{21} = 0$, $\eta_{00} = 1$, $\eta_{11} = -1$. The field ψ has two complex components,

$$\psi = \begin{pmatrix} \psi_+ \\ \psi_- \end{pmatrix}.$$

We use the c-number version of the classical Dirac field. In the case of $(1+1)$-dimensional space-time we have two Dirac matrices γ^0, γ^1, and $\gamma_5 = \gamma^0\gamma^1$. We take the following representation for them

$$\gamma^0 = \sigma_1, \;\; \gamma^1 = -i\sigma_2, \;\; \gamma_5 = \sigma_3, \tag{14.1}$$

where σ_i are Pauli matrices. The Dirac matrices and γ_5 have the usual properties:

$$\gamma^\mu\gamma^\nu + \gamma^\nu\gamma^\mu = 2\eta^{\mu\nu} I_2, \quad \gamma_5\gamma^\mu + \gamma^\mu\gamma_5 = 0, \quad \gamma_5^2 = I_2, \quad \gamma_5^\dagger = \gamma_5.$$

As the Lagrangian for the free, massless, classical Dirac field we take

$$\mathcal{L}_0 = \frac{i}{2}(\overline{\psi}\gamma^\mu\partial_\mu\psi - \partial_\mu\overline{\psi}\gamma^\mu\psi), \tag{14.2}$$

where $\overline{\psi} = \psi^\dagger\gamma^0$. The Dirac equation, that follows from this Lagrangian as the Euler–Lagrange equation, has the form

$$\gamma^\mu\partial_\mu\psi = 0. \tag{14.3}$$

In the absence of the mass term $m_0\overline{\psi}\psi$, the Lagrangian $\mathcal{L}_0$ can be split into two independent parts, $\mathcal{L}_0 = \mathcal{L}_1 + \mathcal{L}_2$, where

$$\mathcal{L}_1 = \frac{i}{2}\,(\psi_+^*\,\partial_0\psi_+ + \psi_+^*\,\partial_1\psi_+ - \partial_0\psi_+^*\,\psi_+ - \partial_1\psi_+^*\,\psi_+), \tag{14.4}$$

$$\mathcal{L}_2 = \frac{i}{2}\,(\psi_-^*\,\partial_0\psi_- - \psi_-^*\,\partial_1\psi_- - \partial_0\psi_-^*\,\psi_- + \partial_1\psi_-^*\,\psi_-). \tag{14.5}$$

The Dirac equation (14.3) is equivalent to the following simple equations

$$\partial_0\psi_+ + \partial_1\psi_+ = 0, \tag{14.6}$$

$$\partial_0\psi_- - \partial_1\psi_- = 0. \tag{14.7}$$

The general solutions of (14.6), (14.7) have the form $\psi_+(t,x) = f(t-x)$ and $\psi_-(t,x) = h(t+x)$, where f and h are arbitrary differentiable functions. For this reason ψ_+ is called the right-mover field, and ψ_- the left-mover field.

We see that the fields ψ_+ and ψ_- are independent of each other. The split of ψ into ψ_+ and ψ_- is analogous to the decomposition of the (3+1)-dimensional Dirac field into right- and left-handed components ψ_R and ψ_L, as discussed in Chap. 5. Such a decomposition is Poincaré invariant. This can be seen from the formulas

$$\begin{pmatrix} \psi_+ \\ 0 \end{pmatrix} = \frac{1}{2}\,(I_2 + \gamma_5)\psi, \quad \begin{pmatrix} 0 \\ \psi_- \end{pmatrix} = \frac{1}{2}\,(I_2 - \gamma_5)\psi,$$

and the fact that γ_5 is invariant with respect to proper Lorentz transformations in $(1+1)$-dimensional space-time (Exercise 14.1). From now on we will consider ψ_+ and ψ_- separately, so we have two models: one with ψ_+, and the other with ψ_-. We shall return to the Dirac field at the end of this section.

Let us generalize our two models by including interactions with classical external gauge fields: $B_\mu(t,x)$ in the case of the right-mover field ψ_+, and $C_\mu(t,x)$ in the case of ψ_-. We apply the minimal coupling rule, i.e., the only change in the pertinent Lagrangian is

$$\partial_\mu\psi_+ \to D_\mu(B)\psi_+ = \partial_\mu\psi_+ - iB_\mu\psi_+, \quad \partial_\mu\psi_- \to D_\mu(C)\psi_- = \partial_\mu\psi_- - iC_\mu\psi_-. \tag{14.8}$$

Here B_μ and C_μ are arbitrary, but fixed—there is no evolution equation for them. Models which differ only by the form of the external field should in general be regarded as different (an exception to this is discussed below). Thus, we are led to consider two independent classes of models: one class with Lagrangians of the form

$$\mathcal{L}_+ = \frac{i}{2}[\psi_+^*\, D_0(B)\psi_+ + \psi_+^*\, D_1(B)\psi_+ - (D_0(B)\psi_+)^*\psi_+ - (D_1(B)\psi_+)^*\psi_+] \tag{14.9}$$

and the Euler-Lagrange equations

$$D_0(B)\psi_+ + D_1(B)\psi_+ = 0, \tag{14.10}$$

and the other class with

$$\mathcal{L}_- = \frac{i}{2}[\psi_-^*\, D_0(C)\psi_- - \psi_-^*\, D_1(C)\psi_- - (D_0(C)\psi_-)^*\psi_- + (D_1(C)\psi_-)^*\psi_-] \tag{14.11}$$

and the Euler-Lagrange equations

$$D_0(C)\psi_- - D_1(C)\psi_- = 0. \tag{14.12}$$

The models within one such class differ by the form of the external field.

The Lagrangian $\mathcal{L}_+$ is invariant with respect to the $U_+(1)$ group of transformations of the form

$$\psi_+(t,x) \to e^{i\chi}\psi_+(t,x), \tag{14.13}$$

where χ is a real parameter. As a consequence of this symmetry we have the conserved current density

$$j_+^\mu(t,x) = \overline{\begin{pmatrix}\psi_+\\0\end{pmatrix}}\gamma^\mu\begin{pmatrix}\psi_+\\0\end{pmatrix} \tag{14.14}$$

that obeys the continuity equation

$$\partial_\mu j_+^\mu(t,x) = 0,$$

provided that ψ_+ obeys (14.10). Formula (14.14) gives the two-vector of the form

$$(j_+^\mu) = \begin{pmatrix}\psi_+^*\psi_+\\ \psi_+^*\psi_+\end{pmatrix}. \tag{14.15}$$

Similarly, $\mathcal{L}_-$ is invariant with respect to the transformations

$$\psi_-(t,x) \to e^{i\eta}\psi_-(t,x) \tag{14.16}$$

with arbitrary real η. These transformations form the group $U_-(1)$. The components of the corresponding conserved current density are given by the formula

$$j_-^\mu(t,x) = \overline{\begin{pmatrix}0\\ \psi_-\end{pmatrix}}\gamma^\mu\begin{pmatrix}0\\ \psi_-\end{pmatrix},$$

or in the two-vector form

$$(j_-^\mu) = \begin{pmatrix}\psi_-^*\psi_-\\ -\psi_-^*\psi_-\end{pmatrix}. \tag{14.17}$$

Similarly as in the previous case,

$$\partial_\mu j_-^\mu(t,x) = 0,$$

provided that ψ_- obeys (14.12).

Because B_μ and C_μ are fixed functions and not dynamical fields, there is no gauge invariance of the type discussed in Chap. 4. Nevertheless, the particular form of the coupling implies that models with various choices of these functions can be equivalent to each other. Specifically, the model (14.9) with the external field B_μ and the field ψ_+ is equivalent to the model with the external field B'_μ and the field ψ'_+, if

$$B'_\mu(t,x) = B_\mu(t,x) + \partial_\mu\chi(t,x), \quad \psi'_+(t,x) = e^{i\chi(t,x)}\psi_+(t,x), \tag{14.18}$$

where $\chi(t, x)$ is an arbitrary differentiable function that vanishes together with its derivatives in the limit $|x| \to \infty$. The conserved current j_+^μ and all the other physical quantities have exactly the same values in all equivalent models. The equivalence transformation in the models of the type (14.11) has the form

$$C'_\mu(t, x) = C_\mu(t, x) + \partial_\mu \eta(t, x), \quad \psi'_-(t, x) = e^{i\eta(t,x)}\psi_-(t, x), \tag{14.19}$$

where $\eta(t, x)$ has the same properties as the function $\chi(t, x)$ above. Of course, these equivalence transformations are akin to the gauge transformations of the type $U(1)_{\pm,loc}$ of the classical Schwinger model, in which B_μ and C_μ are also dynamical fields. The Lagrangian of the Schwinger model contains the standard kinetic terms for these fields, i.e.,

$$-\frac{1}{4}F_{\mu\nu}(B)F^{\mu\nu}(B) - \frac{1}{4}F_{\mu\nu}(C)F^{\mu\nu}(C),$$

where $F_{\mu\nu}(B) = \partial_\mu B_\nu - \partial_\nu B_\mu$, $F_{\mu\nu}(C) = \partial_\mu C_\nu - \partial_\nu C_\mu$. Because of that relationship, we will use the more popular name 'gauge transformations' also for the transformations (14.18), (14.19), instead of the more precise 'equivalence transformations'.

Using gauge transformations (14.18), (14.19) we can eliminate B_0, C_0:

$$B_0 = 0, \quad C_0 = 0.$$

These conditions, called the temporal gauge conditions, do not eliminate the gauge transformations, but restrict them to the functions χ and η that do not depend on t, $\chi = \chi(x)$ and $\eta = \eta(x)$. In all our considerations below, we assume that the external fields have been transformed to the temporal gauge.

Now let us construct quantum versions of our models. By analogy with the previously discussed (in Sect. 6.2) free Dirac field on the space R^3, we postulate the equal time anticommutation relations for the fields $\psi_\pm$, namely

$$\left\{\hat{\psi}_+(t, x), \hat{\psi}_+(t, x')\right\} = 0, \quad \left\{\hat{\psi}_+^\dagger(t, x), \hat{\psi}_+(t, x')\right\} = \delta(x - x')\, I \tag{14.20}$$

and

$$\left\{\hat{\psi}_-(t, x), \hat{\psi}_-(t, x')\right\} = 0, \quad \left\{\hat{\psi}_-^\dagger(t, x), \hat{\psi}_-(t, x')\right\} = \delta(x - x')\, I. \tag{14.21}$$

Note that at this point it is not possible to specify the (anti-)commutation relations between the fields $\hat{\psi}_+$ and $\hat{\psi}_-$ because they act in different Hilbert spaces.

The classical energy density obtained from the Lagrangian $\mathcal{L}_+$ has the form

$$T_{00} = \frac{i}{2}(D_1(B)\psi_+)^*\psi_+ - \frac{i}{2}\psi_+^*\, D_1(B)\psi_+.$$

Therefore, as the quantum Hamiltonian, we would like to take the operator[2]

$$\frac{i}{2}\int dx\ (D_1(B)\hat{\psi}_+(t,x))^\dagger\hat{\psi}_+(t,x) + \text{ h.c.} \tag{14.22}$$

Unfortunately, such an expression is not well-defined. First, we expect that $\hat{\psi}_+(t,x)$ is an operator-valued generalized function of (t,x), and we know that expressions like $(\hat{\psi}_+(t,x))^\dagger\hat{\psi}_+(t,x)$ should be avoided. In fact, the second relation (14.20) suggests that indeed, such a product is ill-defined. Second, assuming that we can 'repair' the products of the generalized functions, the integrand in our candidate formula will be another generalized function, and as such it can only be integrated with a test function. Such a test function is missing from our formula. Therefore, we first consider the well-defined operator

$$\hat{H}_\epsilon[B_1] = -\frac{i}{2}\int dx\ f(x)\,\hat{\psi}_+^\dagger(t,x+\epsilon)\,W[\epsilon;B_1]\,D_1(B)\hat{\psi}_+(t,x) + \text{h.c.}\,, \tag{14.23}$$

where $\epsilon > 0$, and $f(x)$ is a test function (real-valued). The operator $\hat{H}_\epsilon[B_1]$ is the regularized form of operator (14.22).

Similarly, the regularized two-current density $\hat{j}^\mu_{+,\epsilon}(t,x)$ has the form (14.15) with $j^0_+ = \psi^*_+\psi_+ = j^1_+$ replaced by

$$\hat{j}^0_{+,\epsilon}(t,x) = \frac{1}{2}\hat{\psi}^\dagger_+(t,x+\epsilon)\,W[\epsilon;B_1]\,\hat{\psi}_+(t,x) + \text{h.c.} = \hat{j}^1_{+,\epsilon}(t,x). \tag{14.24}$$

The factor

$$W[\epsilon;B_1] = \exp\left(i\int_x^{x+\epsilon} dx'\ B_1(t,x')\right) \tag{14.25}$$

is the parallel transporter from the point (t,x) to the point $(t,x+\epsilon)$ along the rectilinear segment connecting these points. It is analogous to the one considered in Chap. 4. We have included it in order to ensure that $\hat{H}_\epsilon[B_1]$ is gauge invariant. The gauge transformation corresponding to (14.18) in the quantum model (with χ independent of t) has the form (Exercise 14.2)

$$U^{-1}[\chi,t]\ \hat{\psi}_+(t,x)\ U[\chi,t] = \exp(i\chi(x))\ \hat{\psi}_+(t,x), \tag{14.26}$$

where

$$U[\chi,t] = \exp\left(i\int dx\ \chi(x)\hat{j}^0_+(t,x)\right). \tag{14.27}$$

---

[2] Let us recall that a term denoted as 'h.c.' is obtained by the Hermitian conjugation of the preceding term or terms.

Here $\hat{J}^0_+(t,x)$ is the quantum counterpart of the classical charge density $j^0_+(t,x)$—it will be obtained below from $\hat{j}^0_{+,\epsilon}$. Because $\hat{j}^0_+(t,x)$ is a generalized function of x, the function $\chi(x)$ should be a test function. Note that as far as transformation law (14.26) is concerned, we may add to $\hat{J}^0_+$ certain terms proportional to the identity operator—they will cancel out in the product on the l.h.s. of formula (14.26). The gauge invariance of the operator $\hat{H}_\epsilon[B_1]$ means that

$$U^{-1}[\chi,t]\;\hat{H}_\epsilon[B_1+\partial_1\chi]\;U[\chi,t]=\hat{H}_\epsilon[B_1].$$

The regularization employed in formula (14.23) for $\hat{H}_\epsilon[B_1]$ consists of two steps. The first step, that is the introduction of ϵ and $W[\epsilon;\,B_1]$ is called the gauge invariant point splitting. It is applied in order to 'repair' the product of generalized functions without spoiling the gauge invariance. The second step consists in introducing the test function f in order to secure the convergence of the integral over x. This step is sometimes called the regularization in the infrared (because the problem lies at large values of x), while the first step is the regularization in the ultraviolet. After the calculation of $\hat{H}_\epsilon[B_1]$ we shall attempt to take the limit $\epsilon\to 0,\ f(x)\to 1$. It turns out that such a limit exists if we abandon some terms proportional to the identity operator I. Of course, in the case of the current density only the first step—the gauge invariant point splitting—is needed because there is no integration.

Analogously, in the case of the left-mover field $\hat{\psi}_-$ we consider the operators

$$\hat{H}_\epsilon[C_1]=\frac{i}{2}\int dx\;f(x)\,\hat{\psi}^\dagger_-(t,x+\epsilon)\;W[\epsilon;\,C_1]\;D_1(C)\hat{\psi}_-(t,x)+\text{h.c.}\,, \tag{14.28}$$

and

$$\hat{j}^0_{-,\epsilon}(t,x)=\frac{1}{2}\hat{\psi}^\dagger_-(t,x+\epsilon)\;W[\epsilon;\,C_1]\;\hat{\psi}_-(t,x)+\text{h.c.}=-\hat{j}^1_{-,\epsilon}(t,x). \tag{14.29}$$

The change of sign in $\hat{H}_\epsilon[C_1]$, as compared with $\hat{H}_\epsilon[B_1]$, is due to the difference in the signs of the terms containing $D_1(B)\psi_+$ and $D_1(C)\psi_-$ in Lagrangians (14.9), (14.11).

The Heisenberg equations of motion for the fields $\hat{\psi}_+$ and $\hat{\psi}_-$ have the form

$$\partial_0\hat{\psi}_+ + D_1(B)\hat{\psi}_+=0, \tag{14.30}$$

$$\partial_0\hat{\psi}_- - D_1(C)\hat{\psi}_-=0. \tag{14.31}$$

They can easily be solved if B_1 and C_1 do not depend on x, i.e., when

$$B_1=B(t),\quad C_1=C(t). \tag{14.32}$$

In this case

$$\hat{\psi}_+(t,x) = e^{i\int_0^t dt' B(t')}\,\frac{1}{\sqrt{2\pi}}\int dp\; e^{ip(x-t)}\;\hat{c}_+(p), \tag{14.33}$$

$$\hat{\psi}_-(t,x) = e^{-i\int_0^t dt' C(t')}\,\frac{1}{\sqrt{2\pi}}\int dp\; e^{ip(x+t)}\;\hat{c}_-(p). \tag{14.34}$$

The anticommutation relations (14.20), (14.21) are satisfied if

$$\{\hat{c}_+(p),\,\hat{c}_+(p')\} = 0,\quad \{\hat{c}_+^\dagger(p),\,\hat{c}_+(p')\} = \delta(p-p')I, \tag{14.35}$$

$$\{\hat{c}_-(p),\,\hat{c}_-(p')\} = 0,\quad \{\hat{c}_-^\dagger(p),\,\hat{c}_-(p')\} = \delta(p-p')I. \tag{14.36}$$

Let us construct the quantum Hamiltonian for the right-mover field. Note that the integration in the parallel transporters is now trivial, $W[\epsilon;\, B_1] = e^{i\epsilon B(t)}$. Inserting solution (14.33) in formula (14.23) and integrating over x we obtain

$$\hat{H}_\epsilon[B] = \frac{e^{i\epsilon B(t)}}{4\pi}\int dp' \int dp\; \tilde{f}(p'-p)\, e^{it(p'-p)-ip'\epsilon}\,(p-B)\,\hat{c}_+^\dagger(p')\,\hat{c}_+(p)\;+\text{h.c.}\,,$$

where

$$\tilde{f}(p'-p) = \int dx\; e^{-i(p'-p)x}\; f(x),$$

and $B \equiv B(t)$. The limit $f(x) \to 1$ corresponds to $\tilde{f}(p'-p) \to 2\pi\delta(p'-p)$. Note also that $\tilde{f}^*(p'-p) = \tilde{f}(p-p')$. The limit $\epsilon \to 0,\;\; f(x) \to 1$ gives the operator

$$\hat{H}_0[B] = \int dp\,(p-B)\,\hat{c}_+^\dagger(p)\,\hat{c}_+(p),$$

which is well-defined in the Fock space generated by the operators $\hat{c}_+^\dagger(p)$ acting on the vacuum state $|0\rangle$ such that $\hat{c}_+(p)|0\rangle = 0$ for all $p \in R$. It is clear that this operator is not bounded from below. It is the same problem as that encountered in Sect. 6.2 in the case of the free massive Dirac field. We are going to use essentially the same solution, that is, we will use the Dirac vacuum $|0\rangle_D$ instead of $|0\rangle$, and transform the negative energy sector into the sector of antiparticles with positive energy.

Unfortunately, in the case of the massless Dirac field we have to pay more attention to the mathematical side of the theory. The reason is that the positive and negative energy sectors are not well-separated—in fact they touch each other at $p = B$. This has a rather unexpected consequence in that the integral $\int dp\;(...)$ in general cannot be simply written as $\int_{-\infty}^{B} dp(...) + \int_B^{\infty} dp(...)$! This integral is already present in formula (14.33). Because of the presence of the Dirac delta on the r.h.s. of anticommutation relation (14.35) $\hat{c}_+(p)$ is an operator-valued generalized function of p. As explained in the Appendix, for the safe approach to the decomposition of the integral, one should introduce a smooth function $\theta_\kappa(p)$ that represents a smoothed step function $\Theta(p)$. Then,

$$\int dp\,(...) = \int dp\,\theta_\kappa(p-B)(...) + \int dp\,[1-\theta_\kappa(p-B)](...).$$

We shall use a function $\theta_\kappa(p)$ of the form

$$\theta_\kappa(p) = \begin{cases} 1 & \text{when} \quad p \geq \kappa \\ \alpha(p) & \text{when} \ -\kappa < p < \kappa \\ 0 & \text{when} \quad p \leq -\kappa, \end{cases}$$

where $\alpha(p)$ is a smooth monotonic function that interpolates between 0 and 1, and $\kappa > 0$ is a constant. For simplicity, we also assume that $\alpha(p) + \alpha(-p) = 1$ for $p \in (\kappa, -\kappa)$. Then

$$\theta_\kappa(B-p) + \theta_\kappa(p-B) = 1. \tag{14.37}$$

The step function $\Theta(p-B)$ is obtained in the limit $\kappa \to 0_+$. The mathematically correct decomposition of $\hat{\psi}_+$ into the positive and negative energy components has the form

$$\hat{\psi}_+(t,x) = \hat{\psi}_+^{(+)}(t,x) + \hat{\psi}_+^{(-)}(t,x), \tag{14.38}$$

where

$$\hat{\psi}_+^{(+)}(t,x) = e^{i\int_0^t dt' B(t')} \frac{1}{\sqrt{2\pi}} \int dp\,\theta_\kappa(p-B)\,e^{ip(x-t)}\,\hat{c}_+(p),$$

$$\hat{\psi}_+^{(-)}(t,x) = e^{i\int_0^t dt' B(t')} \frac{1}{\sqrt{2\pi}} \int dp\,\theta_\kappa(B-p)\,e^{ip(x-t)}\,\hat{c}_+(p).$$

The first integral extends essentially over the interval $[B-\kappa, \infty)$ and the second over $(-\infty, B+\kappa]$. Such a smoothing of the decomposition is not needed in the case of the massive Dirac field in Sect. 6.2, because there the positive and negative energy sectors are well-separated: the operators $\hat{a}_s^{(+)}(\vec{p})$ and $\hat{a}_s^{(-)}(\vec{p})$, present in formula (6.79) are never equal to each other.

Let us recalculate $\hat{H}_\epsilon[B]$ using decomposition (14.38). Inserting (14.38) into formula (14.23), we obtain terms of the type $\hat{\psi}_+^{(+)\dagger}\hat{\psi}_+^{(+)}$, $\hat{\psi}_+^{(+)\dagger}\hat{\psi}_+^{(-)}$, $\hat{\psi}_+^{(-)\dagger}\hat{\psi}_+^{(+)}$, $\hat{\psi}_+^{(-)\dagger}\hat{\psi}_+^{(-)}$. In the last two we use anticommutation relation (14.35). Next, we take the limits $f(x) \to 1$ and $\epsilon \to 0$ in all the terms except the two that are proportional to the identity operator I. Finally, we apply identity (14.37), and change the integration variable $(p \to -p)$ in the two terms that contain the expression $\hat{c}_+(p)\hat{c}_+^\dagger(p)$. We obtain

$$\begin{aligned} \hat{H}[B] = &\int dp\ \theta_\kappa(p-B)\,(p-B)\,\hat{c}_+^\dagger(p)\hat{c}_+(p) \\ &+ \int dp\ \theta_\kappa(p+B)\,(p+B)\,\hat{c}_+(-p)\hat{c}_+^\dagger(-p) \\ &- \frac{\tilde{f}(0)}{4\pi} \int dp\ \theta_\kappa(B-p)\,(B-p)\,(e^{i\epsilon(B-p)} + e^{i\epsilon(p-B)})\,I. \end{aligned} \tag{14.39}$$

The third term is the consequence of using the second relation (14.35). It is singular in the limit $f \to 1$ because then $\tilde{f}(0) \to \int dx 1$. This singularity can be called the infrared one. There is also the singularity at $\epsilon = 0$ which is of the ultraviolet type. Therefore, we just omit that term, and define the quantum Hamiltonian of the right-mover field as the sum of the first two terms in formula (14.39). In the $\kappa \to 0$ limit

$$\hat{H}_+[B] = \int_B^\infty dp\ (p - B)\ \hat{a}_+^\dagger(p)\hat{a}_+(p) + \int_{-B}^\infty dp\ (p + B)\ \hat{d}_+^\dagger(p)\hat{d}_+(p), \quad (14.40)$$

where

$$\hat{a}_+(p) = \hat{c}_+(p) \ \text{ for } \ p \in [B, \infty), \qquad \hat{d}_+(p) = \hat{c}_+^\dagger(-p) \ \text{ for } \ p \in (-B, \infty).$$

$\hat{a}_+(p)$ and $\hat{d}_+(p)$ are the annihilation operators for the right-mover particle and anti-particle, respectively, $\hat{a}_+^\dagger(p)$ and $\hat{d}_+^\dagger(p)$ are the corresponding creation operators.

Now let us turn to the current $\hat{J}_+^\mu$. Similarly as in the case of the Hamiltonian, our calculations are restricted to the particular case of external fields (14.32). Starting from formula (14.24), we would like to obtain an operator that is well-defined in the Fock space based on the Dirac vacuum. Hence, it should contain, like $\hat{H}[B]$, the normal ordered products $\hat{a}_+^\dagger\hat{a}_+$ and $\hat{d}_+^\dagger\hat{d}_+$. To this end we use decomposition (14.38), and the relation ($\epsilon > 0$)

$$\{\hat{\psi}_+^{(-)\dagger}(t, x + \epsilon),\ \hat{\psi}_+(t, x)\} = \frac{1}{2\pi}\int dp\ \theta_\kappa(B - p)e^{-ip\epsilon}\ I,$$

which follows from anticommutation relation (14.35). The term proportional to I obtained from the anticommutator above is omitted. The resulting expression for the components of the current has the form

$$\begin{aligned}\hat{J}_+^0(t, x) &= \hat{J}_+^1(t, x) \\ &= \hat{\psi}_+^{(+)\dagger}(t, x)\hat{\psi}_+^{(+)}(t, x) + \hat{\psi}_+^{(+)\dagger}(t, x)\hat{\psi}_+^{(-)}(t, x) - \hat{\psi}_+^{(+)}(t, x)\hat{\psi}_+^{(-)\dagger}(t, x) \\ &\qquad - \hat{\psi}_+^{(-)}(t, x)\hat{\psi}_+^{(-)\dagger}(t, x). \quad (14.41)\end{aligned}$$

The transition from $\hat{j}_{+,\epsilon}^\mu$ to $\hat{J}_+^\mu$ is reminiscent of the normal ordering (and in the limit $\kappa \to 0$ it coincides with), hence we may write $\hat{J}_+^\mu = \lim_{\epsilon\to 0} :\hat{j}_{+,\epsilon}^\mu:$.

The total charge operator $\hat{Q}_+$ is obtained in the limit $f \to 1$ of the integral $\hat{Q}_+[f] = \int dx\ f(x)\ \hat{J}_+^0(t, x)$. Simple calculations with the use of identity (14.37) give

$$\hat{Q}_+ = \int dp\ \left[\theta_\kappa(p - B)\ \hat{c}_+^\dagger(p)\hat{c}_+(p) - \theta_\kappa(B - p)\ \hat{c}_+(p)\hat{c}_+^\dagger(p)\right]. \quad (14.42)$$

In the limit $\kappa \to 0$, and after the change $p \to -p$ of the integration variable in the second term,

$$\hat{Q}_+ = \int_B^\infty dp\ \hat{a}_+^\dagger(p)\hat{a}_+(p) - \int_{-B}^\infty dp\ \hat{d}_+^\dagger(p)\hat{d}_+(p). \tag{14.43}$$

This operator has been calculated in the Heisenberg picture, hence it should be constant in time if the charge is conserved. Because the field B can vary with time, one may suspect that this is not the case. Unfortunately, the time derivative of $\hat{Q}_+$ gives the mathematically meaningless (but physically justified, see below) result

$$\dot{\hat{Q}}_+ = -\dot{B} \int dp\ \theta_\kappa'(p-B)\left(\hat{c}_+^\dagger(p)\hat{c}_+(p) + \hat{c}_+(p)\hat{c}_+^\dagger(p)\right) = -\dot{B}\delta(0)I.$$

The last equality is written because of (14.35). Here $\theta_\kappa'(q) = d\theta_\kappa(q)/dq$ and $\int dq\ \theta_\kappa'(q) = 1$. We have also used the equality

$$\theta_\kappa'(p-B) = \theta_\kappa'(B-p)$$

that follows from identity (14.37).

Let us check instead whether $\hat{J}_+^\mu$ obeys the continuity equation. It is convenient to use here the identities

$$\partial_0\hat{\psi}_+^{(\pm)} + (\partial_1 - iB)\,\hat{\psi}_+^{(\pm)} = \mp\frac{\dot{B}}{\sqrt{2\pi}}\, e^{i\int_0^t dt' B(t')} \int_{-\infty}^\infty dp\ \theta_\kappa'(\pm p)\, e^{i(B+p)(x-t)}\ \hat{c}_+(B+p).$$

It turns out that the current (14.41) is not conserved, namely

$$\partial_0\hat{J}_+^0 + \partial_1\hat{J}_+^1 = -\frac{\dot{B}}{2\pi} I \int dp\ \theta_\kappa'(p) = -\frac{\dot{B}}{2\pi} I. \tag{14.44}$$

This formula is called the anomaly equation.

Let us stress that the non-conservation of the current is the result of the very construction of the quantum model. In particular, it is not a dynamical effect related to some peculiar interactions between the particles. To illuminate this point, let us compare the total charge $\hat{Q}_+$ with the operator $\hat{q}_+ = \lim_{\epsilon\to 0}\int dx\ \hat{j}_{+,\epsilon}^0(t,x)$. Using formulas (14.24), (14.33) we obtain $\hat{q}_+ = \int dp\ \hat{c}_+^\dagger(p)\hat{c}_+(p)$. This operator is constant in time, $\dot{\hat{q}}_+ = 0$, but it has an infinite expectation value in the Dirac vacuum, and in every normalized state from the Fock space based on that vacuum. The current $\hat{j}_+^\mu = \lim_{\epsilon\to 0}\hat{j}_{+,\epsilon}^\mu$ is conserved, $\partial_\mu\hat{j}_+^\mu = 0$. Thus, the anomaly is generated by the normal ordering.

Moreover, let us also consider the difference

$$\hat{j}_{+,\epsilon}^0 - \hat{J}_+^0 = \frac{1}{4\pi}\int dp\ \theta_\kappa(B-p)e^{i\epsilon(B-p)} I + \text{h.c.}$$

The expression on the r.h.s. is a generalized function of ϵ; it is the Fourier transform of $\theta_\kappa(B-p)$ (up to a factor). In order to consider the limit $\epsilon \to 0_+$, we have to turn it into an ordinary function of ϵ. To this end, we regularize it by replacing $\theta_\kappa(B-p)$ with $e^{\sigma p}\theta_\kappa(B-p)$, where $\sigma > 0$. At the end of calculations we shall consider the limit $\sigma \to 0_+$. Thus, we now consider the ordinary integral

$$\int_{-\infty}^{\infty} dp\ \theta_\kappa(B-p)e^{(\sigma-i\epsilon)p} = \int_{-\infty}^{B-\kappa} dp\ e^{(\sigma-i\epsilon)p} + \int_{B-\kappa}^{B+\kappa} dp\ \alpha(B-p)e^{(\sigma-i\epsilon)p}$$
$$= \frac{e^{(\sigma-i\epsilon)(B-\kappa)}}{\sigma-i\epsilon} + \int_{-\kappa}^{\kappa} dp\ \alpha(p)e^{(\sigma-i\epsilon)(B-p)}.$$

For $\sigma > 0$, this expression is a regular function of ϵ, and we may put $\epsilon = 0$. Therefore,

$$\left.(\hat{j}^0_{+,\epsilon=0} - \hat{j}^0_+)\right|_{\sigma>0} = \frac{1}{2\pi}\left[\frac{e^{\sigma(B-\kappa)}}{\sigma} + \int_{-\kappa}^{\kappa} dp\ \alpha(p)e^{\sigma(B-p)}\right] I$$
$$= \left[\frac{1}{2\pi\sigma} + \frac{B-\kappa}{2\pi} + \frac{1}{2\pi}\int_{-\kappa}^{\kappa} dp\ \alpha(p) + \mathcal{O}(\sigma)\right] I,$$

where $\mathcal{O}(\sigma)$ denotes the terms which vanish when $\sigma \to 0$. The first term is singular at $\sigma = 0$, but it does not depend on B and κ. We see that the time derivative of $\hat{j}^0_{+,\epsilon=0} - \hat{j}^0_+$ in the limit $\sigma \to 0$ is equal to $\dot{B}I/2\pi$, in agreement with the anomaly equation (14.44).

The correct calculation of $\dot{\hat{Q}}_+$ should start from $\dot{\hat{Q}}_+[f]$. Using the anomaly equation we see that

$$\dot{\hat{Q}}_+[f] = \int dx\ \partial_1 f(x)\ \hat{j}^1_+(t,x) - \frac{\dot{B}}{2\pi} I \int dx\ f(x).$$

In the limit $f \to 1$ the first term on the r.h.s vanishes, while the second becomes proportional to the infinite 'volume' of the one-dimensional space ($\int dx/2\pi$ corresponds to $\delta(0)$ in the meaningless formula for $\dot{\hat{Q}}_+$ shown below (14.43)). This is in fact expected, because the external field (14.32) is constant in x, and therefore the model possesses invariance with respect to spatial translations. For this reason, the production rate for the charge density is the same over the whole space.

The Hamiltonian and the current in the quantum theory of the left-mover field are obtained in a completely analogous manner. The decomposition into the positive and negative energy components reads

$$\hat{\psi}_-(t,x) = \hat{\psi}^{(+)}_-(t,x) + \hat{\psi}^{(-)}_-(t,x), \tag{14.45}$$

where

$$\hat{\psi}^{(+)}_-(t,x) = e^{-i\int_0^t dt' C(t')} \frac{1}{\sqrt{2\pi}} \int dp\ \theta_\kappa(C-p)\ e^{ip(x-t)}\ \hat{c}_-(p),$$

$$\hat{\psi}_-^{(-)}(t,x) = e^{-i\int_0^t dt' C(t')} \frac{1}{\sqrt{2\pi}} \int dp\, \theta_\kappa(p-C)\, e^{ip(x-t)}\, \hat{c}_-(p).$$

The quantum Hamiltonian has the form

$$\hat{H}_-[C] = \int_{-\infty}^{C} dp\, (C-p)\, \hat{a}_-^\dagger(p)\hat{a}_-(p) + \int_{-\infty}^{-C} dp\, (-p-C)\, \hat{d}_-^\dagger(p)\hat{d}_-(p), \tag{14.46}$$

where

$$\hat{a}_-(p) = \hat{c}_-(p) \text{ for } p \in (-\infty, C], \quad \hat{d}_-(p) = \hat{c}_-^\dagger(-p) \text{ for } p \in (-\infty, -C).$$

$\hat{a}_-(p)$ and $\hat{d}_-(p)$ are the annihilation operators for the left-mover particle and antiparticle, respectively. Note that the momenta of the left-mover particles (anti-particles) are restricted from above by C $(-C)$. Formula (14.46) has been obtained from (14.28) in the limits $f(x) \to 1$, $\epsilon \to 0$, $\kappa \to 0$, taken after omitting a singular term proportional to the identity operator I.

The total charge of the left-mover field is given by the formula

$$\hat{Q}_- = \int_{-\infty}^{C} dp\, \hat{a}_-^\dagger(p)\hat{a}_-(p) - \int_{-\infty}^{-C} dp\, \hat{d}_-^\dagger(p)\hat{d}_-(p). \tag{14.47}$$

As the anomaly equation we obtain

$$\partial_0 \hat{J}_-^0 + \partial_1 \hat{J}_-^1 = \frac{\dot{C}}{2\pi} I. \tag{14.48}$$

The anomaly equations (14.44), (14.48) have very similar structures. This fact suggests that one can combine the two models, and the currents, in order to obtain a conserved current. One possibility of such a cancelation of the anomaly is obtained by considering the current $\hat{J}^\mu = \hat{J}_+^\mu + \hat{J}_-^\mu$. Then $\partial_\mu \hat{J}^\mu = (\dot{C} - \dot{B})I/2\pi = 0$ if we assume that $B(t) = C(t)$. The current $\hat{J}_5^\mu = \hat{J}_+^\mu - \hat{J}_-^\mu$ remains not conserved, $\partial_\mu \hat{J}_5^\mu = -\dot{B}I/\pi$. In this case the name 'axial anomaly' is used. The corresponding classical currents j^μ and j_5^μ, and the condition $B(t) = C(t)$, appear automatically if we consider the classical Lagrangian

$$\mathcal{L} = (\mathcal{L}_+ + \mathcal{L}_-)|_{B=C} = \frac{i}{2}[\overline{\psi}\gamma^\mu D_\mu(A)\psi - (D_\mu(A)\overline{\psi})\gamma^\mu\psi],$$

where

$$D_\mu(A)\psi = \partial_\mu\psi - iA_\mu\psi, \quad D_\mu(A)\overline{\psi} = \partial_\mu\overline{\psi} + iA_\mu\overline{\psi}.$$

Here we have changed the notation: for clarity the field $B_\mu = C_\mu$ is denoted as A_μ. The condition $B_\mu = C_\mu$ is only compatible with the subset of gauge transformations (14.18), (14.19) that is obtained by imposing the condition $\chi(t,x) = \eta(t,x)$. Such restricted gauge transformations act on the Dirac field ψ as local $U(1)$ transformations of the form

$$\psi'(t,x) = e^{i\eta(t,x)}\ \psi(t,x). \tag{14.49}$$

The classical counterparts of the currents $\hat{J}^\mu$ and $\hat{J}_5^\mu$ have the form $j^\mu = \overline{\psi}\gamma^\mu\psi$ (the vector current), $j_5^\mu = \overline{\psi}\gamma^\mu\gamma_5\psi$ (the axial vector current). Both classical currents are conserved in the classical theory because they are the Noether currents corresponding to the global symmetries of the Lagrangian $\mathcal{L}$:

$$\psi'(t,x) = e^{i\alpha}\ \psi(t,x), \qquad \alpha \in [0, 2\pi), \tag{14.50}$$

in the case of the vector current, and

$$\psi'(t,x) = e^{i\beta\gamma_5}\ \psi(t,x), \qquad \beta \in [0, 2\pi), \tag{14.51}$$

for j_5^μ.

One can also have $\hat{J}_5^\mu$ as the conserved current. In this case we assume that $B(t) = -C(t)$. Then, the current $\hat{J}^\mu$ is not conserved, $\partial_\mu \hat{J}^\mu = -\dot{B}I/\pi$. The condition $B(t) = -C(t)$ is automatically satisfied if we consider the classical Lagrangian $\mathcal{L}_5 = (\mathcal{L}_+ + \mathcal{L}_-)|_{B=-C}$ with the gauge field $B_\mu(t,x) = -C_\mu(t,x) = A_{5\mu}(t,x)$. This last condition is compatible with the gauge transformations restricted by the condition $\eta(t,x) = -\chi(t,x)$ – then $\psi'(t,x) = e^{i\eta(t,x)\gamma_5}\ \psi(t,x)$. The Lagrangian $\mathcal{L}_5$ can be written in the form

$$\mathcal{L}_5 = \frac{i}{2}[\overline{\psi}\gamma^\mu D_\mu(A_5)\psi - (D_\mu(A_5)\overline{\psi})\gamma^\mu\psi],$$

where

$$D_\mu(A_5)\psi = \partial_\mu\psi - iA_{5\mu}\gamma_5\psi, \quad D_\mu(A_5)\overline{\psi} = \partial_\mu\overline{\psi} + iA_{5\mu}\overline{\psi}\gamma_5.$$

This Lagrangian is also invariant under global $U(1)$ transformations (14.50), (14.51), but only the axial vector current $j_5^\mu = \overline{\psi}\gamma^\mu\gamma_5\psi$ remains conserved after passing to the quantum theory.

In the explicit construction of the quantum theory presented above we have considered external gauge fields of a particular form (14.32). It turns out that the results remain similar if we consider more general gauge fields. Then the r.h.s.'s of the pertinent anomaly equations have a more general form, e.g., $\dot{B}$ is replaced by $\epsilon^{\mu\nu}F_{\mu\nu}/2$, where $F_{\mu\nu} = \partial_\mu B_\nu - \partial_\nu B_\mu$ and $\epsilon^{\mu\nu}$ is the antisymmetric symbol ($\epsilon^{01} = +1$).

14.2 Anomalies and the Path Integral

The derivation of anomalies presented above relies heavily on the operator formalism of quantum field theory. One may be puzzled about the source of anomalies when we use the path integral formulation of the quantum theory. After all, in the path integral there is the classical action which has the relevant symmetries, and there are no operator-valued generalized functions (field operators). We address this question in a model which is akin to the one discussed above, but the fields are now considered in four-dimensional Euclidean space E with the metric $(\delta_{\mu\nu}) = \mathrm{diag}(+1, +1, +1, +1)$, where $\delta_{\mu\nu}$ is the Kronecker delta.

Let us consider the massless Dirac field $\psi(x)$ interacting with a fixed external, classical, electromagnetic field $A_\mu(x), x \in E$. The classical Lagrangian has the form

$$\mathcal{L} = \frac{1}{2} i[\overline{\psi}(x)\gamma^\mu D_\mu\psi - D_\mu\overline{\psi}\gamma^\mu\psi], \tag{14.52}$$

where

$$D_\mu\psi = \left(\partial_\mu - iA_\mu(x)\right)\psi, \quad D_\mu\overline{\psi} = \left(\partial_\mu + iA_\mu(x)\right)\overline{\psi}.$$

and ψ is Euclidean anticommuting field, that is $\psi(x)$ at each point $x \in E$ is a Grassmann element.

In the Euclidean case the Dirac matrices γ^μ are Hermitian, and the Dirac relations (5.2) are replaced by $\{\gamma^\mu, \gamma^\nu\} = 2\delta_{\mu\nu} I_4$. Such matrices can be obtained, e.g., from the matrices given by formulas (5.3) by multiplying γ_D^1, γ_D^2 and γ_D^3 by i, and leaving γ_D^0 unchanged. The Greek indices still have the values 0, 1, 2 and 3, as in the Minkowski case. The matrix γ_5 is defined as $\gamma_5 = \gamma^0\gamma^1\gamma^2\gamma^3$. It is Hermitian, $\gamma_5^2 = I_4$, and it anticommutes with the matrices γ^μ. The conjugation (6.65) is replaced with a simpler one, namely

$$(\overline{\psi}_\alpha(x))^* = \psi^\alpha(x), \quad (\psi^\alpha(x))^* = \overline{\psi}_\alpha(x).$$

The definition of conjugation is such that the physical quantities like energy, momentum, current density, etc., are selfconjugate, e.g., $\mathcal{L}^* = \mathcal{L}$. It matters here whether the γ^μ matrices are Hermitian or not. The conjugation (6.65) involves the γ^0 matrix precisely because in that case the matrices γ^i are anti-Hermitian, see Exercise 6.5.

Lagrangian (14.52) and the action $S = \int d^4x\, \mathcal{L}$ are invariant under the chiral transformations

$$\psi(x) \to \psi'(x) = e^{i\alpha\gamma_5}\psi(x), \qquad \overline{\psi}(x) \to \overline{\psi}'(x) = \overline{\psi}(x)e^{i\alpha\gamma_5}, \tag{14.53}$$

with constant α. The corresponding classical conserved current reads

$$j_5^\mu(x) = i\,\overline{\psi}(x)\gamma^\mu\gamma_5\psi(x).$$

It is selfconjugate with respect to the above introduced Euclidean conjugation of the Dirac field.

Note that under an infinitesimal form of (14.53) with space-time dependent $\alpha = \alpha(x)$,

$$\psi'(x) = (1 + i\alpha(x)\gamma_5)\psi(x) + \mathcal{O}\left(\alpha^2\right), \qquad \overline{\psi}'(x) = \overline{\psi}(x)(1 + i\alpha(x)\gamma_5) + \mathcal{O}\left(\alpha^2\right),$$

Lagrangian (14.52) transforms as

$$\mathcal{L}(x) \to \mathcal{L}'(x) = \mathcal{L}(x) - \left(\overline{\psi}(x)\gamma^\mu\gamma_5\psi(x)\right)\partial_\mu\alpha(x) + \mathcal{O}\left(\alpha^2\right). \tag{14.54}$$

In order to prove that the current j_5^μ is not conserved on the quantum level we shall consider the so called quantum effective action $W[A_\mu]$, defined in the path integral approach as

$$e^{-W[A_\mu]} = \int [d\psi d\bar{\psi}]\, e^{-S}.$$

The crucial observation is that the path integral measure $[d\psi d\bar{\psi}]$ is not invariant under the chiral rotations with α dependent on x.

First, we need to define the measure more precisely. Let $\phi_n(x)$ denote the normalized eigenfunctions and λ_n the corresponding eigenvalues of the operator $\hat{D} = \gamma^\mu(i\partial_\mu + A_\mu(x))$:

$$\hat{D}\phi_n(x) = \lambda_n\phi_n(x), \qquad \int d^4x\; \phi_n^\dagger(x)\phi_m(x) = \delta_{nm}.$$

The completeness relation has the form

$$\sum_n \phi_n(x)\phi_n^\dagger(y) = I_4\delta(x - y).$$

The eigenvalues λ_n are real. Operator $\hat{D}$ is Hermitian in a Hilbert space $L^2(E, C^4)$ of functions on E with values in a complex 4-dimensional space C^4. The eigenvalues λ_n are gauge invariant in the sense that they do not change when we replace A_μ with $A_\mu + \partial_\mu\chi(x)$ (Exercise 14.4). We can expand $\psi(x)$ and $\overline{\psi}(x)$ in the basis formed by $\phi_n(x)$ and $\phi_n^\dagger(x)$:

$$\psi(x) = \sum_n a_n\phi_n(x), \qquad \overline{\psi}(x) = \sum_n \phi_n^\dagger(x)\bar{a}_n,$$

where a_n and $\bar{a}_n$ are independent Grassmann variables. Then

$$[d\psi d\overline{\psi}] = \prod_m\prod_n da_m d\bar{a}_n.$$

Further, let a'_n denote the coefficients of the decomposition of the chirally rotated spinor $\psi'(x) = \psi(x) + i\alpha(x)\gamma_5\psi(x) + \mathcal{O}(\alpha^2)$ in this basis,

$$\psi'(x) = \sum_n a'_n \phi_n(x).$$

Using the orthogonality relation for the eigenfunctions $\phi_n(x)$ we have:

$$a_n = \int d^4x\ \phi_n^\dagger(x)\psi(x)$$

and

$$\begin{aligned}
a'_n &= \int d^4x\ \phi_n^\dagger(x)\psi'(x) \\
&= \int d^4x\ \phi_n^\dagger(x)\psi(x) + i\int d^4x\ \alpha(x)\phi_n^\dagger(x)\gamma_5\psi(x) + \mathcal{O}(\alpha^2) \\
&= a_n + i\int d^4x\ \alpha(x)\phi_n^\dagger(x)\gamma_5\left(\sum_m a_m\phi_m(x)\right) + \mathcal{O}(\alpha^2) \\
&= \left(\delta_{nm} + i\int d^4x\ \alpha(x)\phi_n^\dagger(x)\gamma_5\phi_m(x)\right) a_m + \mathcal{O}(\alpha^2) \equiv \sum_m C_{nm}a_m + \mathcal{O}(\alpha^2).
\end{aligned}$$

Therefore, after neglecting the $\mathcal{O}(\alpha^2)$ terms,

$$\prod_m da'_m = \left(\det \hat{C}\right)^{-1} \prod_m da_m,$$

where $\hat{C}$ denotes the infinite matrix $[C_{nm}]$. The inverse determinant appears since a_m and a'_m are Grassmann variables, see Exercise 11.2 (b).

Let us write $\hat{C}$ in the form $\hat{C} = I + \hat{\epsilon}$, where

$$\epsilon_{nm} = i\int d^4x\ \alpha(x)\phi_n^\dagger(x)\gamma_5\phi_m(x).$$

Using the formula

$$\ln\det\hat{C} = \mathrm{tr}\ln\hat{C} = \mathrm{tr}\hat{\epsilon} + \mathcal{O}(\hat{\epsilon}^2)$$

we have

$$\begin{aligned}
(\det\hat{C})^{-1} &= \exp\left\{-\mathrm{tr}\ln\hat{C}\right\} = \exp(-\sum_n \epsilon_{nn})\left(1 + \mathcal{O}(\alpha^2)\right) \\
&= \exp\left\{-i\int d^4x\ \alpha(x)\left(\sum_n \phi_n^\dagger(x)\gamma_5\phi_n(x)\right)\right\}\left(1 + \mathcal{O}(\alpha^2)\right).
\end{aligned} \tag{14.55}$$

Since

$$B(x) \equiv \sum_n \phi_n^\dagger(x)\gamma_5\phi_n(x)$$
$$= \mathrm{tr}_4\left(\gamma_5 \sum_n \phi_n(x)\phi_n^\dagger(x)\right) = \mathrm{tr}_4\gamma_5 \cdot \delta(0) \text{ “=” } 0\cdot\infty,$$

where the trace tr_4 is over the bispinor indices, we see that $B(x)$ is not a well defined quantity. In order to obtain a meaningful expression for $B(x)$ we shall introduce a Gaussian regularization and define

$$B(x) = \lim_{M\to\infty}\lim_{x'\to x}\sum_n \mathrm{tr}_4\left(\gamma_5\, e^{-\left(\frac{\lambda_n}{M}\right)^2}\phi_n(x)\phi_n^\dagger(x')\right)$$
$$= \lim_{M\to\infty}\lim_{x'\to x}\mathrm{tr}_4\left(\gamma_5\, e^{-\frac{\hat{D}^2}{M^2}}\sum_n \phi_n(x)\phi_n^\dagger(x')\right)$$
$$= \lim_{M\to\infty}\lim_{x'\to x}\mathrm{tr}_4\left(\gamma_5\, e^{-\frac{\hat{D}^2}{M^2}}\right)\delta(x-x').$$

Using the representation

$$\delta(x-x') = \int \frac{d^4k}{(2\pi)^4}\, e^{-ik(x-x')},$$

we get

$$B(x) = \lim_{M\to\infty}\lim_{x'\to x}\int \frac{d^4k}{(2\pi)^4}\,\mathrm{tr}_4\left(\gamma_5\, e^{-\frac{\hat{D}^2}{M^2}}\right)e^{-ik(x-x')}$$
$$= \lim_{M\to\infty}\int \frac{d^4k}{(2\pi)^4}\,\mathrm{tr}_4\left(\gamma_5\, e^{ikx}e^{-\frac{\hat{D}^2}{M^2}}e^{-ikx}\right)$$
$$= \lim_{M\to\infty}\int \frac{d^4k}{(2\pi)^4}\,\mathrm{tr}_4\left(\gamma_5 \exp\left\{-\frac{1}{M^2}e^{ikx}\hat{D}^2e^{-ikx}\right\}\right).$$

Now, let us introduce the operator $d_\mu = i\partial_\mu + A_\mu$ (it coincides with iD_μ, where D_μ is the covariant derivative of the field ψ). We have

$$\hat{D}^2 = \gamma^\mu\gamma^\nu d_\mu d_\nu = \frac{1}{2}\left(\{\gamma^\mu,\gamma^\nu\} + [\gamma^\mu,\gamma^\nu]\right)d_\mu d_\nu$$
$$= \delta_{\mu\nu}d_\mu d_\nu I_4 + \frac{1}{2}\gamma^\mu\gamma^\nu[d_\mu, d_\nu] = d_\mu d_\mu I_4 + \frac{i}{2}\gamma^\mu\gamma^\nu F_{\mu\nu},$$

where the identity

$$[d_\mu, d_\nu] = iF_{\mu\nu}$$

has been used. Consequently,

$$\hat{D}^2\, e^{-ikx} = e^{-ikx}\left[(k+A)^2 + i\partial_\mu A_\mu + \frac{i}{2}\gamma^\mu\gamma^\nu F_{\mu\nu}\right],$$

and, changing the integration variable from k to $q = \frac{1}{M}(k+A)$, we get

$$B(x) = \lim_{M\to\infty}\int \frac{d^4k}{(2\pi)^4}\mathrm{tr}_4\left(\gamma_5\, e^{-\frac{i}{M^2}\left[(k+A)^2 + i\partial_\mu A_\mu + \frac{i}{2}\gamma^\mu\gamma^\nu F_{\mu\nu}\right]}\right)$$
$$= \lim_{M\to\infty} M^4 \int \frac{d^4q}{(2\pi)^4}\mathrm{tr}_4\left(\gamma_5\, e^{-q^2} e^{-\frac{i}{M^2}\left[\partial_\mu A_\mu + \frac{1}{2}\gamma^\mu\gamma^\nu F_{\mu\nu}\right]}\right).$$

Expanding the exponent in powers of M^{-2}, and using the facts that

$$\mathrm{tr}_4\gamma_5 = \mathrm{tr}_4\left(\gamma_5\gamma^\mu\gamma^\nu\right) = 0, \qquad \mathrm{tr}_4\left(\gamma_5\gamma^\mu\gamma^\nu\gamma^\rho\gamma^\lambda\right) = 4\epsilon_{\mu\nu\rho\lambda},$$

and

$$\int \frac{d^4q}{(2\pi)^4}\, e^{-q^2} = \frac{1}{16\pi^2},$$

we get

$$B(x) = \lim_{M\to\infty}\frac{iM^4}{16\pi^2}\left[\frac{1}{2}\left(\frac{-i}{2M^2}\right)^2 \mathrm{tr}_4\left(\gamma_5\gamma^\mu\gamma^\nu\gamma^\rho\gamma^\lambda\right) F_{\mu\nu}F_{\rho\lambda} + \mathcal{O}\left(M^{-6}\right)\right]$$
$$= -\frac{1}{16\pi^2}F_{\mu\nu}\tilde{F}_{\mu\nu},$$

where

$$\tilde{F}_{\mu\nu} = \frac{1}{2}\epsilon_{\mu\nu\rho\lambda}F_{\rho\lambda}.$$

Inserting this back into (14.55), we see that under the infinitesimal chiral transformation the integration measure changes according to the formula

$$[d\psi'] = \exp\left\{\frac{i}{16\pi^2}\int d^4x\, \alpha(x) F_{\mu\nu}(x)\tilde{F}_{\mu\nu}(x)\right\}[d\psi] + \mathcal{O}(\alpha^2). \tag{14.56}$$

Analogous calculation shows that a chiral transformation of the measure $[d\overline{\psi}]$ is given also by formula (14.56).

The chiral rotation can be viewed as a change of integration variables, which does not influence the value of the integral,

$$\int [d\psi d\overline{\psi}]\, e^{-S[A_\mu,\psi,\overline{\psi}]} = \int [d\psi' d\overline{\psi}']\, e^{-S[A_\mu,\psi',\overline{\psi}']}.$$

Neglecting the terms of order α^2, and with the help of (14.54) and (14.56), we get the identity

$$\begin{aligned}0 &= \int [d\psi' d\overline{\psi}'] \, e^{-S[A_\mu,\psi',\overline{\psi}']} - \int [d\psi d\overline{\psi}] \, e^{-S[A_\mu,\psi,\overline{\psi}]} \\ &= \int [d\psi d\overline{\psi}] \, e^{-S[A_\mu,\psi,\overline{\psi}]} \int d^4x \left(\frac{i}{8\pi^2} \alpha(x) F_{\mu\nu} \tilde{F}_{\mu\nu} + \overline{\psi} \gamma^\mu \gamma_5 \psi \partial_\mu \alpha(x) \right) \\ &= \int [d\psi d\overline{\psi}] \, e^{-S[A_\mu,\psi,\overline{\psi}]} \int d^4x \, \alpha(x) \left(\frac{i}{8\pi^2} F_{\mu\nu} \tilde{F}_{\mu\nu} - \partial_\mu (\overline{\psi} \gamma^\mu \gamma_5 \psi) \right).\end{aligned}$$

The vacuum expectation value of the quantum chiral current is given by

$$\langle \hat{J}_5^\mu(x) \rangle = \mathcal{N}^{-1} \int [d\psi d\overline{\psi}] \, e^{-S[A_\mu,\psi,\overline{\psi}]} \, i\overline{\psi}(x) \gamma^\mu \gamma_5 \psi(x),$$

where $\mathcal{N} = \int [d\psi d\overline{\psi}] \, e^{-S}$. Because $\alpha(x)$ is arbitrary, the identity obtained above implies an axial anomaly equation of the form

$$\partial_\mu \langle \hat{J}_5^\mu(x) \rangle = -\frac{1}{8\pi^2} F_{\mu\nu}(x) \tilde{F}_{\mu\nu}(x). \tag{14.57}$$

Let us end this section with a comment. Suppose that $\phi_n(x)$ is an eigenfunction of the Hermitian operator $\hat{D}$ with a non-zero eigenvalue λ_n,

$$\hat{D}\phi_n(x) = \lambda_n \phi_n(x).$$

Since

$$\left\{ \gamma_5, \, \hat{D} \right\} = 0,$$

the function $\gamma_5 \phi_n(x)$ is an eigenfunction of $\hat{D}$ with the eigenvalue $-\lambda_n$. Eigenfunctions corresponding to different eigenvalues are orthogonal, therefore

$$\int d^4x \, \phi_n^\dagger(x) \gamma_5 e^{-\frac{\hat{D}^2}{M^2}} \phi_n(x) = e^{-\frac{\lambda_n^2}{M^2}} \int d^4x \, \phi_n^\dagger(x) \gamma_5 \phi_n(x) = 0.$$

This implies that for a constant parameter of the chiral rotation α, only the functions $\phi_n(x)$ corresponding to the zero eigenvalues—the so called zero modes of the Dirac operator $\hat{D}$—contribute to the chiral variation of the path integral measure. These eigenfunctions are denoted as $\phi_i^{(0)}(x), \; i = 1, 2, \ldots, n_0$. Since $\gamma_5 \phi_i^{(0)}(x)$ also is a zero mode of $\hat{D}$ and γ_5 is Hermitian, we can split the set of zero modes into mutually orthogonal subsets of eigenfunctions of γ_5 with the eigenvalues $+1$ and -1, denoted by $\phi_{i,+}^{(0)}(x), \; i = 1, 2, \ldots, \nu_+$ and $\phi_{i,-}^{(0)}(x), \; i = 1, 2, \ldots, \nu_-$, correspondingly,

$$\gamma_5 \phi_{i,\pm}^{(0)}(x) = \pm \phi_{i,\pm}^{(0)}(x).$$

With this notation

$$\int d^4x\, B(x) = \int d^4x \sum_{i=1}^{n_0} \phi_i^{(0)}(x)^\dagger \gamma_5 \phi_i^{(0)}(x)$$
$$= \sum_{i=1}^{\nu_+} \int d^4x\, \phi_{i,+}^{(0)}(x)^\dagger \gamma_5 \phi_{i,+}^{(0)}(x) + \sum_{i=1}^{\nu_-} \int d^4x\, \phi_{i,-}^{(0)}(x)^\dagger \gamma_5 \phi_{i,-}^{(0)}(x) = \nu_+ - \nu_-.$$

The quantity

$$\nu_+ - \nu_- \equiv \mathrm{ind}(\hat{D}_+)$$

is called the index of the projected Dirac operator $\hat{D}_+ = \hat{D}(1+\gamma_5)/2$. From the calculations above we see that the index determines the chiral anomaly.

Exercises

14.1 The Lorentz group in $(1+1)$-dimensional space-time consists of real, two by two matrices $(L^\mu_{\ \nu})$, where $L^0_{\ 0} = L^1_{\ 1} = \cosh u$ and $L^1_{\ 0} = L^0_{\ 1} = \sinh u$. The parameter u (rapidity) can have an arbitrary real value. The corresponding transformation law for the Dirac field has the form $\psi'(x') = S(L)\psi(x)$, where $x' = Lx$, and the matrix $S(L)$ obeys the conditions $S(L_1)S(L_2) = S(L_1L_2)$ and $S^{-1}(L)\gamma^\mu S(L) = L^\mu_{\ \nu}\gamma^\nu$.
(**a**) Check that

$$S(L) = \begin{pmatrix} e^{\frac{u}{2}} & 0 \\ 0 & e^{-\frac{u}{2}} \end{pmatrix}.$$

(**b**) Check that the Lagrangians $\mathcal{L}_\pm$ (given by formulas (14.10), (14.12)), as well as γ_5, are invariant with respect to the Lorentz transformations.

14.2 Derive formula (14.26).
Hints: Instead of $\hat{J}^0_+(t,x)$ we may insert $\hat{j}^0_{+,\epsilon}(t,x)$ given by formula (14.24), because the two charge densities differ from each other by a term proportional to the identity operator I. Next, introduce the operators

$$U(s) = \exp\left(is \int dx'\, \chi(x')\hat{j}^0_{+,\epsilon}(t,x')\right), \quad \hat{\psi}_s(t,x) = U^{-1}(s)\hat{\psi}_+(t,x)U(s),$$

where s is a real parameter. Using the anticommutation relations (14.20) obtain, in the limit $\epsilon \to 0$, the equation

$$\frac{d\hat{\psi}_s(t,x)}{ds} = i\chi(x)\hat{\psi}_s(t,x).$$

Find its solution such that $\hat{\psi}_s(t,x)|_{s=0} = \hat{\psi}_+(t,x)$ and put $s = 1$.

14.3 Derive the anomaly equation (14.48) for the left-mover current.

14.4 Prove that the eigevalues λ_n of the operator $\hat{D}$ are gauge invariant.

14.5 Check that the quantum effective action $W[A_\mu]$ is not invariant under the gauge transformations $A_\mu(x) \to A_\mu(x) + \partial_\mu \alpha(x)$.
Hint: Consider $\delta W[A_\mu + \partial_\mu \alpha]/\delta \alpha(x)$.

Appendix: Some Facts About Generalized Functions

Here we recall some basic facts and formulas from the theory of generalized functions. There are many mathematical textbooks on this subject. Physicists may find useful concise texts, for example Chaps. 2–4 in [12], or Chaps. 2 and 3 in [8]. A comprehensive introduction to the subject can be found in [2].

The generalized functions that appear in field theory are of the so called Schwartz class, denoted as $S'(R^n)$ or $S^*(R^n)$. The reason is that all such generalized functions have a Fourier transform. A generalized function of the Schwartz class[1] (g. f.) is, by definition, a linear and continuous functional on the Schwartz space of functions, denoted by $S(R^n)$. Elements of $S(R^n)$ are called test functions. They are complex valued functions on R^n of the C^∞ class. Moreover, it is assumed that such functions, and all their derivatives, vanish in the limit $|x| = \sqrt{(x^1)^2 + (x^2)^2 + \ldots + (x^n)^2} \to \infty$, also when multiplied by any finite order polynomial in the variables $x^1, \ldots, x^n$. Here x denotes arbitrary point in R^n and $x^1, \ldots, x^n$ are its Cartesian coordinates. The space $S(R^n)$ is endowed with a topology, but we shall not describe it here. Examples of test functions from the space $S(R^1)$ include e^{-ax^2} and $1/\cosh(ax)$, where $a > 0$ is a real constant. On the other hand, $(1 + x^2)^{-1}$ is not test function from $S(R^1)$.

The value of a generalized function $F \in S^*(R^n)$ on a test function $f \in S(R^n)$ is denoted in mathematical literature as $\langle F(x), f(x)\rangle$, but in physics the most popular is the misleading notation $\int d^n x\, F(x) f(x)$, for example, $\int d^n x\, \delta(x) f(x)$ in the case of the Dirac delta. One should keep in mind that the integral here is merely a symbol that replaces $\langle\ ,\ \rangle$ from the mathematical notation—it is not a true integral. It may happen however, that a generalized function is represented by an ordinary function $F(x)$ such that the true integral $\int d^n x\, F(x) f(x)$ exists for all $f \in S(R^n)$. Such an F is called a regular g.f. For example, the step function $\Theta(x)$, $x \in R^1$, is a regular generalized function from $S^*(R^1)$, because the integral $\int_{-\infty}^{\infty} dx\, \Theta(x) f(x) = \int_0^\infty dx\, f(x)$ exists for every $f \in S(R^1)$. We show the integration range when we deal with a true integral. The Dirac delta $\delta(x)$ is the prominent example of a non regular g.f. In the mathematical notation its definition has the form $\langle \delta(x), f(x)\rangle = f(0)$.

[1] Other names are also used: distribution for generalized function, and tempered distribution for generalized functions from the Schwartz class.

H. Arodź and L. Hadasz, *Lectures on Classical and Quantum Theory of Fields*, Graduate Texts in Physics, DOI 10.1007/978-3-319-55619-2

The derivative $\partial_i F$ of g.f. F is defined as follows:

$$\int d^n x \, \partial_i F(x) f(x) = -\int d^n x \, F(x) \partial_i f(x)$$

for all test functions f. One should remember that this is the definition, and not the formula of integration by parts. For example,

$$\int dx \, \frac{d\Theta(x)}{dx} f(x) = -\int_{-\infty}^{\infty} dx \, \Theta(x) \frac{df(x)}{dx} = -\int_{0}^{\infty} dx \, \frac{df(x)}{dx} = f(0),$$

hence $\Theta'(x) = \delta(x)$. The derivative of g.f. always exists and is a generalized function.

The Fourier transform $\tilde{f}(k) = (2\pi)^{n/2} \int_{R^n} d^n x \exp(ikx) f(x)$ of a test function f is also a test function from the space $S(R^n)$. The g.f. $\tilde{F}(x) \in S^*(R^n)$ such that for every $f \in S(R^n)$

$$\int d^n x \, \tilde{F}(x) f(x) = \int d^n k \, F(k) \tilde{f}(k),$$

is called the Fourier transform of the g.f. F. It exists for any $F \in S^*(R^n)$. The operation of taking the Fourier transform is continuous with respect to F. This property is used in order to facilitate the computation of the Fourier transform of $\Theta(x)$—we first compute the Fourier transform of $e^{-\epsilon x}\Theta(x)$, where $\epsilon > 0$, and take the limit $\epsilon \to 0_+$ at the end.[2] Thus,

$$\int dx \, \tilde{\Theta}(x) f(x) = \lim_{\epsilon \to 0_+} \int_{-\infty}^{\infty} dk e^{-\epsilon k} \Theta(k) \tilde{f}(k)$$
$$= \lim_{\epsilon \to 0_+} \frac{1}{\sqrt{2\pi}} \int_{0}^{\infty} dk \int_{-\infty}^{\infty} dx \, e^{ikx - \epsilon k} f(x) = \lim_{\epsilon \to 0_+} \frac{i}{\sqrt{2\pi}} \int_{-\infty}^{\infty} dx \, \frac{1}{x + i\epsilon} f(x).$$

The r.h.s. of this formula defines the g.f. denoted as $\frac{i}{\sqrt{2\pi}} \frac{1}{x+i0_+}$. Therefore,

$$\tilde{\Theta}(x) = \frac{i}{\sqrt{2\pi}} \frac{1}{x + i0_+}.$$

Note that the g.f. $\frac{1}{x+i\epsilon}$ is regular if $\epsilon > 0$.

One can prove that

$$\frac{1}{x + i0_+} = P\frac{1}{x} - i\pi\delta(x),$$

where the principal value distribution $P\frac{1}{x}$ is defined as

$$\int dx \, P\frac{1}{x} \, f(x) = \lim_{\epsilon \to 0_+} \left(\int_{-\infty}^{-\epsilon} dx \, \frac{f(x)}{x} + \int_{\epsilon}^{\infty} dx \, \frac{f(x)}{x} \right)$$

[2]The notation $\epsilon \to 0_+$ means that $\epsilon = 0$ is approached from the side $\epsilon > 0$.

(it is not regular). The result for $\tilde{\Theta}(x)$ obtained above is often written in the form

$$\int_0^\infty dp\, e^{ipx} = iP\frac{1}{x} + \pi\delta(x).$$

The form of a generalized function can be probed only with test functions. Because there is no test function with support consisting of just a single-point, it is not possible to tell what is the value of the g.f. at the given point. One can however check whether a g.f., say $F(x)$, is constant in a vicinity V_{x_0} of a point $x_0 \in R^n$—it is sufficient to show that the first derivatives of $F(x)$ vanish in that vicinity, i.e., that

$$\int \partial_i F(x)\, f(x) = 0$$

for every test function that has its support in $V_{x_0} \subset R^n$. For example, for $\delta(x) \in S^*(R^1)$ one may say that $\delta(x) = 0$ on every interval (a, b) that does not contain 0, and that $\delta(x) \neq 0$ at $x = 0$, but not that $\delta(1) = 0$.

A consequence of the lack of definite value at a single point is that there is no general definition of a product of generalized functions. We know the generalized functions $F_1(x) \in S^*(R^n)$ and $F_2(x) \in S^*(R^n)$ if we know the values of $\int d^n x\, F_1(x) f(x)$ and $\int d^n x\, F_2(x)\, f(x)$ for every $f \in S(R^n)$. It is not possible to infer from this what values $\int d^n x\, F_1(x) F_2(x)\, f(x)$ should have. Only in some special cases, e.g., for certain regular generalized functions, can such product be defined. In particular, there is no problem with multiplication by an ordinary function $\psi(x)$, provided that $\psi(x) f(x) \in S(R^n)$ for every $f \in S(R^n)$. Then, the product $\psi(x)F(x)$ is the g.f. defined by the formula

$$\int d^n x\, (\psi(x)F(x))\, f(x) = \int d^n x F(x)\, (\psi(x) f(x)).$$

For example, if k is a fixed real number, $e^{ikx}\delta(x)$ is a generalized function, while $x^a\delta(x)$ with non integer constant $a > 0$ is not (not all functions $x^a f(x)$ belong to $S(R^1)$ because of the problem with derivatives at $x = 0$).

On the other hand, there is no difficulty with a product of generalized functions with different arguments. If $F(x) \in S^*(R^n)$ and $G(y) \in S^*(R^m)$, then we know $\int d^n x\; F(x) f(x)$ and $\int d^m y\; G(y) g(y)$ for all $f \in S(R^n)$ and $g \in S(R^m)$. The generalized function $H(x, y) = F(x)G(y) \in S^*(R^{n+m})$ is defined by its action on the test functions $h(x, y) \in S(R^{n+m})$ of the form $h(x, y) = f(x)g(y)$, namely

$$\int d^n x d^m y\; H(x, y) h(x, y) = \int d^n x\; F(x) f(x) \int d^m y\; G(y) g(y).$$

Such factorized test functions $f(x)g(y)$ form a subset of $S(R^{n+m})$ that is sufficiently large to uniquely determine $H(x, y)$ on the whole space $S(R^{n+m})$. An example: if $x, y \in R^1$ are independent variables, then $\delta(x)\delta(y) \in S^*(R^2)$.

Finally, let us consider the question whether

$$\int dx\ \delta(x) \stackrel{?}{=} \int_0^\infty dx\ \delta(x) + \int_{-\infty}^0 dx\ \delta(x),$$

or, in a more meaningful form, whether

$$\delta(x) = 1\delta(x) = (\Theta(x) + \Theta(-x))\delta(x) \stackrel{?}{=} \Theta(x)\delta(x) + \Theta(-x)\delta(x).$$

The answer is that such a formula is wrong, because the products $\Theta(x)\delta(x)$, $\Theta(-x)\delta(x)$ are not defined. The way to correct the splitting consists in replacing $\Theta(\pm x)$ with two smooth functions $\theta_1(x)$ and $\theta_2(x)$ which obey the condition $\theta_1(x) + \theta_2(x) = 1$, and resemble $\Theta(x)$ and $\Theta(-x)$, respectively. Moreover, these functions should be such that $\theta_i(x) f(x) \in S(R^1)$, $i = 1, 2$, for every $f \in S(R^1)$. Then we may safely write

$$F(x) = \theta_1(x)F(x) + \theta_2(x)F(x)$$

for any $F(x) \in S^*(R^1)$.

Bibliography

1. Abramowitz, M., Stegun, I. (eds.): Handbook of Mathematical Functions. Applied Mathematics Series 55. National Bureau of Standards (1964)
2. Vladimirov, V.S.: Methods of the Theory of Generalized Functions. CRC Press, Boca Raton, FL (2002)
3. Bogoliubov, N.N., Shirkov, D.V.: Introduction to the Theory of Quantized Fields. Wiley-Interscience, New York, London, Sydney, Toronto (1980)
4. Jackiw, R., Manton, N.: Symmetries and conservation laws in gauge theories. Ann. Phys. **127**, 257 (1980)
5. Schweber, S.S.: An Introduction to Relativistic Quantum Field Theory. Peterson and Co., Evanston, Ill., Elmsford, N.Y., Row (1961)
6. Henneaux, M., Teitelboim, C.: Quantization of Gauge Systems. Princeton University Press, Princeton, New Jersey (1992)
7. Faddeev, L.D., Jackiw, R.: Hamiltonian reduction of unconstrained and constrained systems. Phys. Rev. Lett. **60**, 1692 (1988)
8. Bogoliubov, N.N., Logunov, A.A., Oksak, A.I., Todorov, I.T.: General Principles of Quantum Field Theory. Kluwer, Dordrecht (1990)
9. Weinberg, S.: The Quantum Theory of Fields, vol. I. Cambridge University Press. Cambridge, New York, Melbourne (1995)
10. Sakita, B.: Quantum Theory of Many-Variable Systems and Fields. World Scientific, Singapore (1985)
11. Sterman, G.: An Introduction to Quantum Field Theory. Cambridge University Press, Cambridge, New York, Melbourne (1993)
12. Richtmyer, R.D.: Principles of Advanced Mathematical Physics, vol. 1. Springer, New York, Heidelberg, Berlin (1978)

H. Arodź and L. Hadasz, *Lectures on Classical and Quantum Theory of Fields*,
Graduate Texts in Physics, DOI 10.1007/978-3-319-55619-2

Index

H. Arodź and L. Hadasz, *Lectures on Classical and Quantum Theory of Fields*, Graduate Texts in Physics, DOI 10.1007/978-3-319-55619-2

Printed in the United States
By Bookmasters

Printed in the United States
By Bookmasters